Prediction of Reservoir Quality Through Chemical Modeling

Edited by
Indu D. Meshri
and
Peter J. Ortoleva

AAPG Memoir 49

Published by
The American Association of Petroleum Geologists
Tulsa, Oklahoma 74101, U.S.A.

Copyright ©1990
The American Association of Petroleum Geologists
All Rights Reserved
Published November 1990

ISBN: 0-89181-327-6
ISSN: 0065-731X

AAPG grants permission for a single photocopy of any article herein for research or noncommercial educational purposes. Other photocopying not covered by the copyright law as Fair Use is prohibited. AAPG participates in the Copyright Clearance Center. For permission to photocopy more than one copy of any article, or parts thereof, contact: AAPG Permissions Editor, P.O. Box 979, Tulsa, Oklahoma 74101-0979.

Association Editor: Susan A. Longacre
Science Director: Gary D. Howell
Publications Manager: Cathleen P. Williams
Special Projects Editor: Anne H. Thomas
Copy Editor: Robert J. Floyd
Production: Custom Editorial Productions, Inc., Cincinnati, Ohio

PREFACE

Prediction of reservoir quality ahead of the drill is one of the most complex problems facing exploration geologists today, especially when they are exploring in frontier basins, where rock and water data are minimal or nonexistent. Although useful descriptive models of diagenesis have existed in the past, they cannot be applied in the areas where rock and water data do not exist.

Therefore, in June 1987, the American Association of Petroleum Geologists sponsored an international research conference on the subject of "Prediction of Reservoir Quality Through Chemical Modeling," held in Park City, Utah. At this conference, petrologists, paleohydrologists, chemical modelers, kinetic and thermodynamic experts, and formation water geochemists came together and presented a total of 44 papers on various aspects of rock-water interaction. Out of these papers emerged a strategy for modeling reservoir quality ahead of the drill, and it was then presented in a condensed form in a half-day research symposium at AAPG's 1989 Annual Meeting held in San Antonio. I believe that this book, which evolved from that condensed symposium, documents the substantial progress made toward the goal of modeling reservoir quality.

The first two chapters bring to our attention those aspects of geology that directly point the way toward the chemical controls on porosity. Shanmugam documents the secondary porosity associated with erosional unconformities in hydrocarbon-bearing sandstones in Algeria, Australia, China, Libya, the Netherlands, the Norwegian North Sea, and various parts of the United States. For the existence of this secondary (reaction-induced) porosity at erosional unconformities, he proposes the model of a hydrologically and geochemically "open" system operating as a result of the influx of fresh (under-saturated) meteoric waters. Similarly, in Chapter 2 Stonecipher and May document the influence of the depositional environment and, therefore, the effect of early pore water composition on the early diagenetic facies.

Before we plunge into various applications of chemical models in hydrologically and geochemically open and closed systems, Chapter 3 provides a brief overview of chemical models. The models are classified and a brief discussion of theory, the inputs, the outputs, and the use and application of each type, is provided, in language that we hope is palatable to exploration geologists. In the chapter I co-authored with Walker, we present an example of the simulation of a diagenetic sequence in the upper Almond Sandstone of the Red Desert and Washakie basins.

Because we provide the reader with a discussion of the models that utilize ideas of thermodynamic equilibrium and the models that use a pseudo-kinetic approach, and we point out how disequilibrium is handled, it is appropriate to follow with Machel's ideas on bulk solution disequilibrium in diagenetic carbonate minerals (Chapter 5). Over the years, precipitation and recrystallization of diagenetic carbonates have been assumed to take place in thermodynamic equilibrium with the bulk pore solutions. However, Machel, basing his argument on trace element and isotopic composition data, demonstrates that the precipitation of diagenetic carbonates occurs under a state of disequilibrium.

Moore and Ortoleva (in Chapter 8) not only agree that disequilibrium exists, but have written a coupled geochemical modeling code wherein the kinetics of mineral dissolution/precipitation and the thermodynamics of fast aqueous reactions are coupled to the transport mechanisms. This chapter applies the one-dimensional reaction-transport code, REACTRAN, to the modeling of clean and dirty arkosic rock compositions and shows how evolution of diagenetic minerals, and hence porosity, can take place in time and one-dimensional space as a function of assumed reaction rates and flow velocities.

The data on mineral dissolution/precipitation rates are sparse when compared with the data on thermodynamic stabilities of minerals. The kinetics of mineral reaction are extremely important when modeling porosity evaluation in a reservoir. Nagy et al. (Chapter 6) describe the dissolution precipitation kinetics of kaolinite and show the results of an isothermal simulation of diagenesis in a subarkosic sandstone, where the volume fraction of minerals varies as a result of precipitation and dissolution reactions driven by the influx of reactive fluids.

In Chapter 7, Chen et al. use mathematical reaction-transport modeling to evaluate the coupling of diagenetic reaction-transport processes. Through several examples, the authors illustrate how reaction-transport coupling can generate spatial patterns of petroleum and minerals that can play a central role in determining reservoir quality and the diagenetic influences on petroleum distribution in the basin.

To complete the story of predicting reservoir quality, we must discuss pressure-solution-induced porosity changes. Dewers and Ortoleva provide us with an overview of their recent attempts to quantify the phenomenon of pressure solution. Chapter 9 examines the two mechanisms of mass transfer, such as water-film diffusion (WFD) and free-face pressure solution (FFPS), in the context of reaction-transport modeling.

One facet of chemical modeling, namely porosity prediction, is the thrust of this book. However, chemical modeling has contributed heavily in the field of environmental geochemistry, nuclear waste disposal, and in the thermal recovery of heavy oil and the like. We include one such chapter by Hutcheon and Abercrombie. Although the original, major goal of this paper was to assist in thermal recovery of heavy oil deposits of Alberta, it documents the importance of silicate hydrolysis as the controlling mechanism in natural, as well as experimental, rock-water interactions.

Testing via application of chemical models in natural geologic settings and via experimental rock-water interactions must continue. The phenomenal improvements in the size and rate of computations, and improved numerical techniques such as adoptive gridding, finite element and finite difference techniques, tempt one to believe that prediction of porosity evolution on geologically reasonable scales of time and space is just around the corner. However, to achieve this the multidisciplinary approach adopted in the book in hand must continue.

I want to express my special gratitude to Amoco Production Company for their generous support of the 1987 research conference, and to the AAPG for its sponsorship. I am also greatly indebted to my colleagues G. Garven, R. W. Scott, J. S. Bradley, P. D. Wagner, J. B. Fisher, R. E. Larese, J. M. Walker, L. C. Babcock, and D. R. Prezbindowski for their help in reviewing papers for this volume. I am thankful to the staff of AAPG for their assistance in making the publication of this volume an easier task.

Indu D. Meshri

ABOUT THE EDITORS

Indu D. Meshri

Indu Meshri was born in Sind, a province of British India that is now in Pakistan. She received her B.Sc. (with honors) degree in chemistry and physics and her M.Sc. in Physical Chemistry from the University of Bombay while she was a fellow of the Royal Institute of Science, Bombay. Her first geological work was at the University of Idaho, where she developed a K-Ar isotope method of age dating of rocks. She then found an opportunity to work on RNA synthesis with the Nobel Laureate Professor R. W. Holley at Cornell University.

After moving to Tulsa she joined Amoco and received her Ph.D. from the University of Tulsa. Her current research interests include basin chemical modeling, prediction of the thermal structure of basins, pore pressure, fluid flow and porosity evolution, genesis of diagenetic traps, coal degasification, petrophysical rock properties, and nuclear magnetic resonance.

Dr. Meshri has organized an international research conference on porosity prediction and has chaired many symposia and research conference sessions. She is the author and coauthor of many technical papers and numerous Amoco reports and is a coeditor of a monograph on "Geochemical Self-Organization."

Peter J. Ortoleva

Peter Ortoleva was born in Brooklyn, New York, in 1942. He received his B.S. in physics from Rensselaer Polytechnic Institute; he earned his Ph.D. in applied physics at Cornell University, under Professor Mark Nelkin, on the statistical mechanics of simple fluids. Dr. Ortoleva carried out postdoctoral research in theoretical physical chemistry with Professor John Ross at Massachusetts Institute of Technology. In 1975 he joined the physical chemistry faculty at Indiana University, where he is presently Professor of Chemistry and of Geology. He has received Sloan and Guggenheim awards, is the author of more than 100 scientific publications, and is author or editor of several monographs.

His main research interests are in the theory of nonlinear phenomena in far-from-equilibrium reaction-transport systems. Areas of specialization include nonlinear waves and pattern formation, mechano-chemical coupling, and electrical aspects of reaction-transport systems. Over the past ten years he has focused on problems in geochemistry, including the physico-chemical dynamics of a sedimentary basin, metamorphic differentiation, zoned crystal growth from magmas and aqueous solutions and reaction fronts. He has developed a number of general reaction-transport codes to simulate these phenomena for academic and industrial research.

Table of Contents

Chapter 1
Porosity Prediction in Sandstones Using Erosional Unconformities
G. Shanmugam 1

Chapter 2
Facies Controls on Early Diagenesis: Wilcox Group, Texas Gulf Coast
S. A. Stonecipher and J. A. May 25

Chapter 3
An Overview of Chemical Models and Their Relationship to Porosity Prediction in the Subsurface
Indu D. Meshri 45

Chapter 4
A Study of Rock-Water Interaction and Simulation of Diagenesis in the Upper Almond Sandstones of the Red Desert and Washakie Basins, Wyoming
Indu D. Meshri and Jana M. Walker 55

Chapter 5
Bulk Solution Disequilibrium in Aqueous Fluids as Exemplified by Diagenetic Carbonates
Hans-G. Machel 71

Chapter 6
Dissolution and Precipitation Kinetics of Kaolinite: Initial Results at 80°C with Application to Porosity Evolution in a Sandstone
K. L. Nagy, C. I. Steefel, A. E. Blum, and A. C. Lasaga 85

Chapter 7
Diagenesis Through Coupled Processes: Modeling Approach, Self-Organization, and Implications for Exploration
W. Chen, A. Ghaith, A. Park, and P. Ortoleva 103

Chapter 8
Effects of Fluid and Rock Compositions on Diagenesis: A Modeling Investigation
Craig H. Moore and Peter J. Ortoleva 131

Chapter 9

Interaction of Reaction, Mass Transport, and Rock Deformation During Diagenesis: Mathematical Modeling of Intergranular Pressure Solution, Stylolites, and Differential Compaction/Cementation
Thomas Dewers and Peter J. Ortoleva 147

Chapter 10

Fluid-Rock Interactions in Thermal Recovery of Bitumen, Tucker Lake Pilot, Cold Lake, Alberta
Ian Hutcheon and Hugh J. Abercrombie 161

Index 171

Porosity Prediction in Sandstones Using Erosional Unconformities

G. Shanmugam
Mobil Research and Development Corporation
Dallas, Texas, U.S.A.

Erosional unconformities of subaerial origin are created by tectonic uplifts and eustatic sea-level falls. Porosity increases below most erosional unconformities developed on sandstones, because uplifted sandstones are exposed to undersaturated CO_2-charged meteoric waters, resulting in dissolution of unstable framework grains and cements. The chemical weathering of sandstones is intensified in humid regions by the heavy rainfall, soil zones, lush vegetation, and the accompanying voluminous production of organic and inorganic acids. Erosional unconformities are considered hydrologically and geochemically "open" systems because of the abundant supply of fresh meteoric water and relatively unrestricted transport of dissolved constituents away from the site of dissolution, causing a net gain in porosity near unconformities. Consequently, porosity in sandstones tends to increase toward overlying unconformities. Such porosity trends have been observed in hydrocarbon-bearing sandstone reservoirs in Alaska, Algeria, Australia, China, Libya, the Netherlands, Norwegian North Sea, Norwegian Sea, and Texas. A common attribute of these reservoirs is that they were all subaerially exposed under warm and heavy rainfall conditions.

An empirical model has been developed for the Triassic and Jurassic sandstone reservoirs in the Norwegian North Sea on the basis of the observed relationship that shows an increase in porosity in these reservoirs with increasing proximity to the overlying Base-Cretaceous unconformity. An important practical attribute of this model is its capability of predicting porosity in the neighboring undrilled areas by recognizing the Base-Cretaceous unconformity in seismic reflection profiles and by constructing subcrop maps of stratigraphic units susceptible to porosity formation by meteoric waters. Caution must be exercised in developing predictive models using unconformities, because porosity reduction due to cementation may also occur beneath some erosional unconformities.

INTRODUCTION

Prediction of porosity in frontier or undrilled areas is a formidable task in hydrocarbon exploration. Although predictive models were proposed in the past (Scherer, 1987), their numerous constraints often render these models impractical (Shanmugam and Alhilali, 1988). In order for a model to be practical in frontier areas of exploration, predictive models of porosity should be tied to seismic reflection profiles that are most commonly available from areas unexplored by drilling. In seismic profiles, erosional unconformities are recognized routinely. Therefore, a predictive model of porosity using erosional unconformities will be of practical value to explorationists. The objectives of this paper are: (1) to summarize factors that promote mineral dissolution when rocks are subaerially exposed, (2) to document evidence for dissolution and porosity enhancement during subaerial exposure of rocks with selected modern and ancient examples, and (3) to discuss a method of predicting porosity in sandstones using erosional unconformities.

Origin, recognition, and importance of erosional unconformities in sedimentary basins have recently been discussed by Shanmugam (1988). Erosional unconformities are created by allocyclic and autocyclic processes in both subaerial and submarine environments. This study deals only with erosional unconformities of subaerial origin. Subaerial unconformities are believed to have been caused by tectonic uplifts and by eustatic sea-level falls. Subaerial unconformities are recognized by: (1) discordance of dip in outcrops and seismic sections,

(2) erosional surfaces and karst facies, (3) basal conglomerates, (4) weathered cherts, (5) abrupt faunal breaks, and (6) paleosol horizons.

FACTORS THAT CONTROL ROCK DISSOLUTION DURING SUBAERIAL EXPOSURE

The connection between subaerial unconformities (subaerial exposure) and the process of rock dissolution was first pointed out by Hutton (1788). Although numerous studies have focused on weathering of rocks (such as granites and carbonates), the aspect of sandstone dissolution during weathering has received only sporadic attention in the literature (Levorsen, 1934; Hrabar and Potter, 1969; Selley, 1985; Shanmugam, 1985a, b, 1988, 1989). Subaerial exposure of sandstones brings them in direct contact with meteoric waters. During subaerial exposure, unstable cements (such as calcite) and framework grains (such as feldspar and rock fragments) are susceptible to dissolution by invading meteoric waters at near-surface conditions (Figure 1). Many factors exert control over dissolution of siliciclastic, carbonate, and even igneous rocks during subaerial exposure. These factors are discussed below.

Meteoric Waters

Rainwater (meteoric water) that is slightly acidic due to dissolved carbon dioxide from the atmosphere is available in large quantities to many sandstones exposed in humid climatic regions. These sandstones are in direct contact with a constantly renewed supply of fresh undersaturated rainwater that can react with unstable minerals.

Giles and Marshall (1986), using theoretical considerations, concluded that leaching of minerals by meteoric water is the most viable mechanism for creating significant volumes of secondary porosity in the shallow subsurface. Unlike other mechanisms advocated for the formation of secondary porosity (such as dissolution by acidic fluids and carboxylic acids generated during the thermal maturation of organic matter), meteoric-water leaching does not suffer from a shortage of the dissolving agent. Meteoric-water leaching can occur under two different scenarios: (1) at subaerial depositional sites where sands are exposed to acidic meteoric water during and immediately after deposition, and (2) during unconformity development when uplifted sandstones are invaded by meteoric waters. All sandstones deposited in the subaerial environment undergo meteoric-water leaching to some degree; however, when some of these sandstones are exposed to meteoric water during later uplift and erosion, they may be further leached and thereby gain additional porosity.

Soil Zones

Soil zones play a vital role in the total weathering of near-surface rock strata. Subaerial erosion exposes new surfaces for leaching, commonly over wide areas. During subsequent weathering, extensive veneers of organic-rich soils can develop on erosional surfaces. These soils can help promote mineral dissolution. The soil zone can be considered an acid pump because of its unique ability to increase the acidity of percolating meteoric waters (Freeze and Cherry, 1979). The partial pressure of CO_2 in the soil atmosphere may be as much as 10 times that of the air (Keller, 1955). Slightly acidic rainwater (pH 5–6) becomes even more acidic as it passes through the soil zone in temperate and humid climates. This is because large amounts of carbon dioxide, generated from the decay of organic matter and the respiration of plant roots, react with water to produce carbonic acid (Figure 1). In arid climates, however, soil zones are usually alkaline, and would not produce carbonic acids. In addition to inorganic acid, organic acids such as humic and fulvic acids are produced in the soil zone by plant roots via biochemical processes. Organic acids are powerful agents in the dissolution of rock-forming silicate minerals (Huang and Keller, 1970). Organic acids break down mineral constituents through the introduction of H^+ into the system and the formation of organometal complexes (Schalscha et al., 1967). These chelating agents react with mineral constituents and accelerate chemical weathering in the near-surface environment (Kononova et al., 1966).

Bennett and Siegel (1987) reported increased solubility of quartz at near-surface conditions due to complexing by organic compounds. The percolation of meteoric waters downward through decaying organic matter in the soil zone should thus provide a suitable environment for the dissolution of minerals and for the transportation of dissolved mineral species away from the site of dissolution.

Open System

The unsaturated zone (also known as the zone of aeration or the vadose zone), which occurs above the water table, is an open system in geochemical terms (Figure 1). The continuous influx of undersaturated meteoric water and constant removal of dissolved material away from the site of dissolution result in a net loss of dissolved constituents. Low amounts of total dissolved solids in fresh meteoric water (<1000 mg/L), which are common in unsaturated zones, favor dissolution. In unsaturated zones, the water/rock equilibrium is either attained slowly or not at all. Such disequilibrium conditions are conducive to transport-controlled dissolution (Berner, 1978). The unsaturated zone may range in thickness from zero to hundreds of meters (Bloom, 1978).

The saturated (or phreatic) zone occurs below the water table. Here, meteoric waters may lose their

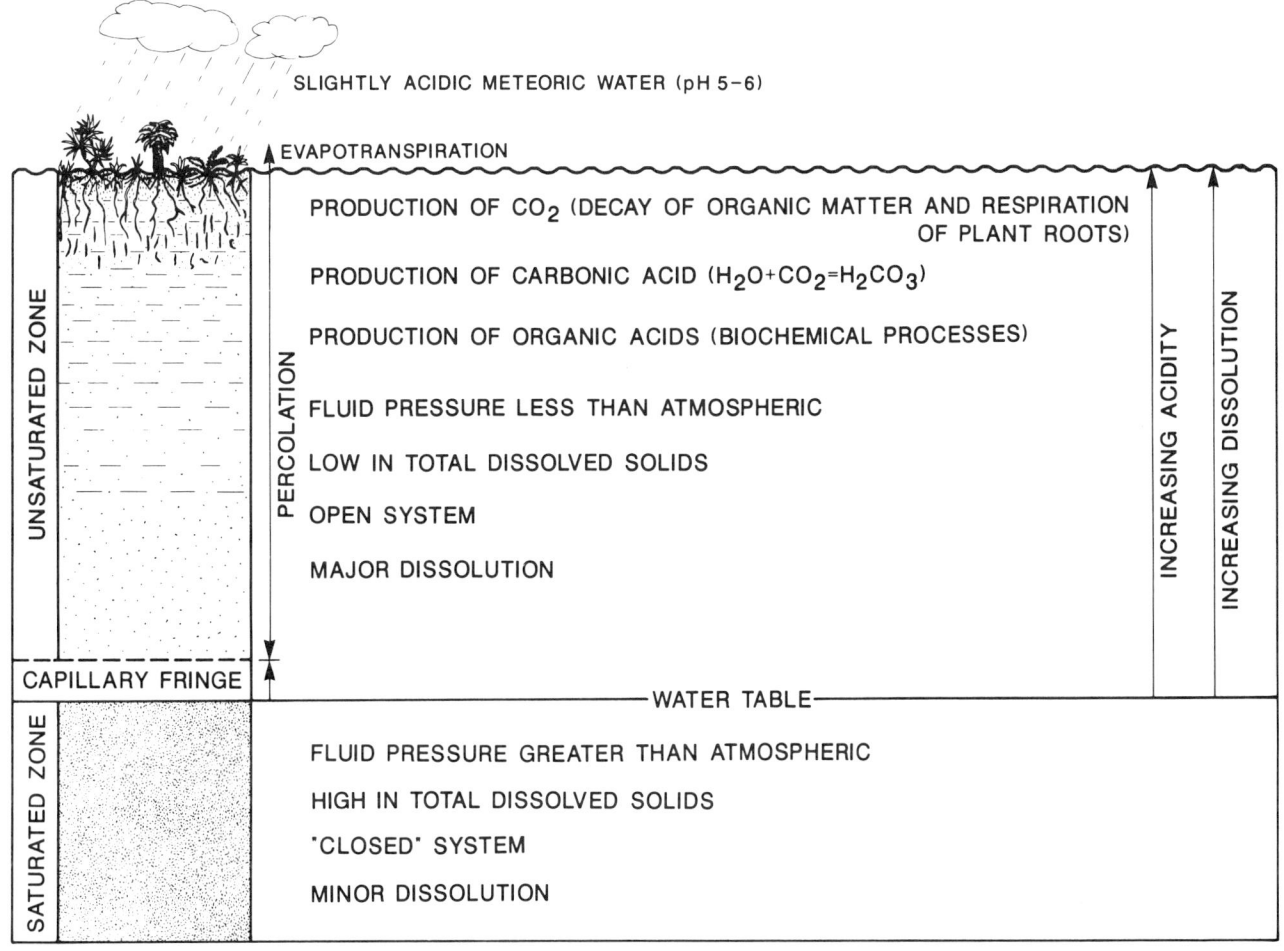

Figure 1. Schematic view of major hydrochemical processes in the soil zone (mostly after Freeze and Cherry, 1979).

potential to dissolve minerals because of poor replenishment of carbon dioxide, high amounts of total dissolved solids, and sluggish circulation.

Depth of Weathering

Meteoric waters in basins may extend to depths approaching 3000 m (Galloway and Hobday, 1983); however, the depth of intense dissolution may be controlled by water-table depth, which is a function of climate, topography, and other factors. Meteoric waters have been reported to reach 1000 m in the Great Basin of Nevada (Winograd and Robertson, 1982), more than 1000 m in the Anadarko basin of Texas (Dutton and Land, 1985), and nearly 5000 m in Bangladesh (Dutta and Suttner, 1987). In uplifted sequences, the ubiquitous presence of fractures and faults would further increase depth of penetration of meteoric waters.

As percolating acidic waters emerge from the soil zone, they may quickly become saturated and lose their aggressiveness within a few decimeters of the soil zone (James and Choquette, 1984). However, a continued supply of fresh meteoric water increases the effective depth of leaching. Bloom (1978) suggested a maximum depth limit of 1000 m for chemical weathering. In unsaturated or vadose zones, meteoric water may be aggressive to depths of 100 m or so (Thrailkill, 1968). Even beneath flat surfaces with only moderate rainfall, the zone of weathering (that is, the zone beneath the B soil horizon) may be as much as 100 m thick (Rose et al., 1979). Thus, it is not surprising that high porosity values (more than 20%) have been observed in sandstones that occur as much as 100 m from an overlying unconformity in sandstones, such as the Ivishak (Alaska) and Brent (North Sea).

Fractures

Fractures and faults, ranging in length from a few millimeters to thousands of meters, and oriented both vertically and horizontally, are characteristic of tectonically uplifted sandstones. In addition to tectonic fractures, near-surface sheeting joints may develop upon unloading during mechanical weathering (Bloom, 1978). The importance of vertical fractures to regolith weathering is that they can increase vertical hydraulic conductivity. Fractures, for example, often play a major role in developing karst topography by serving as conduits for meteoric waters. Fractures and faults tend to create hydrologically open systems.

Topography

Rocks in topographically high and well-drained areas are more prone to leaching than sandstones in topographic lows. The depth of the weathered zone in the tropics is commonly equal to the local relief on hills. Thus the weathered zone may be quite thick and irregular, rather than a uniform thin rind (Bloom, 1978).

Climate

Climatic conditions have a large effect on the extent of chemical weathering in uplifted, subaerially exposed rocks. A preliminary study of selected sandstone and carbonate units from both modern and ancient settings suggests that a warm and humid climate is essential for developing dissolution features (Table 1). In humid regions, an abundant supply of meteoric water facilitates a net movement of fluid through the exposed sandstone; this is a necessary condition for pervasive dissolution. Under warm and humid climatic conditions, the rate of chemical reactions is greatly increased by increased temperature. Furthermore, the lush vegetative cover of warm, humid regions promotes the production of inorganic and organic acids and thereby enhances soil development. Chemical weathering is maximized in the tropics as a product of high meteoric water availability, high temperature, and a high biological activity (Bloom, 1978). Soil zones in warm and humid regions are thick, acidic, and devoid of most chemically unstable minerals because of extensive dissolution. By contrast, soil zones in arid regions are thin, alkaline, and may be rich in unstable minerals in the absence of extensive dissolution. Most examples discussed in this paper are from regions with a warm and humid climate at the time of subaerial exposure.

Framework Composition

Framework composition is an important controlling factor on dissolution. For example, sandstones enriched in unstable rock fragments are more prone to dissolution than sandstones enriched in stable quartz grains. However, other factors such as duration of exposure may facilitate even dissolution of quartz-rich sandstones.

Duration of Exposure

In order for a zone of porosity enhancement to be developed at an unconformity, the favorable conditions discussed above must persist for a considerable length of time (probably millions of years). Erosional unconformities that represent only a short period of subaerial exposure may not be associated with significant zones of weathering.

Repeated flushings by rainwater tend to remove the soluble constituents at the mineral surfaces and transport them downward through the weathering zone. Given sufficient rainfall, permeability, and time, even the most stable minerals, such as quartz, can be destroyed (Loughnan, 1969). Thus, not surprisingly, karst facies, which are characteristic of carbonates, have been reported in quartzites from Australia, South Africa, and Venezuela.

SELECTED MODERN EROSIONAL SURFACES

China

Because of ideal leaching conditions, the world's best developed karst is in southern China (Sweeting, 1978). The karst area covers more than 500,000 km^2 (Figure 2). In places, Devonian to Upper Carboniferous carbonates exhibit well-developed tower karsts with cavernous porosity (Figure 3).

The present climate in southern China, south of the Yangtze River (Figure 2), is warm and humid, with little seasonal variation. Temperatures range from 16 to 24°C, and precipitation averages as much as 200 cm/yr. Rain forests are restricted to the southern part of China (Wang, 1961), where soil types are almost exclusively "red and yellow earths," intensely leached, with moderate to high acidities.

Australia

Orthoquartzites are considered to be extremely stable and unaffected by chemical weathering. However, spectacular tower karsts have been reported in Proterozoic and Paleozoic orthoquartzites (Figure 4) in the east Kimberley region of northwestern Australia (Young, 1987). These sandstones are firmly cemented by quartz (orthoquartzites). Extensive dissolution of quartz and the loss of silica have been attributed to acidic, organic-rich solutions percolating through the sandstone. The karst features in sandstones were primarily developed during the late Mesozoic to late Tertiary, when northwestern Australia was under a warm humid climate (Young, 1987). Quartz is inferred to have been dissolved under acidic, near-surface temperature conditions. This is in conflict with the more traditional view that quartz dissolves only under highly alkaline or high-temperature conditions.

Table 1. Relationship between climate and dissolution

Dissolution	Period of Exposure	Climate
Karst in quartzites, southeastern Venezuela, (White et al., 1966)	? to Present	Hot and humid (T: 40–50°C R: 100–750 cm/yr)*
Tower karst in carbonates, southern China (Sweeting, 1978; Williams, 1978)	Tertiary to Present	Warm and humid (T: 16–24°C R: 100–200 cm/yr)*
Tower karst in quartzose sandstones, northwestern Australia (Young, 1987)	Late Mesozoic to Present	Humid tropical (T: 18–35°C R: 65–75 cm/yr)*
Dolomite, feldspar, and rock fragments in Latrobe Group, Gippsland basin, Australia (Bodard et al., 1984)	Oligocene	Temperate humid (T: 13–16°C R: 100–250 cm/yr) (Shanmugam, 1985c)
Feldspar and rock fragments in Sarir Sandstone, Sirte basin, Libya	Albian to Cenomanian	Warm and semihumid (Sanford, 1970; Al-Shaieb et al., 1981)
Chert in Ivishak Formation, Prudhoe Bay field, Alaska	Early Cretaceous (Neocomian)	Warm and humid (Smiley, 1966)
Plagioclase feldspar in Brent Group, Statfjord field, North Sea	Early Cretaceous	Warm and humid (Hallam, 1984)
Potassium feldspar in Halten Group, Haltenbanken area, Norwegian Sea	Early Cretaceous	Warm and humid (Hallam, 1984)
Chert and carbonates in the Devonian, Permian Basin, Texas	Pre-Pennsylvanian	Humid (Mapel et al., 1979); (Dutton and Land, 1985)**
Karst in carbonates, North China Basin (Guangming and Quanheng, 1982)	Precambrian, Ordovician to Carboniferous, Paleogene	Warm and humid (bauxite/laterite deposits)

T: Temperature
R: Rainfall
* Present climate
** Pennsylvanian climate

Venezuela

In the Gran Sabana, southeastern Venezuela, the Precambrian Roraima quartzarenite contains karstic solution features such as small pits, grooves and ridges (White et al., 1966). Large caves (Figure 5), pillars, and tunnels are common in the quartzarenites (S. K. Ghosh, oral communication, 1989). The present climate here is humid and hot, with alternating wet and dry periods. The annual rainfall is as much as 750 cm at 2000 m elevation. The spectacular development of modern karst in the Precambrian Roraima quartzarenites (George, 1989) attests to quartz dissolution in outcrops by acidic meteoric waters under surface weathering conditions.

South Africa

Another example of karst development is in Precambrian quartzites from the eastern Transvaal, South Africa (Martini, 1981). These quartzites are characterized by large dolines, pinnacles, and caves. The eastern Transvaal region has a hot and cool subtropical climate with heavy rainfall.

POROSITY ENHANCEMENT BENEATH ANCIENT EROSIONAL SURFACES

Many hydrocarbon reservoirs throughout the world exhibit excellent reservoir quality beneath erosional unconformities (Table 2). Selected examples are discussed below.

Alaskan North Slope

In the Prudhoe Bay field, North Slope, Alaska, the Triassic reservoir (Ivishak Formation of the Sadlerochit Group) is truncated by the Lower Cretaceous (Neocomian) unconformity. The Ivishak Formation is composed of a lower deltaic facies (zone 1) and an upper braided fluvial channel facies (zones 2, 3, and 4). Chert is the dominant framework constituent (as much as 70%) in the hydrocarbon-producing fluvial facies but is uncommon in the deltaic facies. Secondary porosity caused by dissolution of framework chert is common in the fluvial facies (Shan-

Figure 2. Karst region of southern China (from Williams, 1978).

mugam and Higgins, 1988). Petrographic evidence for chert dissolution includes: (1) chert weathering rims, (2) completely weathered microporous (tripolitic) chert, (3) clay rims outlining shapes of dissolved chert grains (Figure 6), (4) oversized pores, (5) elongate pores, (6) corroded chert grains, and (7) rounded crystals of microcrystalline quartz in chert.

The dissolution of chert in the Ivishak Formation is primarily attributed to percolating meteoric waters from the Lower Cretaceous (Neocomian) unconformity surface (Shanmugam and Higgins, 1988). Warm and humid climate of present southern China, where karstic solution features are abundant, is considered to be a modern analog of Lower Cretaceous (Neocomian) climate in Alaska. Because of unconformity-related chert dissolution, average core porosity values in the Ivishak increase with increasing stratigraphic proximity to the Lower Cretaceous (Neocomian) unconformity (Figure 7). It must be noted that highest porosity (>25%) occurs in sandstones within 100 m of the unconformity. This is consistent with depths of chemical weathering discussed earlier. It is also important to point out that the porosity increase in the Ivishak Formation occurs within the fluvial facies (zone 4), which has a uniform framework composition, and therefore is not controlled by a difference in depositional facies. Chert dissolution is the major cause of porosity development within fluvial facies, and thus I attribute the increase of porosity toward the unconformity within the fluvial facies to unconformity-related chert dissolution.

Chert dissolution has also been observed in the Mississippian Kekiktuk Conglomerate in the Alaskan North Slope. As in the Ivishak Formation, weathering of chert in the Kekiktuk has also resulted in rounded crystals of microcrystalline quartz (Figure 8). More importantly, dissolution holes have been observed within quartz grains (Figure 8B). The Lower Cretaceous (Neocomian) unconformity is also considered to be responsible for silica dissolution in the Kekiktuk.

Norwegian Sea

Dissolution of feldspar grains can be seen in rocks of the Jurassic Halten Group in the Haltenbanken area of the Norwegian Sea (Figure 9). An important consequence of feldspar dissolution in the Halten Group is the development of high-permeability channels by connecting adjacent pores. These dissolution channels are considered to be responsible for the high permeability measured in many reservoir sandstones (see Table 2). One of the diagnostic diagenetic features in the Halten Group is dissolution of garnet grains (Figure 10). Garnet is stable under deep burial conditions but is unstable under near-surface weathering conditions (Morton, 1984). Dissolution of garnet takes place when fluids are

Figure 3. Development of large caves in Devonian to Upper Carboniferous carbonates, which also form tower karsts near Guilin, southern China. Note man (arrow) standing inside cave.

acidic (Morton, 1984); such conditions would be expected during subaerial exposure. It is proposed that dissolution of feldspar and garnet in the Halten Group was caused by percolating meteoric waters from the Base-Cretaceous unconformity.

The Netherlands

Triassic reservoirs in the Netherlands produce gas in the DeWijk field, where these reservoirs are in contact with the Base-Cretaceous Cimmerian unconformity (Gdula, 1983). Nowhere else in the Netherlands are any of the Triassic reservoirs gas productive (Gdula, 1983). Dissolution of anhydrite cement improved the reservoir quality immediately below the unconformity. Sonic logs show that porosity increases toward the Base-Cretaceous unconformity (Figure 11).

Algeria

In the Hassi Messaoud field of Algeria, the producing reservoir, the Cambrian "Ra" sandstone, directly underlies a post-Hercynian unconformity. In this sequence, permeability increases abruptly below the unconformity. Kaolinite content decreases with increasing depth below the unconformity (Balducchi and Pommier, 1970). This trend has been attributed to the dissolution of feldspar and related precipitation of kaolinite by percolating meteoric waters from the unconformity surface.

Libya

The Augila field in Libya produces oil from fractured and weathered Precambrian granitic basement rocks (Williams, 1972). Reservoir rocks, including weathered granite, granophyre, and devitrified rhyolite, occur beneath a major erosional unconformity. Erosion of the Precambrian basement occurred during Early Paleozoic to Late Cretaceous time. The highest porosities typically occur on the tops of horst blocks, as suggested by higher production of oil from wells located on paleohighs (7627 bbl/day); as compared with production on paleolows (1200 bbl/day). Differential weathering is also indicated by the occurrence of fresh granite in lower areas and weathered granite in higher areas.

Table 2. Reservoir quality of selected sandstones beneath erosional unconformities (Shanmugam, 1988)

Sandstone	Maximum Core Permeability (md)	Dissolved Constituents (secondary porosity)
Latrobe, Cret. - Eocene, Gippsland basin, Australia (Bodard et al., 1984)	3000	Dolomite, feldspar, rock fragments
Sarir, Cretaceous, Sirte basin, Libya (Sanford, 1970; Hea, 1971; Al-Shaieb et al., 1981)	3000	Quartz, feldspar, rock fragments
Halten, Jurassic, Norwegian Sea	22,900	Feldspar, garnet
Brent, Jurassic Statfjord field, North Sea (Sommer, 1978)	6400	Feldspar
Lunde, Triassic, North Sea	8600	Feldspar
Ivishak, Triassic, Prudhoe Bay, Alaska	3000	Chert, siderite
Triassic, DeWijk gas field, The Netherlands (Gdula, 1983)	100	Anhydrite
Mesozoic and Paleozoic, Buried Hills pools, North China Basin (Guangming and Quanheng, 1982)	No data (cavernous porosity in karst facies)	Carbonates
Minnelusa, Permian, Powder River basin, Wyoming and Montana	3200	Anhydrite, carbonates
Kekiktuk, Mississippian, Mikkelsen Bay, Alaska	12,800	Chert, quartz
Devonian, Ector, Crane, and Pecos counties, Permian Basin, Texas (David, 1946; Hanson, 1985)	200	Chert, carbonates

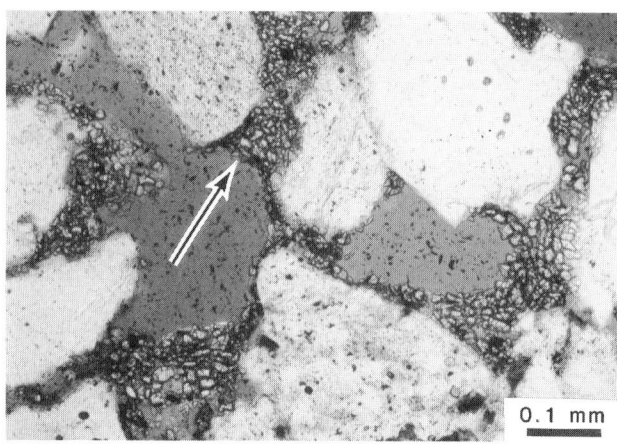

Figure 6. Thin-section photomicrograph showing secondary macroporosity caused by complete dissolution of chert grains; plain light. Note clay rim (arrow) outlining shape of dissolved chert grain. Triassic Ivishak Formation, subsurface sample, Alaskan North Slope.

organic maturation. The increase in porosity toward the unconformity, however, suggests that the unconformity may have played a major role in porosity enhancement. During the Eocene and Oligocene, this area was under a temperate humid climate (Shanmugam, 1985c).

Venezuela

In the B-6 Sands (Eocene) of the Mosoa Formation, Maracaibo basin, the Oligocene regional uplift created extensive secondary porosity through dissolution of framework grains and cements by undersaturated meteoric waters in the vicinity of the truncation surface (Ghosh et al., 1989). Similar to outcrop dissolution of Roraima quartzarenites, these sands exhibit evidence for profuse leaching of quartz. Ghosh et al. (1989) reported that the sands immediately beneath the unconformity have the best reservoir quality.

Texas

The pre-Pennsylvanian unconformity marks a major erosional surface in the Permian basin, Texas. This regional unconformity separates the Devonian chert (marine facies) from the overlying Pennsylvanian chert-pebble conglomerate and sandstone (alluvial plain facies). The Devonian chert is characterized by collapse breccias (Figure 13), which have been interpreted to be products of karst-related processes during subaerial exposure of the Devonian chert.

POROSITY PREDICTION

Norwegian North Sea

In the Statfjord field (Figure 14) and the surrounding area of the Norwegian North Sea, Jurassic and Triassic sandstones occur beneath the Base-Cretaceous unconformity (Figure 15). These feldspathic sandstones exhibit secondary porosity caused by dissolution of feldspar (mainly plagioclase) grains (Figure 16). The dissolution of feldspars in these

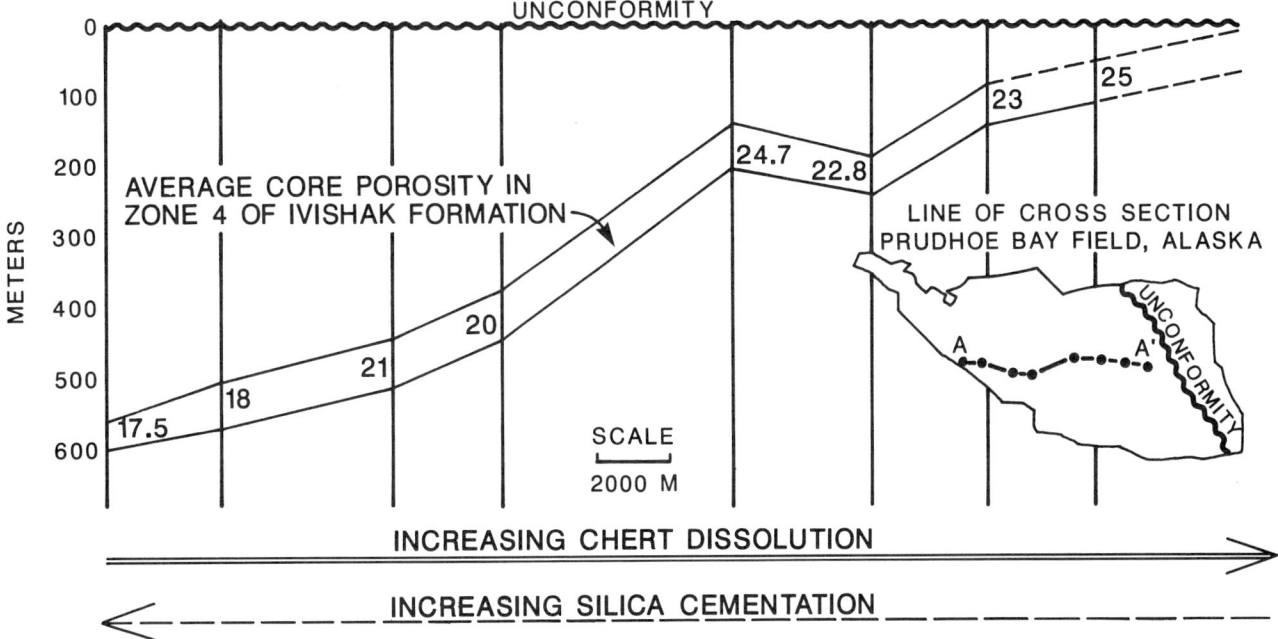

Figure 7. Increase of average core porosity in zone 4 (fluvial facies) of the Triassic Ivishak Formation with increasing stratigraphic proximity to Lower Cretaceous (Neocomian) unconformity (from Shanmugam and Higgins, 1988). Vertical lines represent well locations from which core samples were taken.

reservoirs has been attributed to incursion of meteoric waters during the Early Cretaceous uplift (Hancock and Taylor, 1978; Sommer, 1978; Selley, 1984; Shanmugam, 1988). A major phase of uplift and erosion occurred across Europe during Volgian to Ryazanian time, resulting in the "Late Cimmerian unconformity" (Rawson and Riley, 1982), which is also known as the Base-Cretaceous unconformity (Ziegler, 1975).

Some authors (Bjorlykke, 1983; Bjorlykke and Brendsdal, 1986) proposed two periods of feldspar dissolution by meteoric waters in the Brent Group—initially during Middle Jurassic deposition, and later during Early Cretaceous uplift and erosion. Petrographic evidence and core porosity trends, however, suggest that feldspar dissolution primarily occurred during the uplift. First, the amount of feldspar dissolution increases toward the unconformity, and second, the core porosity values show a general increase toward the unconformity, as illustrated in a well in Figure 17. Core and log porosity values from several wells show a dramatic increase in porosity of the Brent Group toward the unconformity (Figure 18). The role of burial depth on porosity, however, is less striking in the Brent Group (Figure 19).

Kaolinite is a common product of feldspar dissolution along unconformities (Al-Gailani, 1981). Some investigators believe that feldspar dissolution does not enhance reservoir quality, because aluminum is relatively immobile and so the products of feldspar dissolution will invariably precipitate in nearby pores as kaolinite (Thomson and Stoessell, 1985). In the Statfjord field, however, kaolinite is not abundant in zones of extensive feldspar dissolution (Figure 16), suggesting that most dissolved aluminum was transported away from the site of feldspar dissolution (Akselsen, 1985). In other words, there is a net loss of dissolved constituents locally within a sandstone. The dissolved constituents may, however, precipitate some distance away within the same sandstone (Selley, 1984), or they may be lost to adjacent stratigraphic units through fractures and faults.

An empirical model has been developed for predicting porosity in the Triassic and Jurassic reservoirs of the Norwegian North Sea (Figure 20). This model shows that sandstones increase in porosity with increasing proximity to the unconformity (Figures 17, 18, 19). By the empirical model, the Middle Jurassic Brent Group is expected to have porosity values commonly in the range of 25 to 30% in areas where the Base-Cretaceous unconformity is in contact with the Brent Group (Figure 20, second column from left). Seismic reflection profiles across Block 34/8 show areas where the Brent Group is in contact with the Base-Cretaceous unconformity, and these areas of potential porosity development are shown on a subcrop map (Figure 21). The empirical model for the region (Figure 20) and the Base-

Figure 8. SEM photographs showing rounded crystals of microcrystalline quartz in chert grains (arrows in A and B), and a dissolution hole in a quartz grain (B, lower right), Mississippian Kekiktuk Conglomerate, subsurface sample, Alaskan North Slope.

Figure 9. SEM photograph showing etched feldspar, Jurassic Halten Group, Haltenbanken area, Norwegian Sea.

Figure 10. SEM photograph showing dissolved garnet (almandine), Jurassic Halten Group, Haltenbanken area, Norwegian Sea.

Cretaceous subcrop map (Figure 21) have been used to predict that the Brent porosity should range from 25 to 30% in areas where the Brent is in contact with the unconformity in Block 34/8 (Figure 21).

The predictive model has been successfully tested for the Triassic Lunde Formation in Block 34/7 (Figure 22). Core porosities (18–26%) of the Lunde Formation from a well drilled in 1984 were found to be in close agreement with porosity values (15–20%) that were predicted in 1983 using an earlier empirical model for the region.

CAUTION

Although porosity enhancement beneath erosional unconformities is common worldwide, porosity reduction due to cementation can also occur beneath erosional unconformities. An example of cementation (dolomite) beneath an unconformity is found in the Parkman field, Saskatchewan (Miller, 1972). Therefore, caution must be exercised in recognizing these exceptions when developing predictive models by using erosional unconformities.

ACKNOWLEDGMENTS

I am thankful to Indu Meshri for inviting me to submit this paper. I wish to thank L. B. Van Ingen, III, T. B. Christiansen, J. R. Kusiak, J. B. Higgins, M. Venuto, D. Stoneman, and C. M. Wall for their help during this study (1980-1987); R. W. Scott, P. D. Wagner, J. G. McPherson, and D. M. Summers for a critical review of the manuscript; E. L. Jones and R. J. Moiola for managerial support; R. W. Young and S. K. Ghosh for providing photographs of karst in sandstones; and Mobil Research and Development Corporation for granting permission to publish this paper.

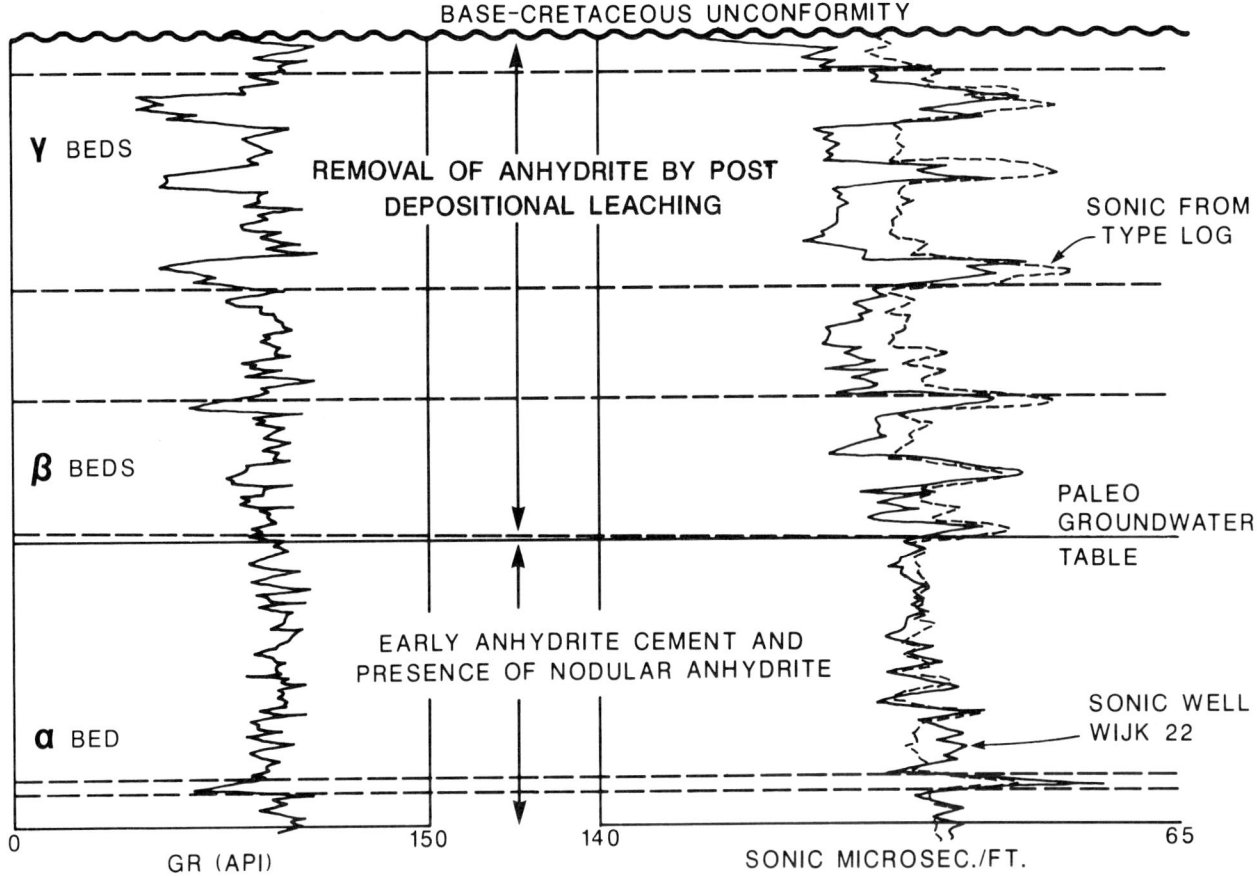

Figure 11. Sonic logs showing anhydrite dissolution in Triassic reservoir beneath Base-Cretaceous unconformity, DeWijk gas field, The Netherlands (Gdula, 1983). Reprinted by permission of Kluwer Academic Publishers.

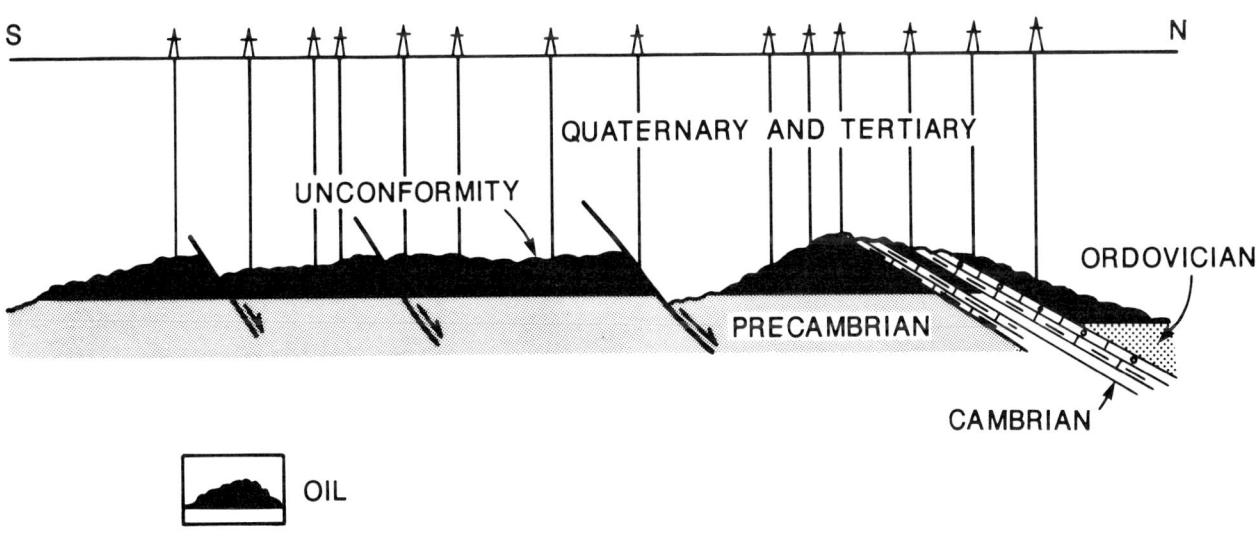

Figure 12. Entrapment of oil in karst facies with large dissolution pores and fractures in Precambrian reservoirs beneath a major erosional unconformity, Renqiu oil field, North China basin. After Guangming and Quanheng (1982).

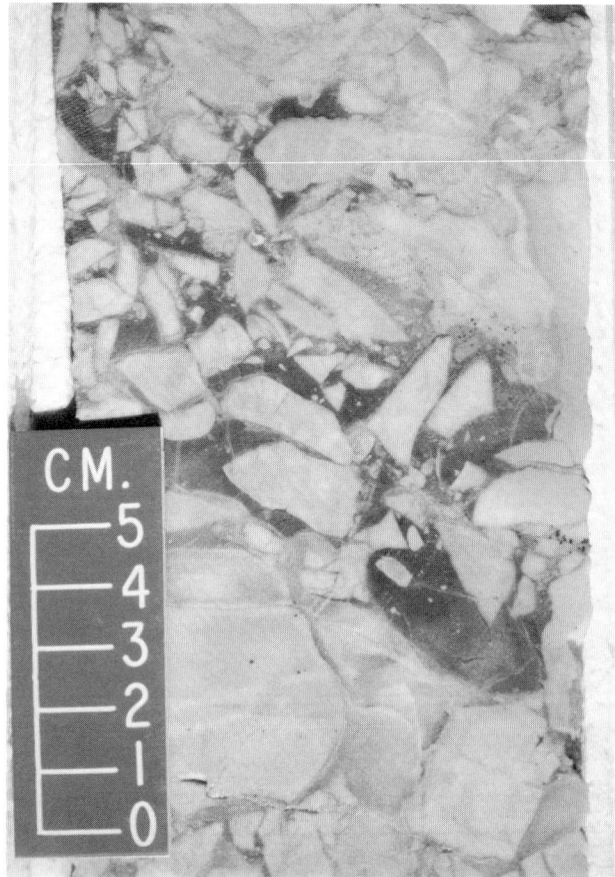

Figure 13. Collapse breccia in Devonian chert beneath the pre-Pennsylvanian unconformity, Permian Basin, Crane County, Texas; subsurface sample at 5671 ft.

REFERENCES CITED

Akselsen, J., 1985, Kaolinite and illite authigenesis control led by fluid flow rates, pore water composition and temperature (abs.): SEPM Annual Midyear Meeting Abstracts, p. 3-4.

Al-Gailani, M. B., 1981, Authigenic mineralizations at unconformities: implication for reservoir characteristics: Sedimentary Geology, v. 29, p. 89-115.

Al-Shaieb, Z., W. C. Ward, and J. W. Shelton, 1981, Diagenesis and secondary porosity evolution of Sarir sandstone, southeastern Sirte basin, Libya (abs.): AAPG Bulletin, v. 65, p. 889-890.

Balducchi, A., and G. Pommier, 1970, Cambrian oil field of Hassi Messaoud, Algeria: AAPG Memoir 14, p. 477-488.

Bennett, P., and D. I. Siegel, 1987, Increased solubility of quartz in water due to complexing by organic compounds: Nature, v. 326, p. 684-686.

Berner, R. A., 1978, Rate control of mineral dissolution under earth surface conditions: American Journal of Science, v. 278, p. 1235-1252.

Bjorlykke, K., 1983, Diagenetic reactions in sandstones, in A. Parker and B. W. Sellwood, eds., Sediment Diagenesis: Dordrecht, D. Reidel Publishing Company, p. 169-213.

Bjorlykke, K., and A. Brensdal, 1986, Diagenesis of the Brent Sandstone in the Statfjord field, North Sea, in D. L. Gautier, ed., Roles of organic matter in sediment diagenesis: SEPM Special Publication 38, p. 157-167.

Bloom, A. L., 1978, Geomorphology: Englewood Cliffs, New Jersey, Prentice-Hall, Inc., 510 p.

Bodard, J. M., V. J. Wall, and R. A. F. Cas, 1984, Diagenesis and the evolution of Gippsland basin reservoirs: Australian Petroleum Exploration Association Journal, v. 24, pt. t, p. 314-335.

David, M., 1946, Devonian (?) producing zone, TXL pool, Ector County, Texas: AAPG Bulletin, v. 30, p. 118-119.

Dutta, P. K., and L. J. Suttner, 1987, Alluvial sandstone composition and paleoclimate, II. Authigenic mineralogy—Reply: Journal of Sedimentary Petrology, v. 57, p. 576.

Dutton, S. P., and L. S. Land, 1985, Meteoric burial diagenesis of Pennsylvanian arkosic sandstones, southwestern Anadarko basin, Texas: AAPG Bulletin, v. 69, p. 22-38.

Freeze, R. A., and J. A. Cherry, 1979, Groundwater: Englewood Cliffs, New Jersey, Prentice Hall, Inc., 604 p.

Galloway, W. E., and D. K. Hobday, 1983, Terrigenous clastic depositional systems: New York, Springer-Verlag, 423 p.

Gdula, J. E., 1983, Reservoir geology, structural framework, and petrophysical aspects of the DeWijk gas field: Geologie en Mijnbouw, v. 62, p. 191-202.

George, U., 1989, Venezuela's islands in time: National Geographic, v. 175, no. 5 (May), p. 526-561.

Ghosh, S. K., J. D. Croce, and A. Isea, 1989, Impact of freshwater diagenesis on reservoir quality: B-6 Sands, Mosoa Formation, Venezuela (abs.): AAPG Bulletin, v. 73, p. 358.

Giles, M. R., and J. D. Marshall, 1986, Constraints on the development of secondary porosity in the subsurface: reevaluation of processes: Marine and Petroleum Geology, v. 3, p. 243-255.

Guangming, Z., and Z. Quanheng, 1982, Buried-hill oil and gas pools in the North China basin, in M. T. Halbouty, ed., The deliberate search for the subtle trap: AAPG Memoir 32, p. 317-335.

Hallam, A., 1984, Continental humid and arid zones during the Jurassic and Cretaceous: Paleogeography, Paleoclimatology, and Paleoecology, v. 47, p. 195-223.

Hancock, N. J., and A. M. Taylor, 1978, Clay mineral diagenesis and oil migration in the Middle Jurassic Brent Sand Formation: Journal of the Geological Society of London, v. 135, p. 69-72.

Hanson, B. M., 1985, Truncated Devonian and Fusselman fields and their relationship to the Permian basin reserve: Trans. Southwest Section AAPG 1985 Convention, Fort Worth, Texas, p. 132.

Hea, J. P., 1971, Petrography of the Paleozoic-Mesozoic sandstones of the southern Sirte basin, Libya, in C. Gray, ed., Symposium on the geology of Libya: University of Libya, Tripoli, p. 100-125.

Hrabar, S. V., and P. E. Potter, 1969, Lower West Baden (Mississippian) Sandstone body of Owen and Greene counties, Indiana: AAPG Bulletin, v. 53, p. 2150-2160.

Huang, W. H., and W. D. Keller, 1970, Dissolution of rock-forming silicate minerals in organic acids: simulated first-stage weathering of fresh mineral surfaces: American Mineralogist, v. 55, p. 2076-2094.

Hutton, J., 1788, Theory of the Earth, or an investigation of the laws observable in the composition, dissolution and restoration of land upon the globe: Royal Society of Edinburgh Transactions, v. 1, p. 109-304.

James, N. P., and P. N. Choquette, 1984, Diagenesis 9—Limestones—The meteoric diagenetic environment: Geoscience Canada, v. 11, p. 161-194.

Keller, W. D., 1955, The principles of chemical weathering: Columbia, Missouri, Lucas Brothers Publishers, 88 p.

Kononova, M. M., T. Z. Nowakowski, and A. C. D. Newman, 1966, Soil Organic Matter: New York, Pergamon Press, 523 p.

Levorsen, A. I., 1934, Relation of oil and gas pools to unconformities in the mid-continent region, in W. E. Wrather and F. H. Lahee, eds., Problems of petroleum geology: AAPG Sidney Powers Memorial Volume, p. 761-784.

Loughnan, F. C., 1969, Chemical weathering of the silicate minerals: New York, Elsevier, 154 p.

Mapel, W. J., et al., 1979, Southern midcontinent and southern Rocky Mountain region: USGS Professional Paper 1010-J, p. 161-187.

Martini, J., 1981, The control of karst development with reference to the formation of caves in poorly soluble rocks in the eastern Transvaal, South Africa: Eighth International Congress of Speleology Proceedings, v. 1, p. 4-5.

Miller, E. G., 1972, Parkman field, Williston basin, Saskatchewan, in R. E. King, ed., Stratigraphic oil and gas fields-classification, exploration methods, and case histories: AAPG Memoir 16, p. 502-510.

Morton, A. C., 1984, Stability of detrital heavy minerals in Tertiary

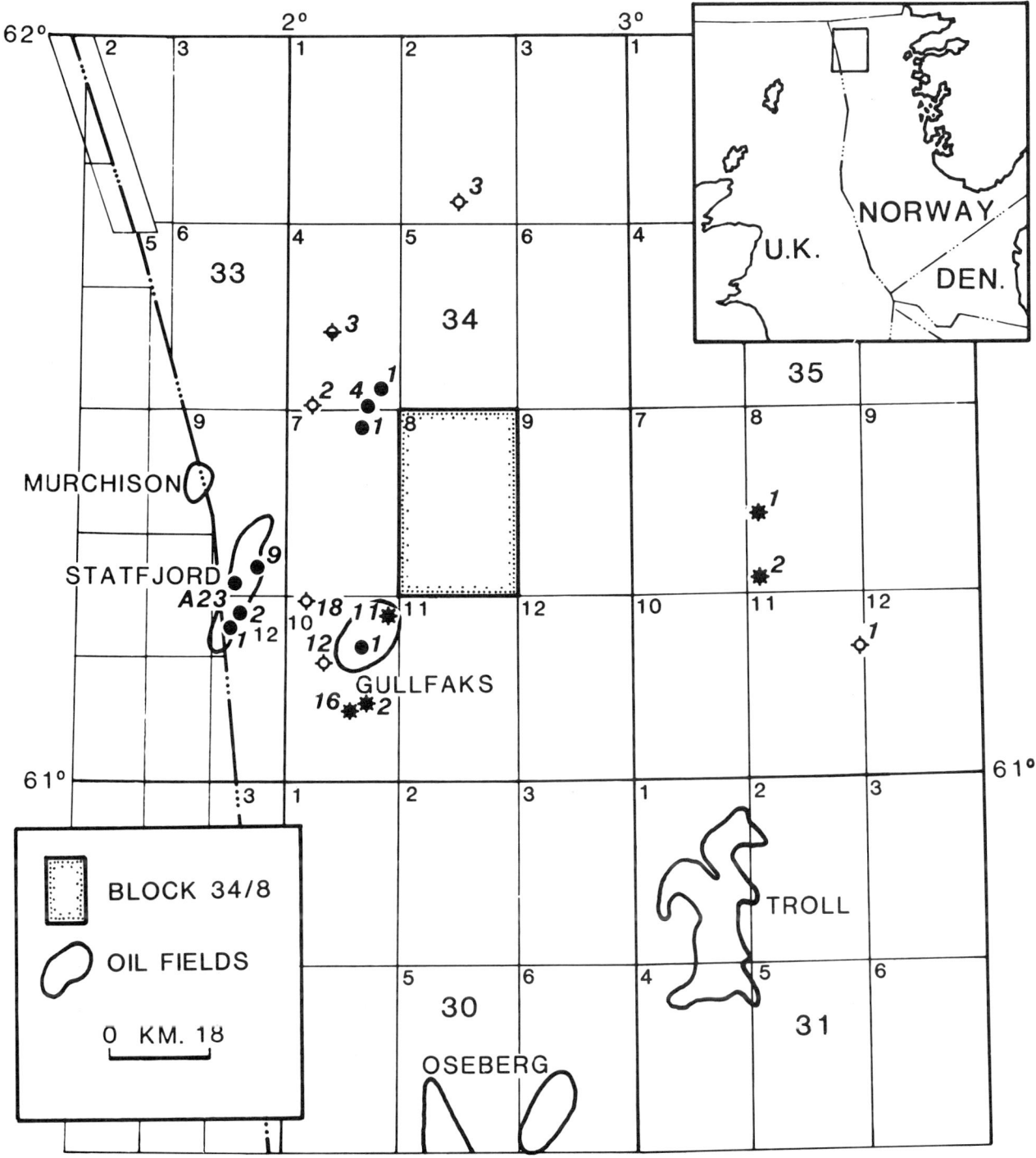

Figure 14. Location map showing wells used in this study, Norwegian North Sea. Reservoirs in Blocks 34/7 and 34/8 were used in porosity predictions.

sandstones from the North Sea Basin: Clay Minerals, v. 19, p. 287-308.

Rawson, P. F., and L. A. Riley, 1982, Late Jurassic-Early Cretaceous events and the "Late Cimmerian unconformity" in North Sea area: AAPG Bulletin, v. 66, p. 2628-2648.

Rose, A. W., H. E. Hawkes, and J. S. Webb, 1979, Geochemistry in mineral exploration: London, Academic Press, 657 p.

Sanford, R. M., 1970, Sarir oil field—Libya desert surprise: AAPG Memoir 14, p. 449-476.

Schalscha, E. B., Appelt, H., and Schatz, A., 1967, Chelation as a weathering mechanism—I. Effect of complexing agents on the solubilization of iron from minerals and granodiorite: Geochimica et Cosmochimica Acta, v. 31, p. 587-596.

Scherer, M., 1987, Parameters influencing porosity in sandstones: a model for sandstone porosity prediction: AAPG Bulletin, v. 71, p. 485-491.

Selley, R. C., 1984, Porosity evolution of truncation traps: diagenetic models and log responses: Stavanger, Norway,

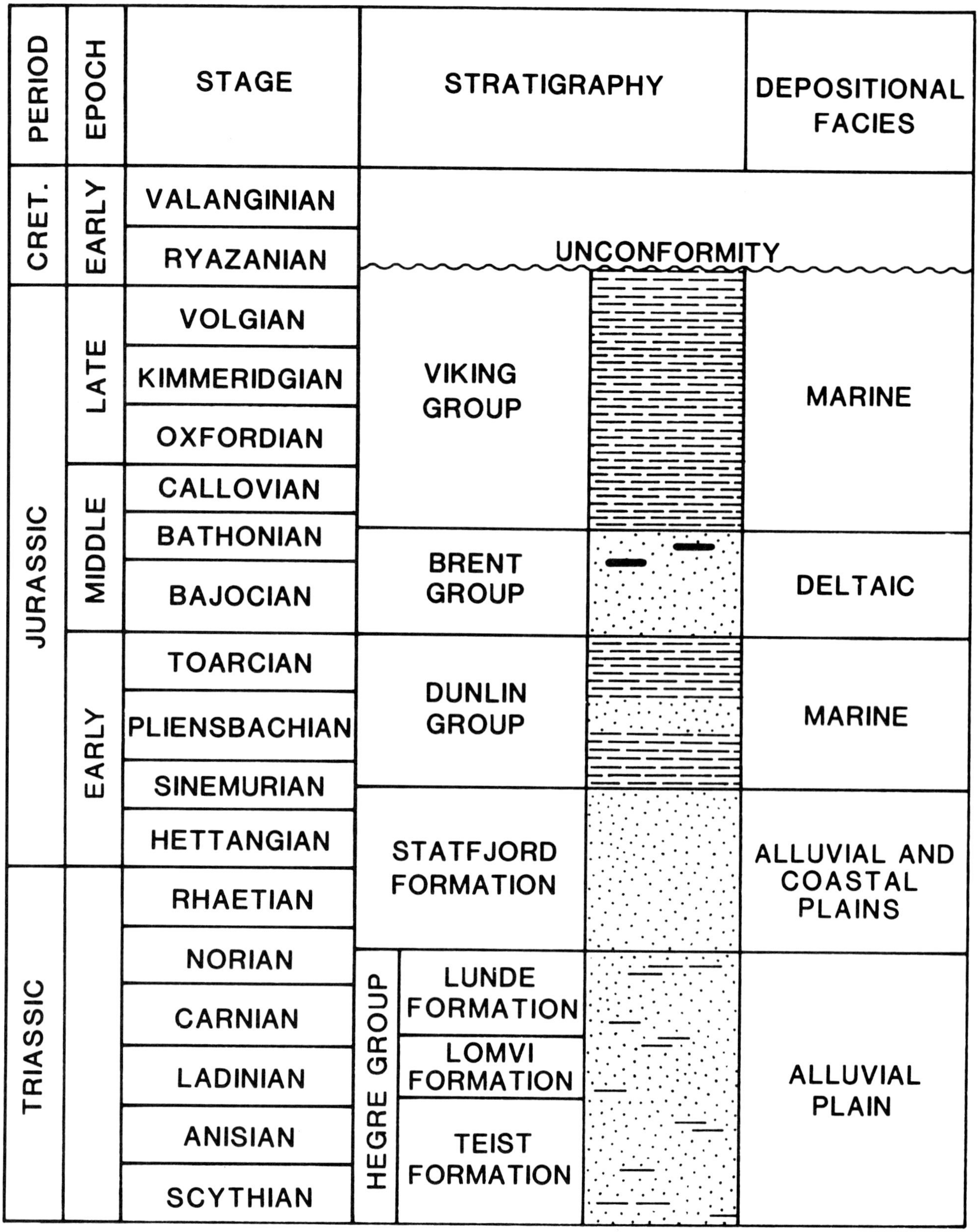

PERIOD	EPOCH	STAGE	STRATIGRAPHY		DEPOSITIONAL FACIES
CRET.	EARLY	VALANGINIAN			
		RYAZANIAN	UNCONFORMITY		
JURASSIC	LATE	VOLGIAN	VIKING GROUP		MARINE
		KIMMERIDGIAN			
		OXFORDIAN			
	MIDDLE	CALLOVIAN	BRENT GROUP		DELTAIC
		BATHONIAN			
		BAJOCIAN			
	EARLY	TOARCIAN	DUNLIN GROUP		MARINE
		PLIENSBACHIAN			
		SINEMURIAN			
		HETTANGIAN	STATFJORD FORMATION		ALLUVIAL AND COASTAL PLAINS
TRIASSIC		RHAETIAN			
		NORIAN	HEGRE GROUP	LUNDE FORMATION	ALLUVIAL PLAIN
		CARNIAN			
		LADINIAN		LOMVI FORMATION	
		ANISIAN		TEIST FORMATION	
		SCYTHIAN			

Figure 15. Generalized stratigraphy, Statfjord field area, Norwegian North Sea. Stratigraphic nomenclature modified after Vollset and Dore (1984).

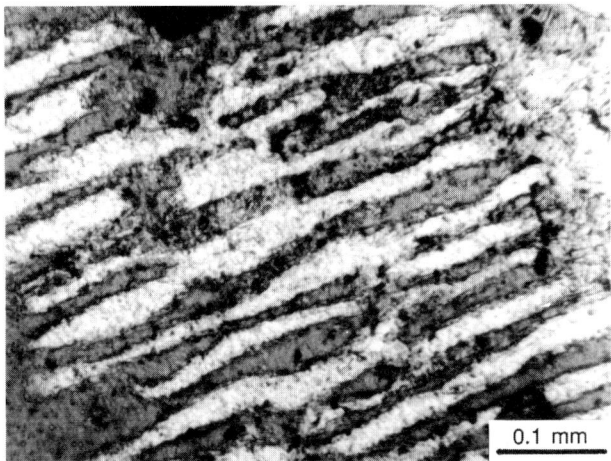

Figure 16. Thin-section photomicrograph showing dissolution of plagioclase feldspar; plain light. Note absence of authigenic clay minerals. Middle Jurassic Brent Group, Norwegian North Sea.

Norwegian Petroleum Society Offshore North Seas Conference, Paper G3, 18 p.
Selley, R. C., 1985, Elements of petroleum geology: New York, W. H. Freeman and Company, 449 p.
Shanmugam, G., 1985a, Significance of secondary porosity in interpreting sandstone composition: AAPG Bulletin, v. 69, p. 378-384.
Shanmugam, G., 1985b, Types of porosity in sandstones and their significance in interpreting provenance, *in* G. G. Zuffa, ed., Provenance of arenites, NATO ASI Series: Dordrecht, D. Reidel Publishing Company, p. 115-137.
Shanmugam, G., 1985c, Significance of coniferous rain forests and related organic matter in generating commercial quantities of oil, Gippsland basin, Australia: AAPG Bulletin, v. 69, p. 1241-1254.
Shanmugam, G., 1988, Origin, recognition and importance of erosional unconformities in sedimentary basins, *in* K. L. Kleinspehn and C. Paola, eds., New perspectives in basin analysis: New York, Springer-Verlag, p. 83-108.
Shanmugam, G., 1989, Porosity development in sandstones beneath erosional unconformities, *in* Search for the subtle trap: West Texas Geological Society Symposium Volume, p. 269-281.
Shanmugam, G., and K. A. Alhilali, 1988, Parameters influencing porosity in sandstones: a model for sandstone porosity prediction: Discussion: AAPG Bulletin, v. 72, p. P,52-853.
Shanmugam, G., and J. B. Higgins, 1988, Porosity enhancement from chert dissolution beneath Neocomian unconformity: Ivishak Formation, North Slope, Alaska: AAPG Bulletin, v. 72, p. 523-535.
Smiley, C. J., 1966, Cretaceous floras from Kuk River area, Alaska: stratigraphic and climatic interpretations: GSA Bulletin, v. 77, p. 1-14.
Sommer, F., 1978, Diagenesis of Jurassic sandstones in the Viking Graben: Journal of the Geological Society of London, v. 135, p. 63-67.
Sweeting, M. M., 1978, The karst of Kweilin, southern China: Geographic Journal, v. 144, p. 199-204.
Thomson, A., and R. K. Stoessell, 1985, Nature of secondary porosity created by dissolution of aluminum silicates (abs.): AAPG Bulletin, v. 69, p. 311.
Thrailkill, J., 1968, Chemical and hydrologic factors in the excavation of limestone caves: GSA Bulletin, v. 79, p. 19-46.
Vollset, J., and A. G. Dore, 1984, A revised Triassic and Jurassic lithostratigraphic nomenclature for the Norwegian North Sea: Norwegian Petroleum Directorate Bulletin 3, 53 p.
Wang, C. W., 1961, The forests of China: Harvard University Press, Maria Moore Cabot Foundation Publication 5, 313 p.
White, W. B., G. L. Jefferson, and J. F. Haman, 1966, Quartzite karst in southeastern Venezuela: International Journal of Speleology, v. 2, p. 309-316.
Williams, J. J., 1972, Augila field, Libya: Depositional environment and diagenesis of sedimentary reservoir and description of igneous reservoir, *in* R. E. King, ed., Stratigraphic oil and gas fields—classification, exploration methods, and case histories: AAPG Memoir 16, p. 623-632.
Williams, P. W., 1978, Karst research in China: British Cave Research Association Transactions, v. 5, p. 29-46.
Winograd, I. J., and F. N. Robertson, 1982, Deeply oxygenated ground water: anomaly or common occurrence?: Science, v. 216, p. 127-130.
Young, R. W., 1987, Sandstone landforms of the tropical east Kimberley region, northwestern Australia: Journal of Geology, v. 95, p. 205-218.
Ziegler, W. H., 1975, Outline of the geological history of the North Sea, *in* A. W. Woodland, ed., Petroleum and the continental shelf of northwest Europe: London, Institute of Petroleum, p. 165-190.

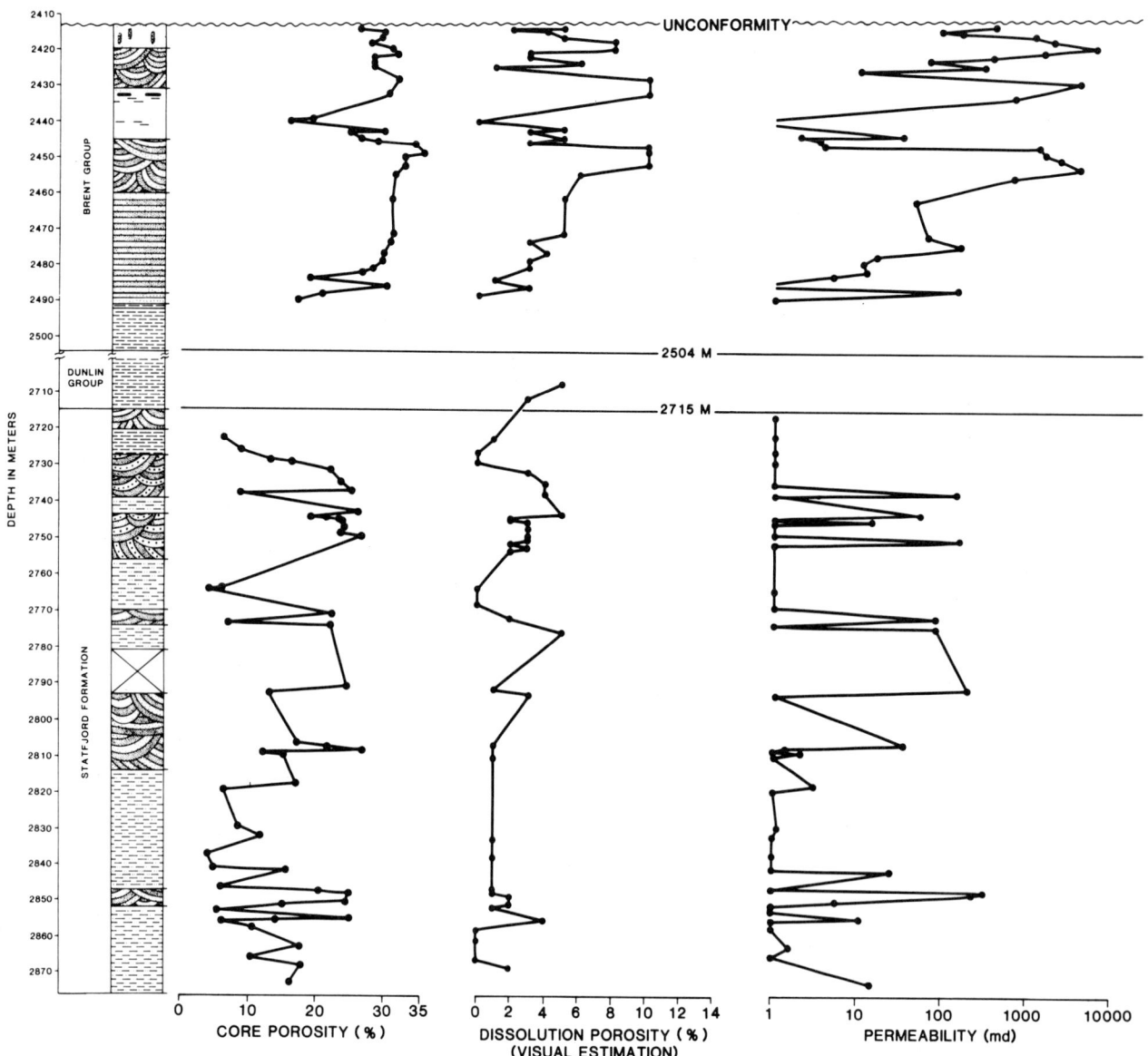

Figure 17. General increase of core porosity, permeability, and dissolution porosity in the Brent Group toward the Base-Cretaceous unconformity. Dissolution porosity represents plagioclase dissolution.

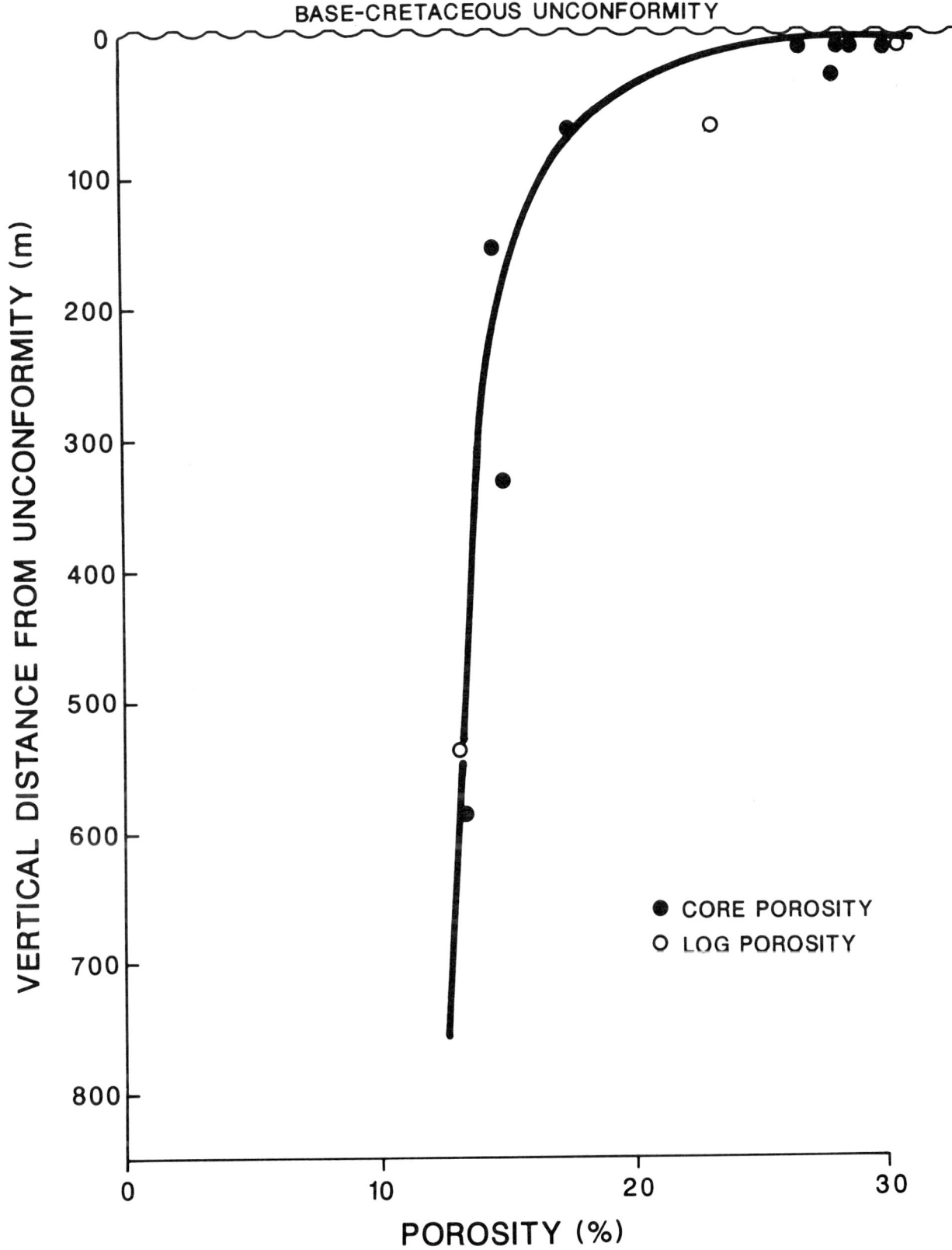

Figure 18. Dramatic increase of core and log porosity in the Brent Group toward the Base-Cretaceous unconformity. Note highest porosity values are in the 100 m immediately beneath the unconformity.

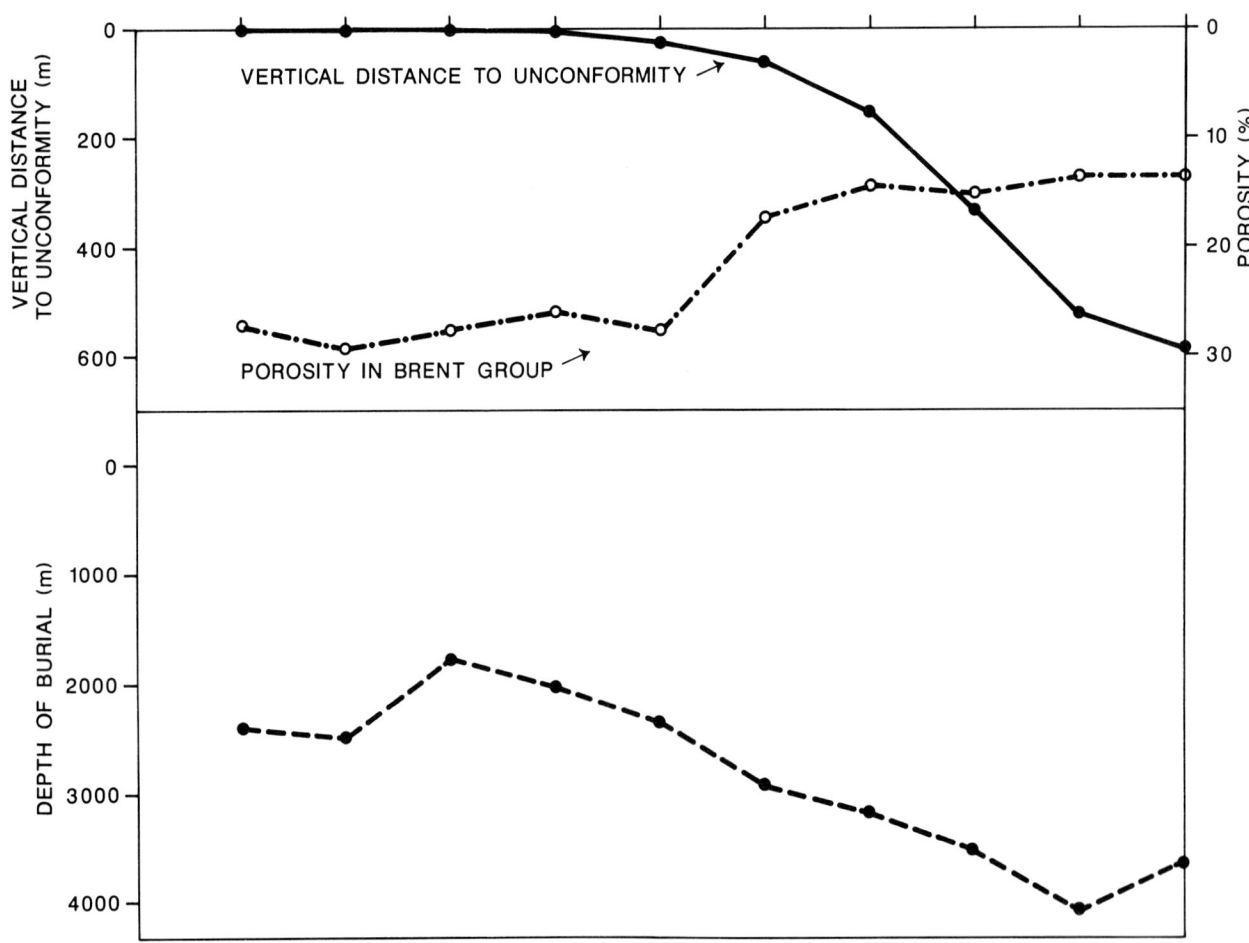

Figure 19. Distribution of porosity in the Brent Group as a function of proximity to unconformity and depth of burial in 10 wells, Statfjord Field area, Norwegian North Sea.

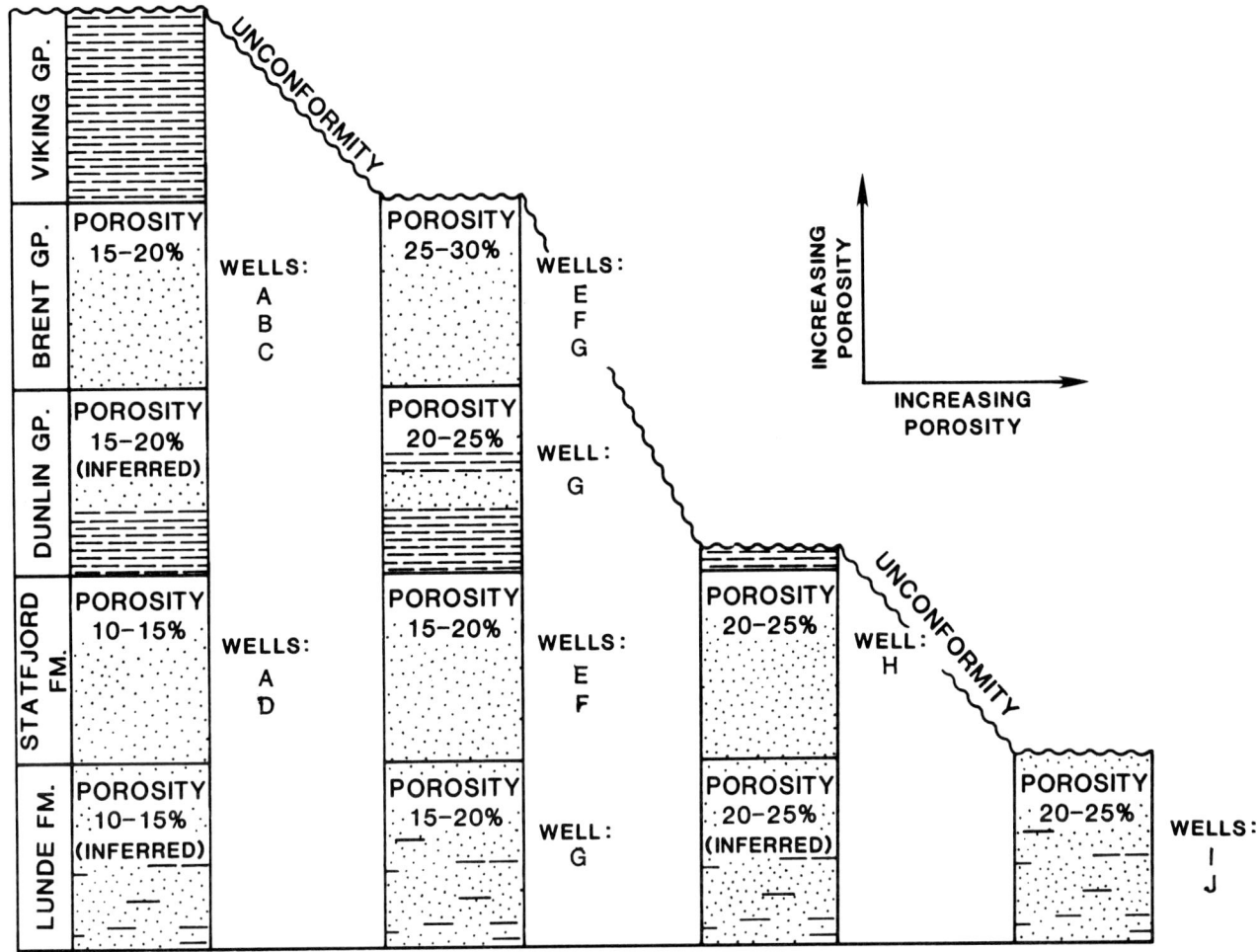

Figure 20. Empirical model showing porosity increase toward the Base-Cretaceous unconformity, Norwegian North Sea. Common range of core porosity is used. Actual well names are not given.

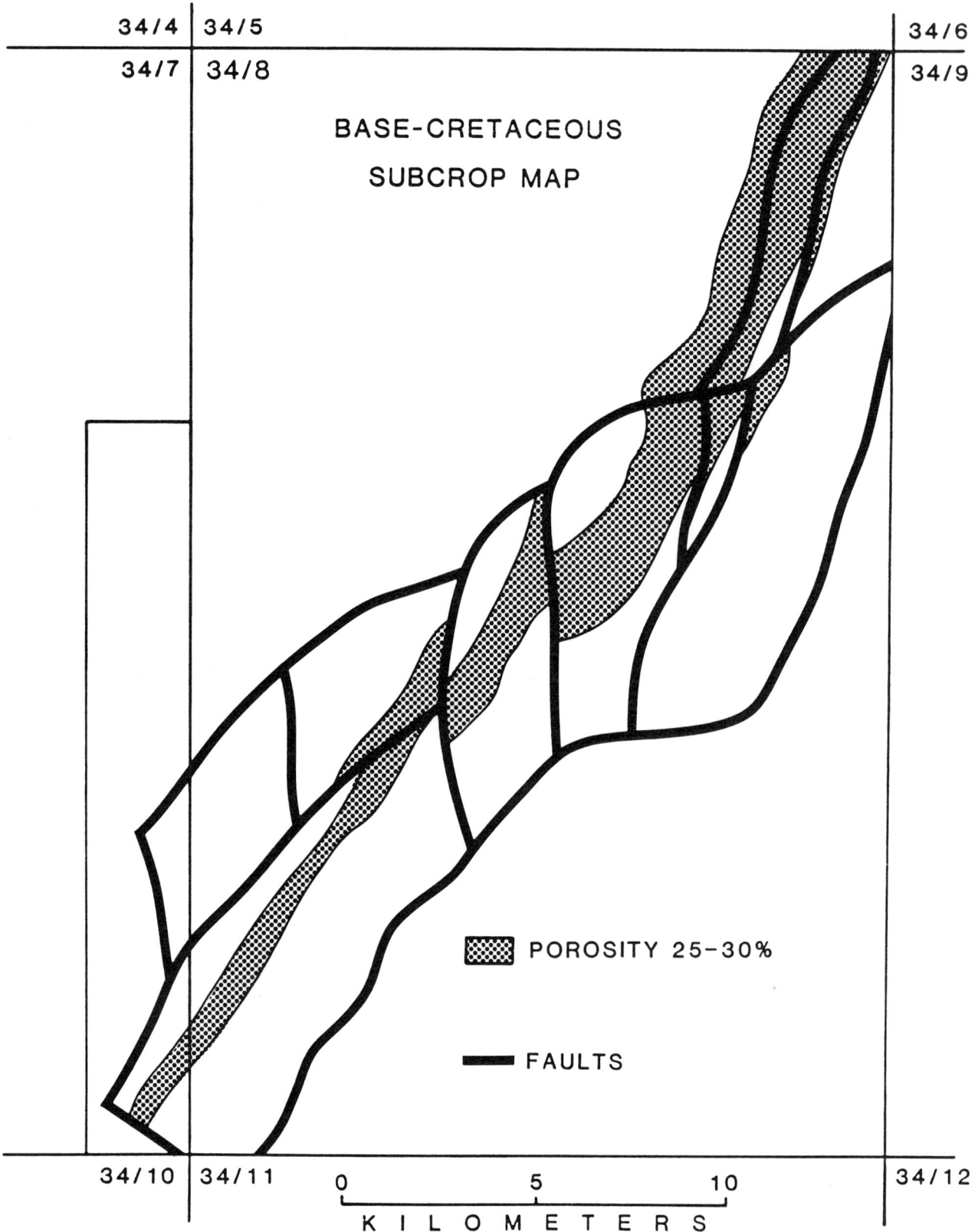

Figure 21. Base-Cretaceous subcrop map showing areas (stippled) where the Middle Jurassic Brent Group is in contact with the unconformity in Block 34/8, Norwegian North Sea. According to the empirical model (Figure 20), the stippled areas should have a porosity range of 25 to 30%. Porosity prediction was made in 1985 before the block was drilled. Subsequently, this block has been drilled, but porosity data are not available to the writer.

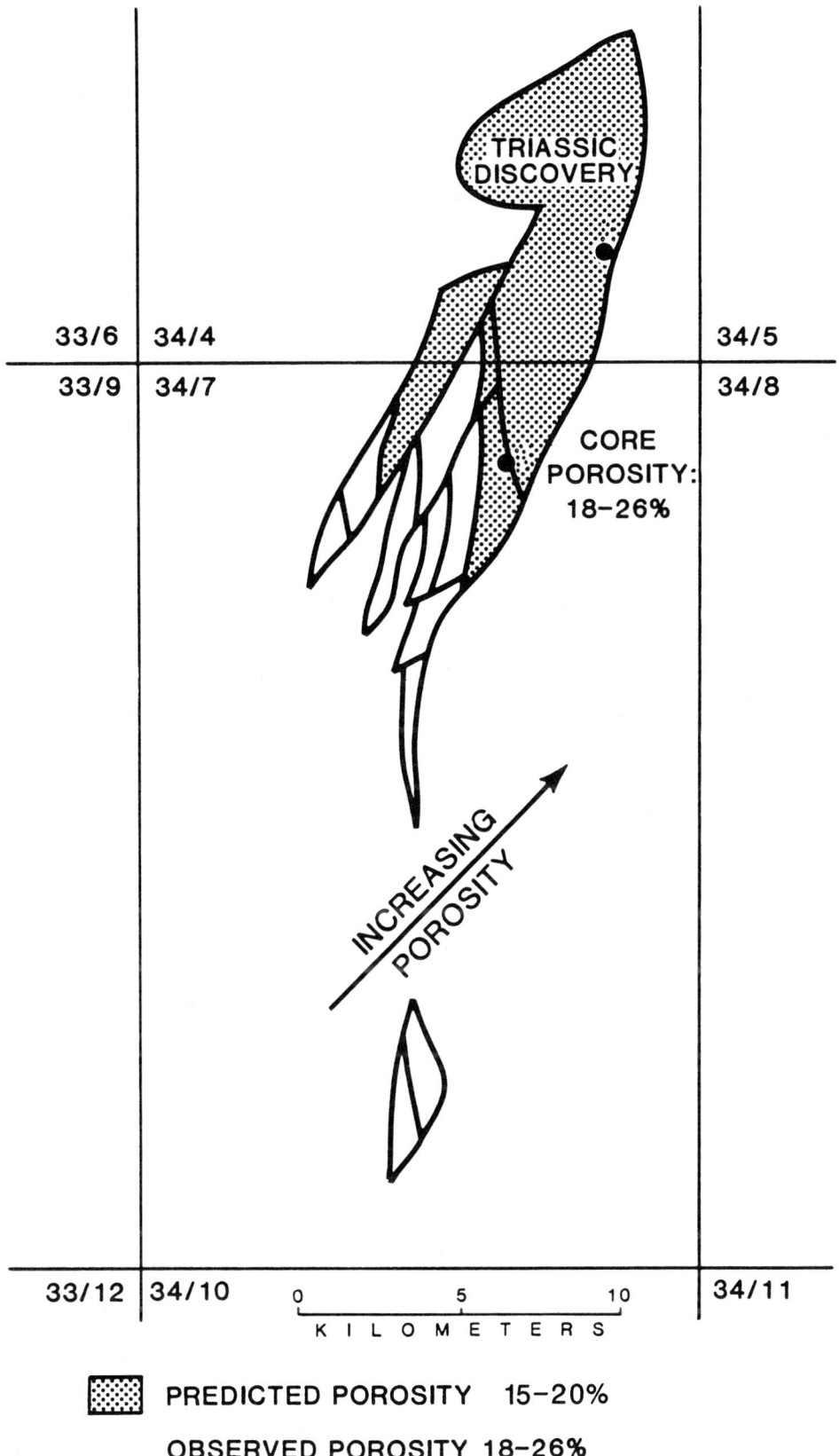

Figure 22. Base-Cretaceous subcrop map of the Triassic Lunde Formation in Block 34/7, Norwegian North Sea. Stippled areas show the regions where the Lunde Formation is in contact with the Base-Cretaceous unconformity. In 1983, predicted porosity range in stippled areas was 15-20%. In 1984, this block was drilled and the observed porosity was 18-26%.

Facies Controls on Early Diagenesis: Wilcox Group, Texas Gulf Coast

S. A. Stonecipher and J. A. May
Marathon Oil Co.
Littleton, Colorado, U.S.A.

In the Paleocene-Eocene Wilcox Group of the Texas Gulf Coast, early diagenetic patterns appear to be related to factors such as original water chemistry, sediment texture, detrital composition, and organic content which were, in turn, related directly or indirectly to depositional environment. Rapid lateral and vertical changes in depositional environments produced markedly different early diagenetic patterns in sand units only a few feet or even inches apart.

One of the most important controls was the chemistry of the original pore waters. Flushing by fresh meteoric water produced early mica-derived kaolinite. Fresh, but anoxic, water resulted in abundant siderite. Saline marine pore waters typically led to the development of chlorite rims. Reducing or mildly reducing marine waters yielded pyrite or glauconite. Mixing of fresh and marine waters resulted in chamosite/berthierine ooids or illite, depending on oxygen content.

These patterns suggest that lateral and vertical variations in diagenetic facies defined on the basis of sediment texture, detrital composition, and early authigenic cements can be used to complement, and test, traditional sedimentologic methods of interpreting depositional environment.

INTRODUCTION

In recent publications, a number of controls on the diagenesis of clastic sediments have been proposed, ranging in scale from local fluctuations in water or sediment composition to patterns of basinal evolution. Suggested controls include provenance (Stalder, 1975), depositional mineralogy (Chan et al., 1981), diagenesis of associated mudrocks (Boles and Franks, 1979), changing pore-fluid chemistry (Kantorowicz, 1985), burial pressures and temperatures (Hutcheon et al., 1980), and plate tectonics (Siever, 1979). These factors influence sediment diagenesis to various degrees at different times during burial as well as during uplift. In the Paleocene-Eocene Wilcox Group of the Texas Gulf Coast, many of the diagenetic patterns seen today appear to be early features related to factors such as original water chemistry, sediment texture, detrital composition, and organic content which, in turn, were directly or indirectly related to depositional environment.

We examined approximately 5000 ft of core from 15 wells to characterize diagenetic trends in the context of depositional environment. Well locations are shown in Figure 1. We compared textural, compositional, and diagenetic characteristics of samples from 12 interpreted depositional facies and developed generalized models for early diagenetic patterns characteristic of each facies. Note that although the relationships between early chemical and physical diagenetic processes and resulting lithofacies may be generalized and applied elsewhere, the specific petrologic characteristics of each lithofacies described below hold true only for the Wilcox Group of Texas. Also note that the following discussion deals only with the very earliest diagenetic reactions. The Wilcox was chosen purposely for this study because it has undergone a relatively simple history of continued subsidence since deposition. In more mature or more complex basins these early patterns are commonly overprinted by later diagenetic patterns caused by changes in structural, stratigraphic, or hydrologic regimes. These later diagenetic patterns are not related to facies, but will cross facies boundaries and will instead reflect fluid migration patterns or basinal "plumbing" systems. But, even in more complex situations, careful delineation of paragenetic sequences should reveal early diagenetic patterns which can be related to depositional facies using the principles described below.

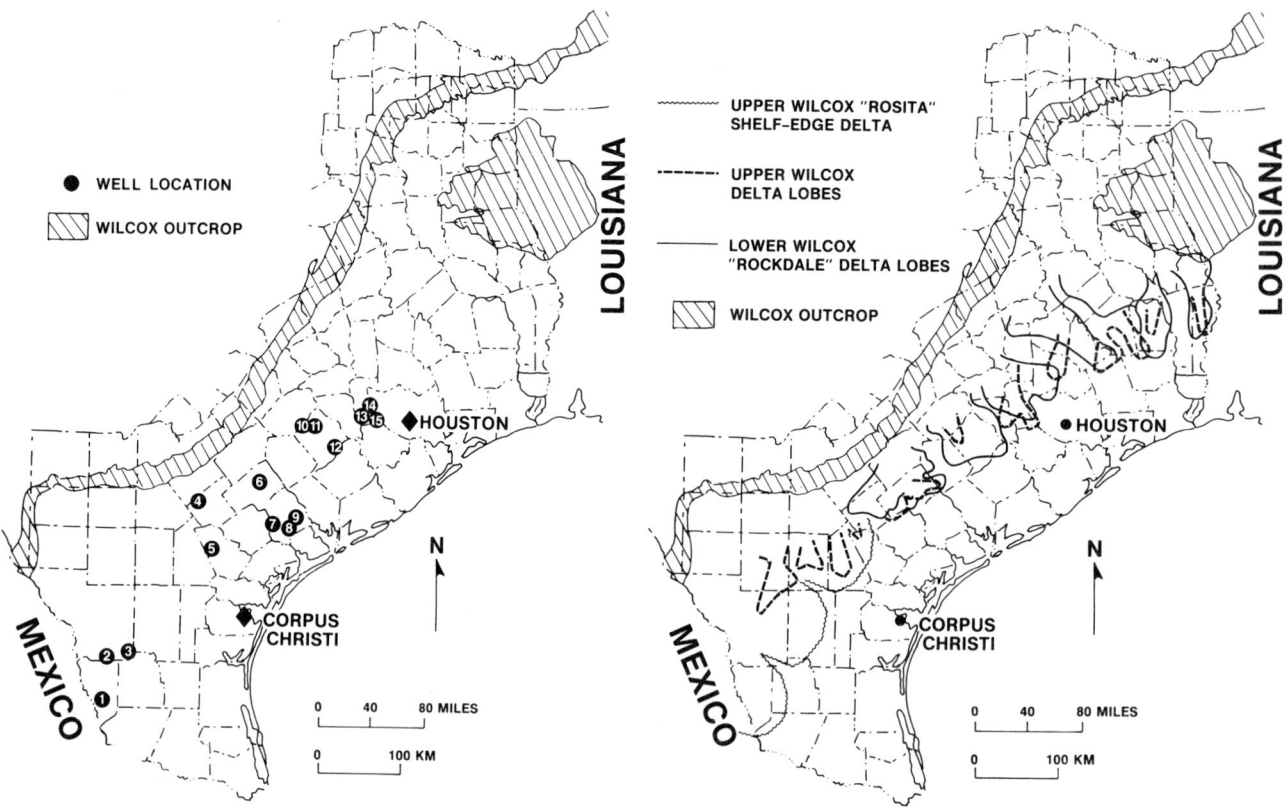

Figure 1. Map of southeastern Texas showing locations of cores utilized for this study. 1 = National #1 Fee, Zapata County. 2 = Forrest #2 Olimitos Ranch, Webb County. 3 = Getty #1 Benavides, Webb County. 4 = Seaboard #1 Kolodziejczyk, Karnes County. 5 = Carl #1 Gillette, Bee County. 6 = Getty #1 Burns, Dewitt County. 7 = Sun #1 Urban, Victoria County. 8 = Amerada #1 Talley, Victoria County. 9 = Amerada #1 Kovar, Victoria County. 10 = Citgo #5 Schobel, Colorado County. 11 = Citgo #16-2 Worthy, Colorado County. 12 = Magnolia #1 Gracey-Wegenhoft, Colorado County. 13 = Humble #W-31 Katy, Waller County. 14 = Humble #W-32 Katy, Waller County. 15 = Humble #W-35 Katy, Waller County.

Figure 2. Map of southeastern Texas showing areas of outcropping Wilcox strata and generalized basinward extent of lower and upper Wilcox delta lobes (modified from Fisher and McGowen, 1967; Williams et al., 1974; Edwards, 1981).

REGIONAL SETTING

The late Paleocene–early Eocene Wilcox Group represents the first major, regional clastic wedge deposited over and beyond the Cretaceous carbonate shelf edges of the Texas Gulf Coast province. The Wilcox produces hydrocarbons along a strip parallel to the coastline from the Mississippi-Alabama state line to the Mexican border. In Texas, the Wilcox Group in outcrop is as much as 1000 ft (300 m) thick (Figure 2), and it thickens basinward to more than 10,000 ft (3000 m) within a distance of 100 mi (160 km).

During Wilcox deposition much of the Gulf Coast was a broad, flat, low-lying coastal plain (Murray and Thomas, 1945; Albach, 1979). Modern drainage systems in this region roughly coincide with those of the Wilcox (Galloway, 1968). The southern Appalachian region supplied detritus to the ancestral Apalachicola, Mississippi, and Sabine rivers of the northern Gulf Coast (Murray, 1961; Mann and Thomas, 1968). The Ouachita Mountains supplanted the Appalachians as a sediment source westward, and the ancestral Brazos and Colorado rivers flowed into the upper and central Texas coast. The ancestral Rio Grande River of the lower Texas coast drained local volcanic and sedimentary sources (Hardin, 1962; Harris, 1962; Loucks et al., 1979; Oliver, 1980). The southern Rocky Mountains were not a major source of Gulf Coast sediments during Wilcox deposition (Mann and Thomas, 1968).

In Louisiana and Mississippi, a relatively straight northeast-trending coastline extended into the Mississippi Embayment during the late Paleocene through early Eocene (Albach, 1979). Concurrently, the large Holly Springs delta system, fed by the ancestral Mississippi River, began filling the Embayment (Galloway, 1968). The Texas coastline was much more irregular; major deposition occurred during two progradational cycles, the lower and upper Wilcox. Fluvial-dominated deltas, of a style and size similar to the Holocene Mississippi deltas, characterized the lower Wilcox "Rockdale System" along the upper Texas Gulf Coast (Figure 2). Longshore currents redistributed sand from this delta system into a strand plain to the southwest (Fisher and McGowen, 1967). In contrast, the upper Wilcox consisted of numerous small, wave-dominated deltas (Figure 2; Fisher, 1969; Williams et al., 1974). Along the lower Texas Gulf Coast, these deltas rapidly prograded basinward, forming the shelf-edge "Rosita System" (Edwards, 1981).

DEPOSITIONAL ENVIRONMENTS AND DIAGENETIC SIGNATURES

An interplay of physical, biological, and chemical processes controlled the development and distribution of the many Wilcox deltaic subenvironments (Figure 3). The purpose of this paper is not to review the different delta models nor to examine their subenvironments in detail. For that, the reader is referred to reviews such as those by Coleman and Prior (1980) or Elliott (1986). Nor is our purpose to describe the Wilcox cores and our stratigraphic facies interpretations in detail (for this the reader is referred to a companion paper in preparation by May and Stonecipher). What follows instead is a breakdown of the Wilcox deltaic systems into their separate components based on sedimentologic and stratigraphic criteria, a brief description of the major diagenetic processes active in each subenvironment, and a discussion of how those processes affected diagenetic signatures.

DELTA-PLAIN ENVIRONMENTS

Distributary Channel

The distributary-channel facies is characterized by a fining-upward sequence of poorly to moderately sorted, coarse- to fine-grained sandstone. Structureless, high-angle cross-beds in sandstones grade upward to horizontal laminae and ripple cross-laminae. Clay-draped laminae and plant debris become more common upward. The part of the sequence representing abandonment is characterized by flaser and wavy bedding, breakup structures, and slumps.

Compositionally, distributary-channel sands are immature to moderately mature (where maturity is proportional to the abundance of chemically and mechanically unstable components such as micas, plagioclase feldspars, and any kind of lithic fragments; Table 1). Organic debris and clay rip-up clasts commonly mark channel bases. Lithic fragments are most common at the base of the sequence; quartz, micas, and detrital clay increase upward.

Upper parts of channel (point-bar) sequences, which are generally very fine to fine grained (Table 1), have higher quartz:lithic ratios than the basal point bar and commonly contain more detrital clay and micas (Figure 4a). In clay-rich units, where the effects of compaction have reduced the initially low permeability even further, the only major diagenetic reactions are illite/smectite clay neomorphosis (recrystallization and/or overgrowth), alteration of micas, and formation of a small amount of authigenic quartz. The source of the silica is likely intergranular pressure solution promoted by the presence of clay (Figure 4a); stylolites are developed in sections where clay laminae are common. Although the amount of silica precipitated in clay-rich zones is inhibited by clay grain-coats, in clay-free sections quartz overgrowth development is commonly extensive (Figure 4b).

In basal point-bar sandstones, on the other hand, the coarser grain size (fine to medium, Table 1) and the lack of clay (Figure 5a) yield better original porosity and permeability and less susceptibility to severe reductions in porosity and permeability due to compaction. Secondary porosity is also best developed in these basal sands where the typically more abundant but chemically less stable lithic fragments are dissolved.

Clay compositions also differ as a function of subenvironment within the distibutary channels. Detrital clay composition is relatively consistent throughout the deltaic system and consists primarily of randomly interstratified illite/smectite mixed-layer clays (MLC). However, the amount of detrital clay ranges from rare to common, depending on the energy characteristics of the part of the distributary-channel sequence in which it occurs. Most of the illite/smectite shows signs of neomorphosis (Figure 5b), but x-ray diffraction data indicate that the proportion of nonexpansible illite layers remains relatively constant (average approximately 80% smectite/20% illite) throughout the section. In contrast, authigenic clay shows a wide range in composition, depending on the depositional environment. Kaolinite is one of the most common authigenic clays in active distributary-channel sands (Figure 6a). It generally occurs as an early alteration product of micas (Figure 7a, b, c) and less commonly of feldspars (Figure 7d, e), and is most abundant in the more porous parts of basal point-bar sands. Abandoned distributary-channel sands (Figure 6b) and distal distributary-channel sands (Figure 6c) both contain more MLC and chlorite than the active channel sands do, presumably because of the lower energy at the time of deposition and the influence of marine pore waters from the surrounding marine shales. Early authigenic chlorite rims are generally patchy and

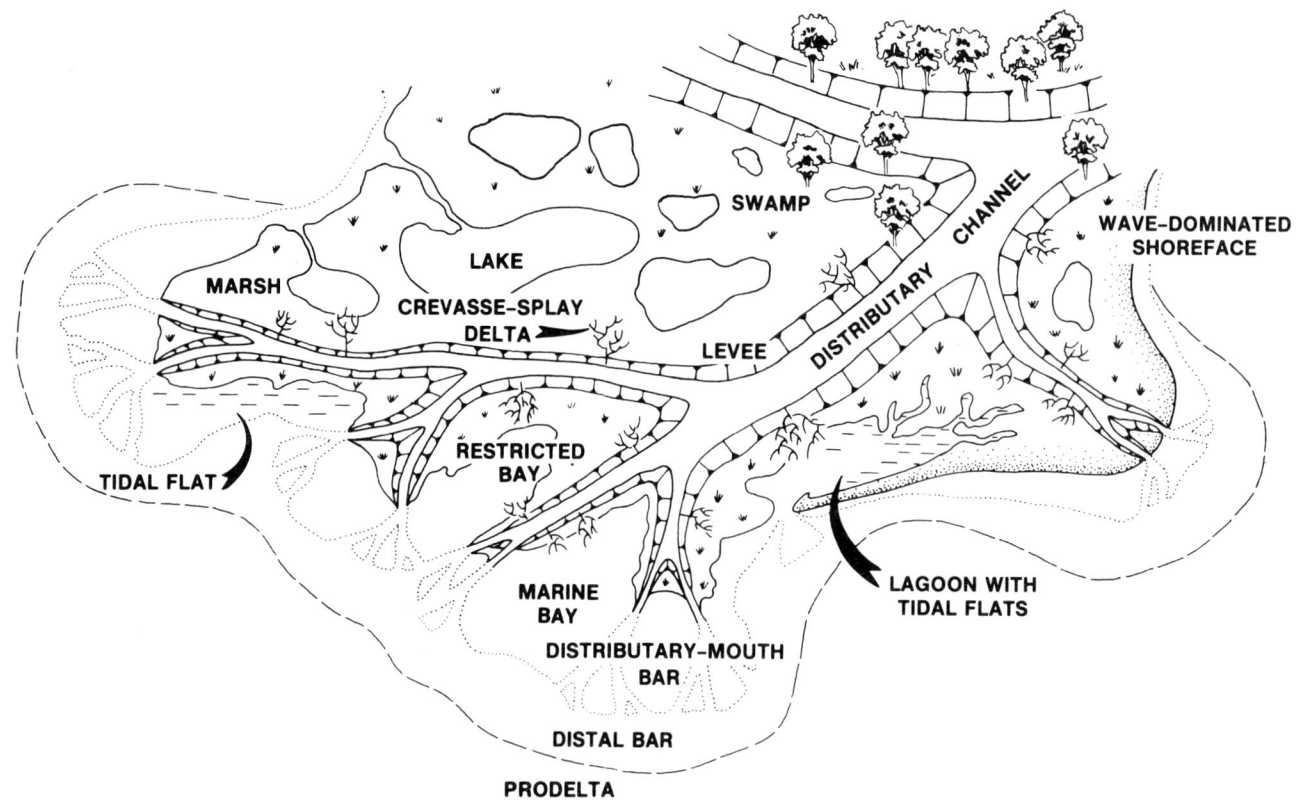

Figure 3. Diagrammatic representation of deltaic environments. Any one specific delta will display various components of this composite system.

incomplete. Carbonate cements are also common, especially in more distal channel sections.

Lake

Lake sediments consist primarily of dark-gray to black claystones which are commonly coaly and laminated. Rare, thin, sandier layers are associated with input from overbanking of levees. Plant debris is common and burrowing is rare. The main diagenetic processes involved formation of early siderite nodules and sideritization of fecal pellets and detrital micas (Figure 8), as well as neomorphosis of abundant detrital MLCs (Figure 6d).

Marine Bay

Interdistributary marine bay deposits range from siltstone and sandy mudstone to fine-grained muddy sandstones which are concentrated in thin horizontal laminae and ripple lenses. In contrast with lake sediments, bioturbation in marine bay deposits was commonly extensive, destroying all primary stratification.

Compositionally, marine bay deposits tend to be fairly quartzose (Table 1) and contain abundant detrital clay, micas, and silt (Figure 9a). Organic material is generally minor. Diagenetic processes are commonly restricted to illite/smectite clay neomorphosis and formation of minor pyrite. Abundant early chlorite (Figure 9b) also formed in some sandier layers but is difficult to distinguish in thin section (Figure 9c) because of the equal abundance of detrital clays (Figure 6e). Minor quartz overgrowths may form where open pore space is available.

Crevasse-Splay Delta

Crevasse-splay deltas, which formed in both lakes and marine bays (Figure 3), generally consist of very-fine-grained, moderately well sorted sandstones with some interlaminated mudstones. Grain size commonly decreases and detrital clay content increases upward. Horizontal and low-angle cross-laminae are common, and tops of beds in many places are capped by ripples. Bed tops may also be rooted or burrowed.

Compositionally, splay delta channel deposits resemble the finer grained parts of distributary channels; they tend to be very quartzose and contain abundant micas and detrital clays (Table 1). Diagenetic patterns depend strongly on whether the crevasse-splay delta extended into a freshwater lake or a marine bay. Splays into lakes generally contain abundant kaolinite (Figure 6f) and authigenic siderite (Figure 10a, Table 1), whereas splays into marine

Table 1. Point count modal analyses[a]

Well Depth	Getty #1 Burns 9271 ft	Getty #1 Burns 9335 ft	Humble Katy #31 10,571.5 ft	Getty #1 Burns 9616.5 ft	Humble Katy #31 10,571.8 ft	Citgo Schobel #5 8378 ft	Citgo Schobel #5 8475 ft	Getty #1 Burns 9346.1 ft	Humble Katy #W-31 12,010.8 ft
Depositional Environment	Distal-Channel-Bar Base	Distal-Channel-Bar Base	Distal-Channel-Bar Base	Distal-Channel-Bar Top	Distal-Channel-Bar Top	Crevasse Splay into Lake	Crevasse Splay into Lake	Crevasse Splay into Bay	Crevasse Splay into Bay
Grain Size	0.28mm ± 0.10	0.26mm ± 0.07	0.35mm ± 0.13	0.14mm ± 0.06	0.19mm ± 0.10	0.07mm ± 0.03	0.13mm ± 0.05	0.17mm ± 0.08	0.10mm ± 0.04
Sorting	0.66ϕ (m-w)	0.46ϕ (w)	0.70ϕ (m-w)	0.87ϕ (m)	1.20ϕ (p)	0.96ϕ (m)	0.73ϕ (m)	1.14ϕ (p)	0.78ϕ (m)
Monocrystalline qtz	39.3	36.3	23.0	41.0	25.0	43.7	41.0	34.7	35.0
Polycrystalline qtz	15.0	14.3	21.7	19.7	22.3	21.3	18.7	18.7	17.7
K-feldspar	1.0	1.3	3.3	1.0	2.0	0.7	1.7	2.0	0.3
Plagioclase	6.0	5.3	8.0	7.3	12.0	4.0	6.0	4.7	10.0
Chert	9.3	5.3	4.7	1.7	3.3	3.7	7.0	6.3	3.3
Sedimentary rock frag	1.3	1.0	2.7	1.3	2.0	4.7	3.3	4.0	4.0
Volcanic rock frag	5.7	4.7	3.0	—	2.3	0.7	—	0.3	5.0
Plutonic rock frag	1.7	1.0	3.3	1.7	2.0	—	1.0	0.7	0.7
Metamorphic rock frag	—	—	2.3	3.0	3.3	0.3	2.0	1.3	3.0
Micas	—	—	1.0	10.0	16.3	2.0	—	—	2.0
Organics	1.0	0.3	—	0.3	0.7	0.3	0.3	—	1.7
Glauconite	0.3	0.3	1.0	2.0	—	0.3	0.3	0.7	—
Heavy minerals	—	—	—	—	—	—	—	—	—
Clay matrix	—	—	—	4.0	5.0	6.0	6.3	14.0[b]	5.7[b]
Chamosite ooid	—	—	—	—	—	—	—	—	—
Quartz cement	8.0	10.7	6.7	4.3	1.7	—	4.3	8.0	6.0
Calcite	—	0.3	0.7	—	—	—	—	—	—
Ankerite	1.0	—	—	—	—	1.0	1.0	—	—
Siderite	—	—	—	—	—	5.7	2.0	—	—
Pyrite	1.0	1.0	—	0.3	0.7	1.3	—	1.3	1.3
Feldspar overgrowths	—	0.3	1.3	—	—	—	—	—	—
Kaolinite	1.3	1.7	2.3	1.0	1.3	0.3	0.3	2.3	0.3
Intergranular porosity	7.3	10.3	12.3	1.3	—	4.3	4.7	—	3.3
Secondary porosity	0.7	6.0	2.7	—	—	—	—	1.0	0.7
-Cem por	20.3%	30.3%	26.0%	7.0%	3.7%	12.7%	12.3%	12.6%	11.7%
QFL[e]	$Q_{68}F_9L_{23}$	$Q_{73}F_{10}L_{17}$	$Q_{62}F_{16}L_{22}$	$Q_{71}F_{10}L_{19}$	$Q_{64}F_{19}L_{17}$	$Q_{78}F_6L_{16}$	$Q_{74}F_{10}L_{16}$	$Q_{73}F_9L_{17}$	$Q_{63}F_{12}L_{25}$

(Continued)

[a]-Point count percentages based on 300 points; grain size analyses based on 100 grains
[b]-Includes authigenic chlorite
[c]-Minimum number; difficult to distinguish from ankerite cement
[d]-Includes reworked sideritized fecal pellets; see text
[e]-After Folk (1980)

Table 1. (Continued)

Well	Getty #1 Burns 10,274 ft	Forrest Olimitos Ranch #2 9189.3 ft	Forrest Olimitos Ranch #2 9211.9 ft	Getty #1 Burns 9282.8 ft	Getty #1 Burns 9572.5 ft	Getty #1 Burns 9353.2 ft
Depositional Environment	Upper Shoreface	Tidal Sand Flat	Tidal Sand Flat	Tidal Channel	Tidal Channel	Distal Mouth Bar Top
Grain Size	0.11 mm ± 0.03	0.08mm ± 0.03	0.06mm ± 0.02	0.19mm ± 0.07	0.12mm ± 0.04	0.30mm ± 0.10
Sorting	0.44ϕ (w)	0.67ϕ (m-w)	0.63ϕ (m-w)	0.67ϕ (m-w)	0.59ϕ (m-w)	0.46ϕ (w)
Monocrystalline qtz	44.0	50.0	47.7	34.0	33.7	34.7
Polycrystalline qtz	17.7	18.7	11.0	15.3	18.3	10.7
K-feldspar	—	—	—	1.3	1.7	5.0
Plagioclase	2.7	2.0	2.3	8.0	9.3	9.3
Chert	3.0	3.7	2.7	8.0	7.0	3.0
Sedimentary rock frag	1.7	3.7	3.7	4.3	8.0	3.3
Volcanic rock frag	—	—	—	3.7	4.3	2.7
Plutonic rock frag	—	—	—	0.7	0.7	3.0
Metamorphic rock frag	0.7	1.0	1.0	2.0	2.7	0.7
Micas	0.7	—	0.3	—	1.0	0.3
Organics	—	0.7	0.3	3.0	1.0	—
Glauconite	—	—	—	2.0	2.3	—
Heavy minerals	0.3	—	—	—	—	—
Clay matrix	—	4.3[c]	4.0[c]	2.3	0.3	—
Chamosite ooid	—	—	—	—	—	2.3
Quartz cement	23.0	0.3	—	9.7	5.7	10.0
Calcite	—	—	26.0	—	—	—
Ankerite	0.7	15.3	—	—	—	0.3
Siderite	—	—	—	—	—	—
Pyrite	—	0.3	1.0	1.0	1.0	—
Feldspar overgrowths	—	—	—	—	—	1.0
Kaolinite	3.3	—	—	3.3	1.7	0.3
Intergranular porosity	2.0	0.3	—	1.0	1.3	9.7
Secondary porosity	0.3	—	—	0.3	—	3.7
-Cem por	29.3%	16.3%	27%	15.3%	9.7%	25.0%
QFL[e]	Q$_{88}$F$_4$L$_8$	Q$_{87}$F$_2$L$_{11}$	Q$_{86}$F$_3$L$_{11}$	Q$_{64}$F$_{12}$L$_{24}$	Q$_{61}$F$_{12}$L$_{27}$	Q$_{63}$F$_{20}$L$_{17}$

(Continued)

Table 1. (Continued)

Well	Humble Katy #W-35	Forrest Olirnitos Ranch #2	Humble Katy #W-31	Sun Urban #1
Depth	10,390.4 ft	9776 ft	13,108 ft	14,108 ft
Depositional Environment	Distributary Mouth Bar Mid	Distal Bar	Prodelta	Shelf
Grain Size	0.22mm ± 0.10	0.11mm ± 0.05	0.15mm ± 0.13	0.08mm ± 0.04
Sorting	1.00ϕ (m)	0.76ϕ (m)	2.29ϕ (vp)	1.50ϕ (vp)
Monocrystalline qtz	29.0	43.3	21.0	33.3
Polycrystalline qtz	10.3	21.3	15.3	14.3
K-feldspar	1.0	0.3	0.7	—
Plagioclase	9.0	7.3	3.3	1.7
Chert	7.0	8.0	4.3	4.3
Sedimentary rock frag	8.0	2.3	42.3^d	3.7
Volcanic rock frag	7.7	4.0	1.0	5.3
Plutonic rock frag	1.7	—	—	—
Metamorphic rock frag	10.7	0.7	0.3	1.3
Micas	1.0	0.3	0.3	2.0
Organics	—	—	—	0.3
Glauconite	—	—	4.0	0.7
Heavy minerals	—	—	—	—
Clay matrix	—	9.7	2.3	21.7
Chamosite ooid	2.0	—	—	—
Quartz cement	10.3	—	—	—
Calcite	—	—	—	—
Ankerite	—	2.0	4.3	10.0
Siderite	—	—	$—^d$	—
Pyrite	—	0.7	1.0	1.0
Feldspar overgrowths	0.7	—	—	—
Kaolinite	—	—	—	0.3
Intergranular porosity	2.7	—	—	—
Secondary porosity	0.3	—	—	—
-Cem por	14.0%	2.7%	5.3%	11.3%
QFL^e	$Q_{47}F_{12}L_{51}$	$Q_{74}F_9L_{17}$	$Q_{41}F_4L_{55}{}^d$	$Q_{74}F_3L_{23}$

Figure 4. Photomicrographs of the upper distributary channel fill. (a) Fine-grained point-bar top is laminated with clays, organic debris, and abundant micas. The abundance of clays prevented the development of extensive quartz overgrowths, but suturing is common. Humble #W-31 Katy, 10571.8 ft; scale bar equals 0.5 mm. (b) Lack of detrital clay in this part of a point bar top allowed extensive quartz overgrowths to develop. Getty #1 Burns, 9597 ft; scale bar equals 0.2 mm.

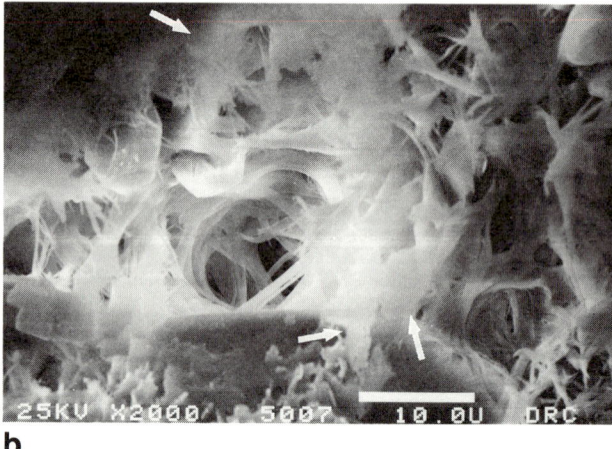

Figure 5. Petrologic features from basal distributary-channel point bars. (a) Point-bar (distributary-channel) base is typified by coarser grain size, lack of detrital clay, greater abundance of lithic fragments, and abundant authigenic kaolinite (k). Patchy quartz overgrowths may be well developed in secondary pores. Getty #1 Burns, 9335 ft; scale bar equals 0.2 mm. (b) Wispy laths of pore-bridging illite appear to grow from the edges of detrital clay flakes which coat the grain surface (arrows). The detrital clay consists of randomly interstratified illite/smectite mixed-layer clay. It appears that neomorphosis (overgrowth and partial transformation) has taken place and that the overgrowths are more illitic than the original detrital core. Citgo #5 Schobel, 10,179.5 ft; 2000 ×.

bays are characterized by minor early pyrite (Figure 10b) and authigenic chlorite (Figure 6g). Like distributary-channel sands, splay delta channel sands also contain abundant kaolinite (Figure 6h), reflecting their connection with the fluvial system. However, they contain more detrital MLC, because of the combination of turbid flood waters pouring through breached levees and the sharp drop in energy as the channelized flows spread out into the lake or marine bay. The upper parts of splay deltas in both environments tended to act as closed hydrologic systems characterized by illite/smectite clay neomorphosis and quartz overgrowth development.

SHORELINE ENVIRONMENTS

Shoreface

The differences in energy levels and depositional processes across the shoreface profile resulted in

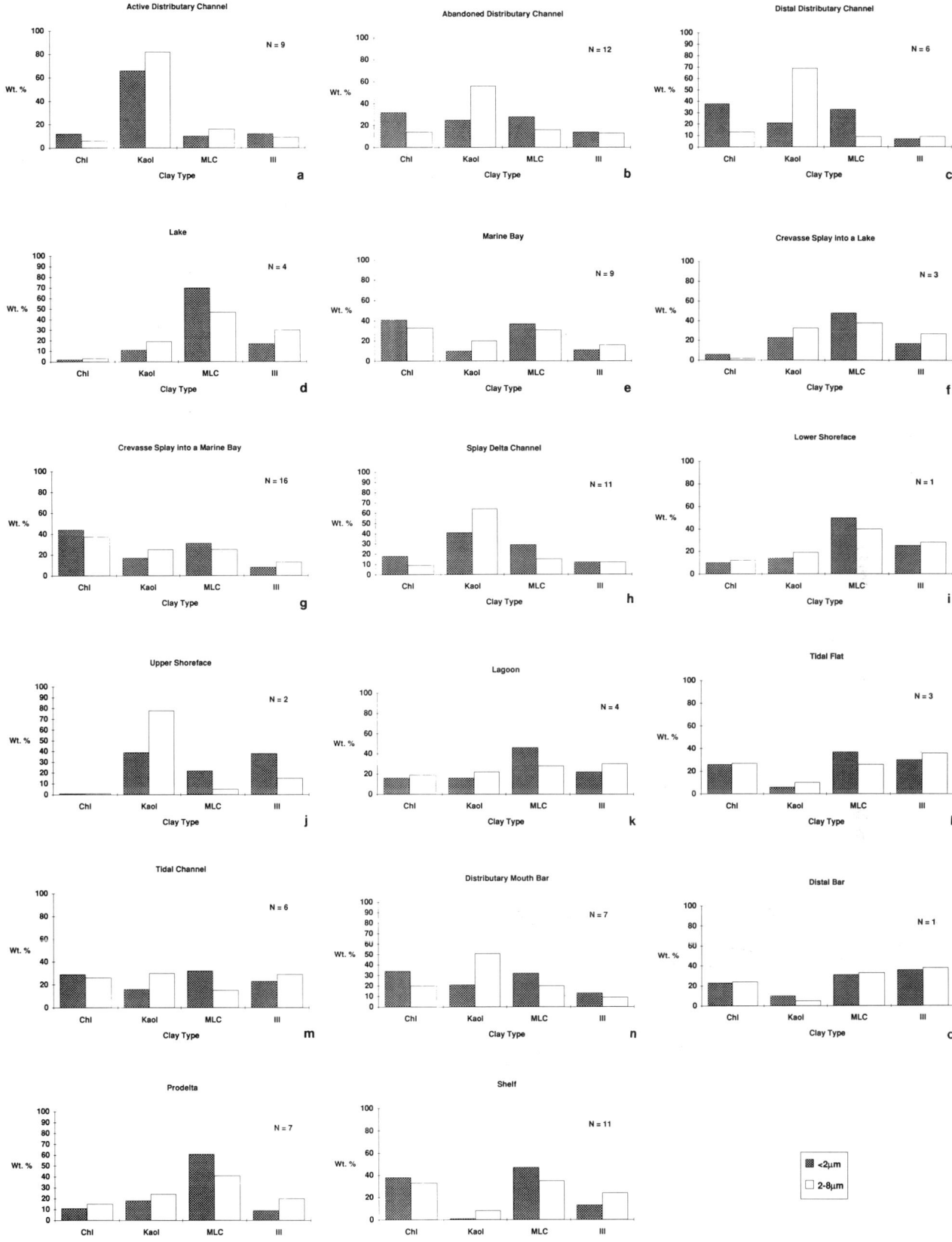

Figure 6. Average clay compositions in various depositional facies based on x-ray diffraction data. Numbers represent the arithmetic mean of the weight percentage of each clay in that respective size fraction. Chl = chlorite; Kaol = kaolinite; MLC = randomly interstratified illite/smectite mixed-layer clay; I I I = illite; N = number of samples averaged.

Prediction of Reservoir Quality Through Chemical Modeling

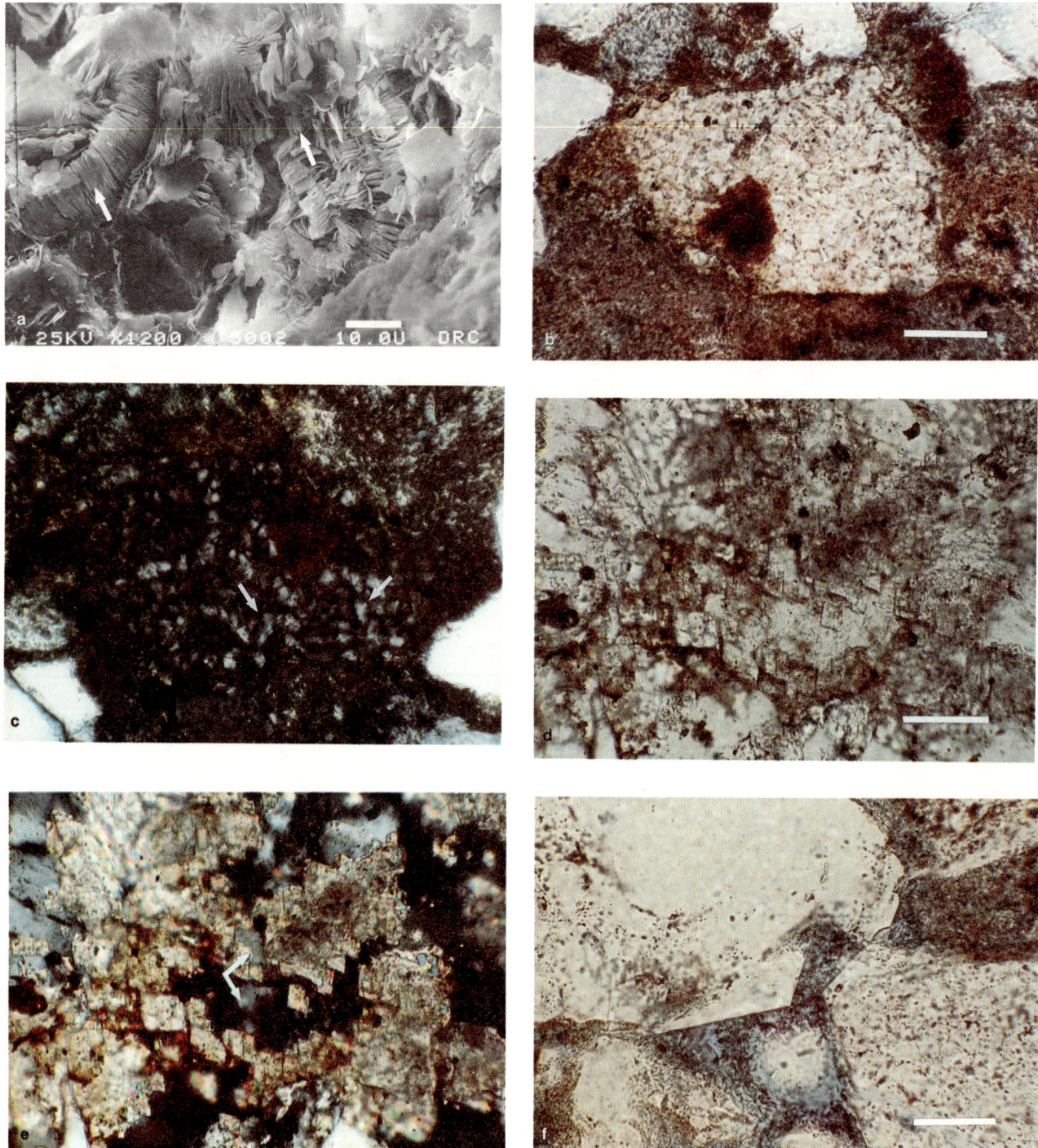

Figure 7. Authigenic clays in basal distributary-channel point bars. (a) Scanning electron photomicrograph of mica flake which has expanded to form a large "accordion." Kaolinite platelets (arrows) fill in between mica flakes to form booklets. Getty #1 Burns, 9266.5 ft; 1200×. (b, c) Thin section photomicrographs of mica-derived kaolinite. Large, well-formed booklets of kaolinite still contain shreds of mica from which they formed (arrows). Note lack of quartz overgrowths on immediately adjacent portions of quartz grains, suggesting early formation of kaolinite. Carl #1 Gillette, 10,177.4 ft; scale bar equals 0.1 mm. (b - plane polarized light; c - crossed nichols). (d, e) This kaolinite appears to have formed from a feldspar, a remnant of which is still visible (arrow). The kaolinite also is believed to have developed later in the diagenetic history, probably after the late-stage ferroan dolomite. Getty #1 Benavides, 9689.1 ft; scale bar equals 0.05 mm. (d - plane polarized light; e - crossed nichols). (f) This pore is filled with authigenic chlorite, most of which appears to have formed after the quartz overgrowths. Getty #1 Burns, 9559.3 ft; scale bar equals 0.05 mm.

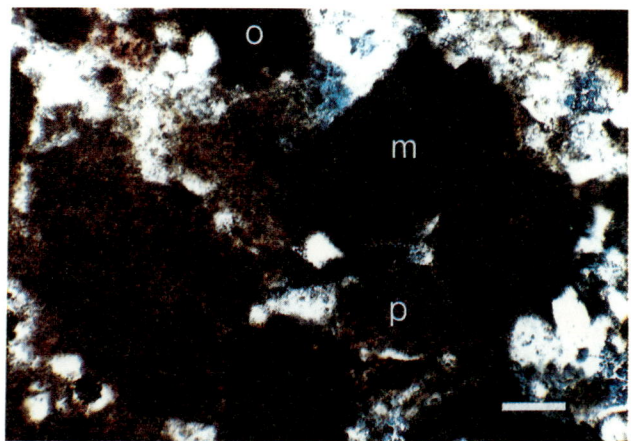

Figure 8. Photomicrograph typical of sandy lake sediment showing organic clots (o), sideritized fecal pellet (p), and sideritized mica (m), as well as sideritic clay matrix. Humble #W-31 Katy, 3108 ft; scale bar equals 0.1 mm.

marked variations in texture. Lower shoreface deposits consist of fine-grained, poorly sorted, clay-rich, horizontal-laminated to hummocky cross-laminated sandstones commonly alternating with burrowed intervals. Upper shoreface deposits, on the other hand, consist of medium- to fine-grained, moderately well sorted, cross-bedded sandstones with rare *Ophiomorpha* burrows. The foreshore is represented by medium- to fine-grained, well sorted to very well sorted, horizontal-laminated sandstone; the backshore is characterized by fine-grained, very well sorted, horizontally laminated and small-scale ripple cross-laminated sandstones. Thus, the vertical shoreface sequence is characterized by a coarsening-upward and then a fining-upward trend.

The differences in energy levels and sorting in shorefaces also caused differences in composition through the profiles. Detrital clays, micas, and fine-grained lithics (including sedimentary, volcanic, plutonic, and metamorphic rock fragments) decrease upward (Table 1). The lower, finer grained, more clay-rich sections of shoreface sequences generally acted as closed hydrogeochemical systems; the dominant diagenetic processes were development of quartz overgrowths and neomorphosis of the abundant detrital illite/smectite clay (Figure 6i). Upper sections are highly quartzose but may also contain shale clasts, fecal pellets, carbonaceous debris, and heavy-mineral stringers, especially in the foreshore. Porosity in this part of the section is commonly occluded by quartz overgrowths (Figure 11). Secondary porosity is generally rare but some has developed locally, probably from dissolution of clay clasts or carbonate debris. Detrital illite/smectite MLCs are less abundant in the upper shoreface than in the lower, but kaolinite and illite are more common (Figure 6j).

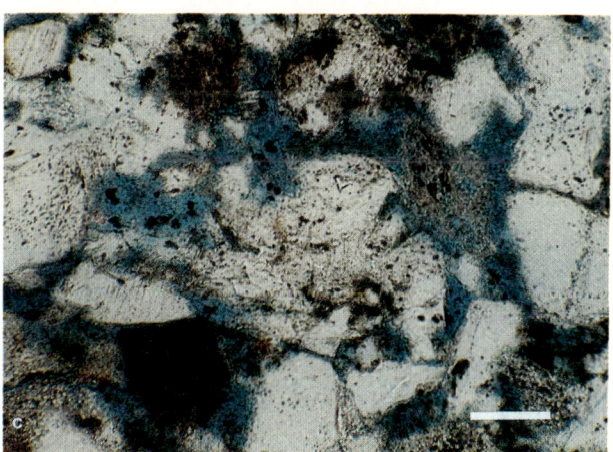

Figure 9. Attributes of marine-bay deposits. (a) This sample represents a thin sandy layer in a marine-bay unit. Note moderately well sorted fine grain size, highly quartzose composition, burrow (brown patch, left side of photo), and pervasive matrix composed of chlorite and illite/smectite mixed-layer clay (I/S MLC). Getty #1 Burns, 9346.1 ft; scale bar equals 0.2 mm. (b) SEM photomicrograph of sediment matrix composed of I/S MLC and chlorite (arrows). Citgo #5 Schobel, 10,252 ft; 2000 ×. (c) Presence of abundant clay matrix and authigenic chlorite has inhibited the formation of quartz overgrowths. Getty #1 Burns, 9346.1 ft; scale bar equals 0.1 mm.

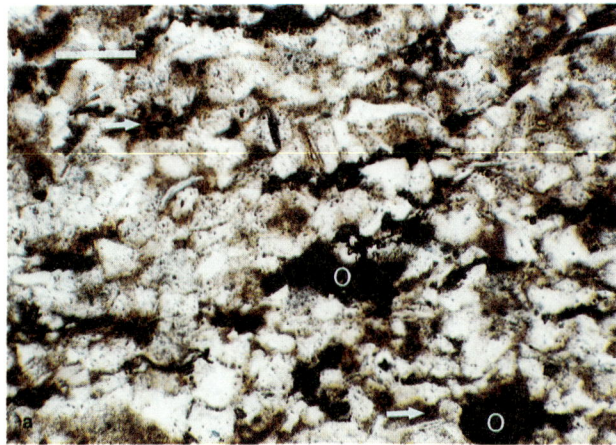 

Figure 10. Characteristics of crevasse-splay deltas. (a) Example of a splay sand from a lacustrine sequence. Note abundant organic clots (o), detrital clays (light-brown patches), and siderite cement (arrows). Humble #W-31 Katy, 13,066.5 ft; scale bar equals 0.1 mm. (b) Example of splay sand from a marine-bay sequence. Note minor pyrite, lack of organics and siderite, and matrix composed of I/S MLC and chlorite. Humble #W-31 Katy, 12,919.2 ft; scale bar equals 0.1 mm.

Lagoon

Lagoonal deposits consist primarily of clayey to limy mudstones interlaminated with siltstone and very-fine-grained sandstone. The coarser sediment forms thin laminae and ripples within the mudstone. Thicker intervals of horizontal-laminated to low-angle cross-bedded siltstones and fine-grained sandstones resulted from washovers. Complete bioturbation is rare but small, silt-filled burrows are common (Figure 12). Sorting is generally moderate to poor.

Compositionally, lagoonal sediments are generally highly quartzose (Table 1) and contain abundant detrital clay (Figure 6k). Shell material is locally common and, depending on productivity and openness of circulation, the deposits may be either limy or dolomitic (see Figure 12). Quartz overgrowths, chlorite rims, and (less commonly) later stage kaolinite developed in the cleaner laminae.

Tidal Flat

Tidal flats formed both in back-barrier settings and along the seaward margins of interdistributary bays open to the ocean. Tidal-flat deposits consist of mudstones interlaminated with fine-grained sandstones and siltstones. The sandier parts of these deposits show moderate to good sorting and a sharp change in clay content across the sand/mud boundaries. Many tidal-flat sequences display an overall slight fining-upward trend. Herringbone cross-laminae, horizontal laminae, and draped ripples (flaser and wavy bedding) are typical sedimentary structures; burrows are rare. In thin section, the most commonly observed features are thin, wavy clay laminae and parallel alignment of elongate grains.

In terms of detrital composition, tidal-flat sands tend to be slightly more quartzose and micaceous than tidal-channel sands (Table 1), primarily because of their finer grain size (very fine to fine). Tidal-flat deposits contain abundant detrital clay, typically as organic-rich fecal pellets (Figure 13), and also locally contain abundant shell debris.

Compaction of the abundant clay laminae and fecal pellets reduced the already low initial permeability. Diagenesis was characterized by the development of common to extensive quartz overgrowths and the alteration of micas to MLCs (Figure 13). Diagenetic clays include minor to common chlorite, minor kaolinite, and minor to common illite (Figure 6l). The dissolution of shell material to form secondary-porosity and/or the development of patchy carbonate cements are also noted.

Tidal Channel

In comparison to tidal flats, tidal-channel deposits contain slightly coarser grained (fine- to medium-grained) sandstones (Table 1). Low- and high-angle cross-bedding also indicates deeper water flow. Clay drapes along cross-laminae and horizontal laminae are again prevalent and are commonly associated with abundant peat and wood debris. Shale rip-ups, shell debris, and lithic fragments are most common at the bases of the channels. Lithics generally decrease upward in contrast with quartz, micas, and detrital clays, which all increase upward (Table 1). Traces to minor amounts of glauconite are commonly found associated with clay laminae (Figure 14a).

In some ways tidal-channel sands are similar to distributary-channel sands; in other ways, they are dissimilar. As in the fluvial point bars, the upward decrease in grain size and increase in clay content

Figure 11. The heavy-mineral stringer running through the center of this photomicrograph formed as a lag in a ripple trough because of extensive reworking and sorting in the swash zone. Another characteristic of the foreshore sands is the highly quartzose composition and the nearly complete cementation by quartz overgrowths. This particular sample represents a Chandaleur-Island-type shoreface in an otherwise marine sequence. Getty Benavides, 10,274 ft; bar equals 0.2 mm.

Figure 13. Typical tidal-flat sand sample is well laminated with organic-rich fecal pellets (arrows) and micas. Small shell fragment (arrow, s) has been replaced by late-stage ferroan dolomite. Forrest #2 Olimitos Ranch, 9189.3 ft; scale bar equals 0.2 mm.

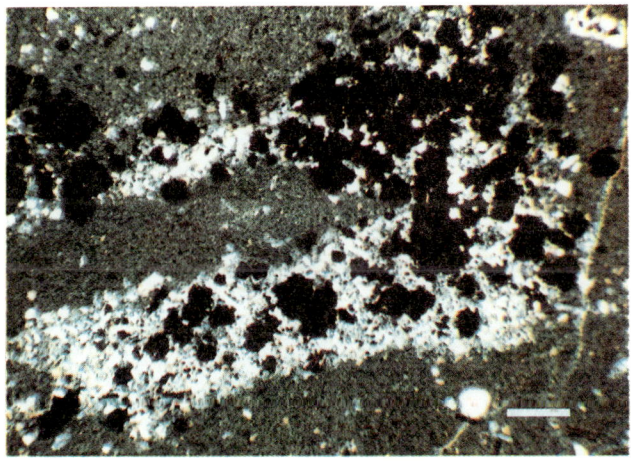

Figure 12. Sand-filled burrow in micritic lagoon deposit. Burrow has been preferentially cemented with pyrite. Forrest #2 Olimitos Ranch, 9451 ft; scale bar equals 0.2 mm.

in tidal-channel point bars produced lower original porosities and permeabilities. Here, quartz cementation was common. The bases of tidal point bars are also similar to their fluvial counterparts in that original porosity and permeability were relatively high because of coarser grain size and better sorting; they also tend to contain more chemically unstable lithic fragments and shell debris (Table 1), which dissolved to produce secondary porosity. As a result, porosity and permeability are better in the base of the tidal point bar than in the upper part. Overall, however, unless shell debris was especially common, or unless early carbonate cements preserved primary intergranular volume, porosity and permeability are generally less in tidal-channel sands than in distributary-channel sands. The abundant clay laminae likely limited and channeled intergranular fluid flow and, as a result, diagenesis commonly varies in adjacent laminae. Quartz overgrowths are common but patchy and incomplete. Early authigenic chlorite rims are generally moderately to well developed (Figure 14b) and common (Figure 6m). Some grains show partial detrital illite/smectite clay rims. Later (postchlorite) kaolinite (Figure 14c) is also prevalent in some samples (Figure 6m).

DELTA-FRONT ENVIRONMENTS

Distributary-Mouth Bar

The distributary-mouth-bar sequence mimics the coarsening-upward lower to upper shoreface succession. Medium- to fine-grained sandstones contain horizontal laminae and cross-beds formed by tractive deposition. Graded beds (turbidites) with structureless bases and rippled tops are also developed in this sequence. *Ophiomorpha* burrows are locally abundant.

Depending on the balance between wave reworking and fluvial sedimentation, distributary-mouth-bar sediments may be slightly more quartzose (Table 1, Figure 15a) or similar in composition to distributary-channel deposits. The lower part of the bar is fine

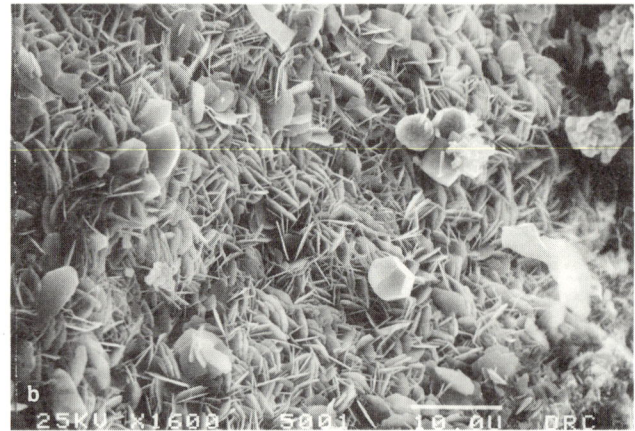
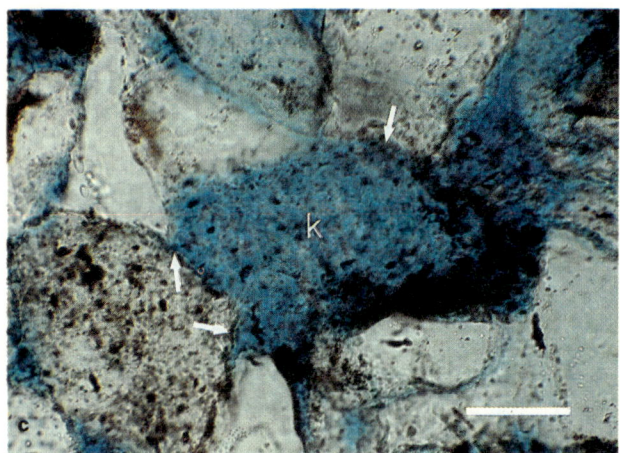

Figure 14. Aspects of tidal-channel deposits. (a) Discontinuous, subparallel to anastomosing clay laminae characteristic of tidal-channel sands are typically organic rich and commonly contain minor glauconite pellets (arrow), apparently swept in from offshore. Getty #1 Burns, 9572 ft; scale bar equals 0.2 mm. (b) Tidal-channel deposits generally have most of the detrital clay concentrated in discrete laminae such as in (a), so the sands themselves tend to be clean and make good conduits for marine pore waters expelled from nearby shales. As a result, most sand grains commonly have well-developed chlorite rims, as shown in this SEM photomicrograph. Getty #1 Burns, 9247.4 ft; 1600 ×. (c) Authigenic kaolinite (k) can also be found in tidal-channel sands, but it is generally later stage (that is, postchlorite, arrows). Forrest #2 Olimitos Ranch, 9670.1 ft; scale bar equals 0.05 mm.

grained, clay rich, and quartzose (Figure 15b), similar to the lower shoreface.

The most characteristic diagenetic aspect of distributary-mouth-bar deposits is the presence of berthierine/chamosite ooids (Figure 15a) or partially chamositized fecal pellets. Early grain-coating chlorite and later clean pore-filling kaolinite are also common (Figure 6n).

Distal Bar

Distal-bar deposits consist of alternations of well-sorted to poorly sorted, fine-grained sandstones, siltstones, and mudstones. Horizontal laminae and clay-draped ripple lenses are common, and many thin beds are graded. Bioturbation ranges from isolated burrows to nearly complete reworking.

Distal-bar sediments are primarily the products of physical and chemical breakdown of existing rocks that contained abundant micas, fine quartz, and detrital clays (Table 1, Figure 16). In intervals where burrowing organisms worked abundant clay into the siltstones and sandstones, compaction reduced porosity and permeability at a very early stage. Neomorphosis of abundant detrital illite/smectite (Figure 6o) and precipitation of chlorite were the dominant processes, and the resulting growth of clays in pores caused a further reduction in permeability. Alteration of micas and development of minor quartz overgrowths were the other two main processes.

Prodelta

The transition from distal bar to prodelta is gradational. These sediments consist primarily of silty to clayey mudstones. Many of the mudstones are burrowed to graded and contain thin laminae and lenses of siltstone and rare very-fine-grained sandstone. Convoluted distributary-mouth-bar and distal-bar deposits were apparently commonly redeposited in prodelta sediments as slumps.

Compositionally, prodelta sediments are very similar to distal-bar sediments, containing abundant micas, fine-grained quartz, and detrital clays (Table 1). Plant debris is also common. One unusual component found in many samples is sideritized fecal pellets. The distribution of the pellets, which commonly are concentrated along bedding planes

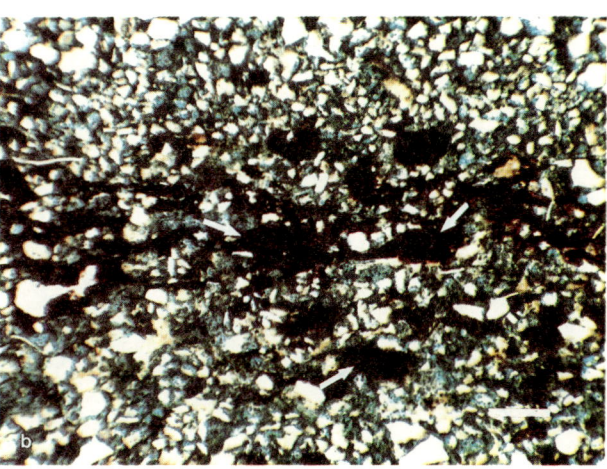

Figure 15. Features of distributary-mouth bars. (a) This distributary-mouth-bar sand looks very much like a distal distributary-channel sand but is slightly more quartzose and contains chamosite ooids (arrow, c). Getty #1 Burns, 9353.2 ft; scale bar equals 0.2 mm. (b) The lower part of a distributary-mouth bar is similar to a lower shoreface, being fine grained, clay rich, and quartzose. However, unlike the lower shoreface, the lower distributary-mouth bar is commonly discontinuously laminated with organic clots (arrows) such as these. Carl #1 Gillette, 10,225.5 ft; scale bar equals 0.2 mm.

Figure 16. Similar to the lower shoreface, distal bar sediments consist primarily of fine quartz, micas, and detrital clays. Burrowing has thoroughly homogenized the sediment. Note common suturing but lack of quartz overgrowths. Forrest #2 Olimitos Ranch, 9776 ft; scale bar equals 0.2 mm.

(Figure 17a), suggests that they probably resulted from reworking of existing sediments, perhaps deposited in tidal (mud) flats or brackish water lagoons.

Clay neomorphosis reduced porosity and permeability still further from the original low values. Authigenic chlorite is present but is difficult to recognize in thin section because of the abundant detrital clays (Figure 6p). Alteration of abundant plant debris resulted in the formation of common early pyrite and the minor late kaolinitization of micas and feldspars (Figure 17b).

Continental Shelf

Continental shelf sediments are primarily dark-gray clayey mudstone, locally containing interlaminated fine-grained sandstone and siltstone. Flaser to wavy bedding (draped ripples) are the two most common bedding types in coarse-grained units, although high- to low-angle cross-beds and hummocky cross-stratification are also preserved. Bioturbation is generally moderate to intense. Rare graded beds with laminated and rippled tops are also present.

As with all the other distal deposits (such as prodelta and lower shoreface), the shelf sediments consist of abundant fine-grained quartz, minor to common micas, abundant detrital clays, and minor lithics (Table 1, Figure 18a). The major diagenetic processes were the development of early authigenic chlorite (Figure 6q), minor quartz overgrowths, and detrital clay neomorphosis. In addition, in the outer parts of the continental shelf, minor to common glauconite formed (Figure 18b).

DISCUSSION

Early diagenesis in the Wilcox was largely controlled by factors such as the original texture of the sediment, mineralogic character of the detritus, amount and type of organic matter, and original water chemistry. These factors, in turn, were directly or indirectly related to depositional environment. Thus, early diagenetic patterns are very much facies related. Later diagenetic patterns tend to cross facies boundaries, and appear to be a function of regional

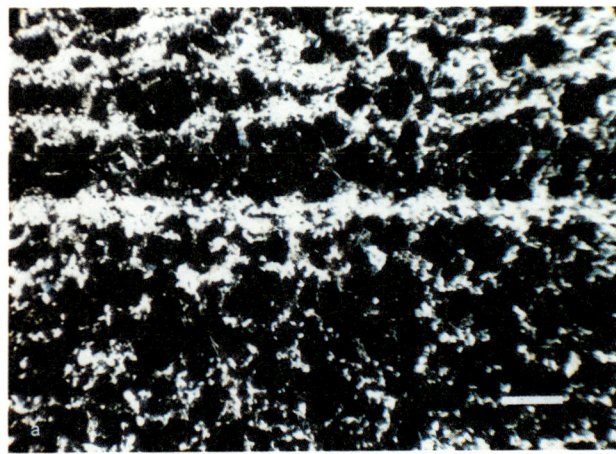 

Figure 17. Characteristics of the prodelta setting. (a) Well-sorted trains of sideritized fecal pellets like these appear to be common in prodelta deposits. Presumably they result from reworking of older delta-plain sediments. Humble #W-31 Katy, 13,108 ft; scale bar equals 0.5 mm. (b) This small patch of kaolinite (arrow) is clean, contains no mica shreds, and likely formed after the surrounding ferroan dolomite. This is typical of late-stage kaolinite, the formation of which apparently is related to organic maturation products. Getty #1 Benavides, 9689.1 ft; scale bar equals 0.1 mm.

Figure 18. Photomicrographs of continental shelf deposits. (a) Typical shelf siltstone showing abundant quartz, micas, and clays. Note how the slightly cleaner, more porous laminae are cemented with late ferroan dolomite. Sun #1 Urban, 14,052 ft; scale bar equals 0.2 mm. (b) Slightly coarser shelf sandstone contains common burrowed-in clay matrix (arrows), clayey fecal pellets (p), and glauconite grain (g). Sun #1 Urban, 14,113.6 ft; scale bar equals 0.1 mm.

fluid migration patterns. The diagenetic history in any sediment package is the result of the balance between these two regimes.

SEDIMENT COMPOSITION

Knowledge of provenance is very important for the interpretation of compositional patterns. In the northern Texas Gulf Coast, Wilcox sediment was derived from older sedimentary and metamorphic rocks in the the Appalachians and was transported through the Mississippi-Apalachicola system. As a result, this material was slightly more mature than sediment derived from the volcanic and sedimentary terranes to the south in Texas and Mexico, which was transported through the shorter Rio Grande system. But even though the source areas differed regionally along the Wilcox depositional trend, detrital composition still appears to have varied in consistent patterns primarily as a function of depositional environment.

As sediment was transported from the provenance area to the final site of deposition, it was acted upon by agents of physical and chemical weathering, which caused changes in the "maturity" of the detritus. Bates and Jackson (1980) define "mature" as "a clastic sediment that has been differentiated or evolved from its parent rock by processes acting over a long time and with a high intensity and that is characterized by stable minerals (such as quartz), deficiency of the more mobile oxides (such as soda), absence of weatherable material (such as clay), and well sorted but sub-angular to angular grains." Thus, Wilcox sediments that underwent more transport or reworking tend to be characterized by a greater proportion of chemically and mechanically stable minerals. Similarly, sorting by size or hydraulic equivalence tended to produce different detrital mineral suites in depositional environments of different energy levels (Davies and Ethridge, 1975; Doe et al., 1976). For example, basal fluvial-point-bar sands are the coarsest Wilcox sands and, regardless of position along the Texas Gulf Coast, contain the highest proportion of nondisaggregated lithic fragments. In more distal settings, such as prodeltas, the sediment contains abundant products of disaggregation—fine quartz, micas, and detrital clays—again regardless of the source area. Shoreline and tidal sands experienced in situ reworking and thus tend to be more quartzose. Therefore, as a result of the combined effects of transport and sorting, different facies have different detrital assemblages.

TEXTURE

Textural characteristics (such as whether a sediment is coarse or fine grained or whether it is predominantly sand or clay or a mixture of the two) are also a function of depositional process and, in turn, strongly affected diagenetic patterns. Detrital clays and sand grains are not hydraulic equivalents, and therefore the many facies containing mixtures of these two sediments must have experienced fluctuations in flow energy during sedimentation, resulting in clay laminae within sand layers. The spacing and distribution of these laminae is a function of how often and in what manner these fluctuations occurred. For instance, in fluvial-point-bar systems, clay laminae indicate waning flow resulting from migration of the channel away from the point of deposition, decrease in flood activity, and/or channel abandonment. In all these cases, the laminae tend to be more common near the top of the unit. In a tidal-channel or tidal-flat setting, however, flow energies varied more often and more regularly, commonly resulting in clay/mica laminae scattered throughout these sand units (Figures 13, 14a). In many other marine environments clay laminae are not commonly preserved as distinctly separate units because of the effects of burrowing organisms. The extent of burrowing ranges from isolated forms to bioturbation that thoroughly homogenized the sediment.

These differences in original Wilcox sediment textures apparently exerted a strong influence on diagenesis by controlling the extent to which water moved through or within a given sediment package. Original porosities and permeabilities of homogeneously burrowed, fine-grained, clay-rich deposits such as those of the lower shoreface, distal bar, and prodelta, were likely low because of the abundance of detrital clay, and were further reduced at an early stage by compaction. Because of the very low permeability, these deposits probably acted as systems which were closed or semiclosed (experiencing expulsion but not input) to migrating pore waters. The dominant diagenetic reactions in these sediments reflected rearrangement of materials in situ and included processes such as clay neomorphosis and minor development of quartz overgrowths (Thomas, 1978).

In coarse-grained, clay-free sediments such as those in the bases of point bars, on the other hand, porosity and permeability were probably relatively high following deposition. Lack of abundant dispersed detrital clay apparently limited the effects of compaction to minor permeability reduction caused by crushing of micas and weathered lithics and squeezing of ductile grains such as mud clasts. Because of the maintenance of good porosity and permeability through early diagenetic stages, these sands probably were preferred conduits for fluid flow (open systems). Fluid throughput appears to have been high both in terms of volume of fluid moved and rate of flow. Previous studies (Stonecipher et al., 1984) have suggested that high fluid fluxes tend to promote dissolution rather than precipitation of new cements, because solutes are flushed out of the system rather than building in concentration locally. Even where fluid fluxes were only moderate, the relative abundance of lithics, feldspars, and other nonmonocrystalline quartz components tended to prevent the occlusion of porosity by quartz overgrowth and to permit the development of secondary porosity through dissolution.

Sediments which had low initial permeabilities due to fine grain size but which lacked appreciable detrital clays, such as sediments at the tops of point bars or in backshore deposits, experienced some fluid throughput, but at very low volumes and rates. The slow movement of water through these sands, combined with the lack of competition from clays for dissolved silica (Mackenzie et al., 1967; Kastner et al., 1977), commonly resulted in cementation of fine-grained clay-free sands by quartz overgrowths.

PORE-WATER CHEMISTRY

Rapid lateral and vertical variations in early diagenetic mineralogy appear also to be related to the chemical character of the original pore water as well as to regional hydrologic regimes (Hurst and

Irwin, 1982) (Figure 19). For example, fluvial and distributary-channel sediments were deposited in highly agitated, oxygenated fresh water; fresh water likely continued to circulate through these sediments throughout shallow burial (approximately 1.5 km; Galloway, 1984; Harrison, 1989). In contrast with the well-oxygenated fluvial system, bottom water in delta-plain lakes, although also generally fresh, tended to be anoxic because of low turnover and high organic content. The low permeability of these shaly lake sediments probably precluded later introduction of more oxygenated waters and, as a result, evidence of anoxic conditions is preserved. Sediments deposited in prodelta, marine-bay, tidal-flat, and lagoonal environments, on the other hand, initially contained seawater in their pores. Compaction of underlying and/or downdip marine shales probably resulted in continued movement of evolved marine fluids through these sediments. Distal sandy parts of the deltaic system, including distal distributary-channel sands, tidal-channel sands, shoreline sands, and distributary-mouth-bar sands, were also apparently influenced by this compactional regime. Early features indicating freshwater influence are commonly overprinted by those typical of marine fluids. Each of these different chemical environments produces a different pattern of early diagenesis in the Wilcox, as discussed below.

Marine vs. Fresh Water

The most common product of early diagenesis in Wilcox sediments deposited in well-oxygenated fresh water (such as fluvial and distributary-channel sands) is kaolinite. Most of this early kaolinite appears to have formed from the alteration of micas. The constant flushing by fresh meteoric water appears to have stripped the mica of interlayer cations and hydroxide-layer components, allowing the layers to fan or spread apart. Silica and aluminum were then reprecipitated as kaolinite between the remnant mica layers (Figure 7a), perhaps using them as a template. Kaolinite formed in this manner tends to occur as large, coarse-grained booklets or fans (Figure 7b, c), which commonly contain remnant shreds of mica between the kaolinite platelets. Mica-derived kaolinite such as this always seems to form very early in the paragenetic sequence and appears to strongly indicate diagenesis in a freshwater environment. In some instances, kaolinite also apparently formed in marine sediments. In these cases, it generally appears to have formed later in the diagenetic history from feldspars, probably associated with the introduction of CO_2 or organic acids generated during the alteration of humic organic matter. Kaolinite produced in this manner typically occurs as small equant booklets without any mica shreds (Figure 7d, e).

In contrast, the most prevalent early diagenetic product of Wilcox sediment deposited under the influence of marine pore waters is authigenic chlorite. Depending on the original porosity and permeability of the sediment and the proximity and thickness of surrounding marine shales, these rims can either be thick and well developed (Figure 14b) or patchy and incomplete (Figure 7f). They generally have inhibited the development of quartz overgrowths, and thereby have preserved large amounts of primary porosity.

One of the most diagenetically interesting areas of the Wilcox deltas is unfortunately also one about which we know the least geochemically, that is, the zone where fresh water flowing through fluvial sands encounters and mixes with marine water. One environment characteristic of this zone is the distributary-mouth bar. An unusual occurrence typical only in distributary-mouth-bar sediment is the presence of chamosite or berthierine (also called guyanite; Odin and Matter, 1981) ooids (Figure 15a). Chemically and structurally chamosite lies somewhere between an iron kaolinite and chlorite, in contrast with glauconite, which is more like an iron-rich mica. So far, chamosite ooids have been found to form at the sediment/water interface only where fresh water from rivers flows into marginal marine areas (Odin and Matter, 1981).

Another environment typical of this boundary zone is the shoreline. Here fresh and marine waters commonly mix to form a brackish water system. Although few samples were taken from Wilcox shoreline sediments, illite is believed to be more common in the lower and upper shoreface, lagoon, tidal-flat, and tidal-channel deposits than in other parts of the delta (Figure 6, i–m). This agrees with the geochemical modeling work of Meshri (personal communication, 1987), which indicates that illite should be the most stable clay mineral in a brackish water system.

Oxic vs. Anoxic Conditions

Oxidizing vs. reducing conditions in the original depositional environment are most clearly shown by the character of authigenic iron minerals in the sediments. Berner (1981) divided sedimentary environments into four basic classifications (Table 2) based on the presence or absence of dissolved oxygen and dissolved sulfate in the sediment at the time of authigenic mineral formation. Iron and manganese mineral associations, which are diagnostic of each of the environments, are also listed in Table 2. In the Wilcox, perhaps the clearest demonstration of diagenetic controls related to oxygenation occurs in crevasse-splay sediments. Splay deposits from freshwater delta-plain lakes likely experienced primarily methanic diagenesis. Bottom waters in such lakes were not only fresh (and, most importantly, sulfate free), but also, because of the high organic input and low turnover, tended to be anoxic. In such an environment siderite should be the most common authigenic iron mineral (Berner, 1981), and, indeed, splay delta sands associated with lakes are commonly heavily cemented with siderite (Figure 10a, Table 1). Splays formed in marine bays,

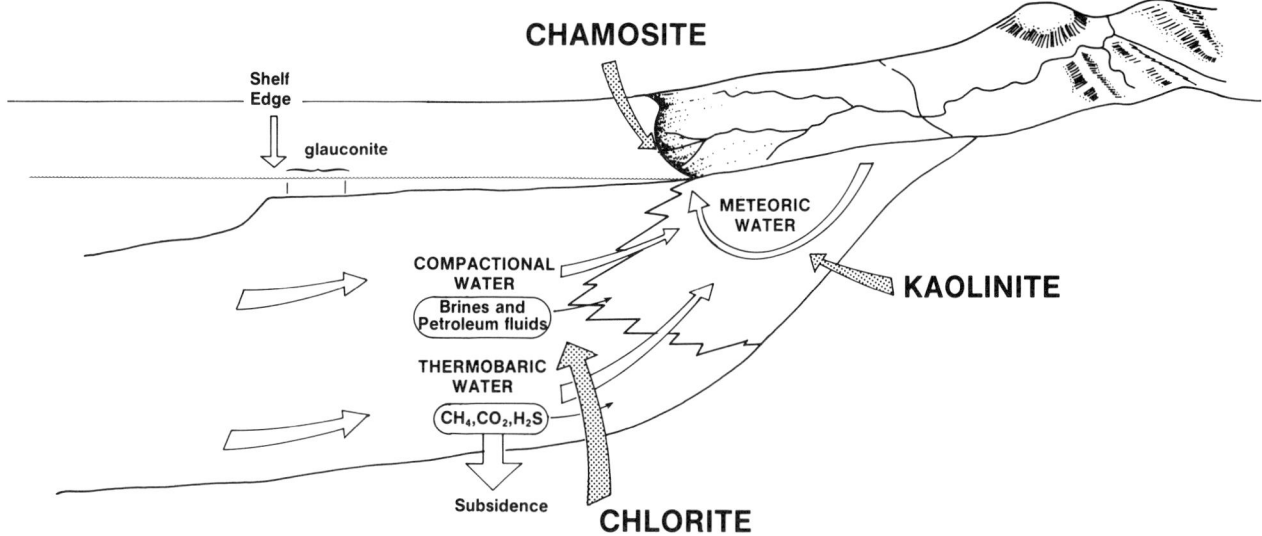

Figure 19. Hydraulic regimes and associated authigenic clays for typical continental margin.

Table 2. Geochemical classification of sedimentary environments[1]

Environment	Characteristic Phases
I. Oxic ($CO_2 \geq 10^{-6}$)	Hematite, goethite, MnO_2-type minerals; no organic matter
II. Anoxic ($CO_2 < 10^{-6}$)	
A. Sulfidic ($CH_2S \geq 10^{-6}$)	Pyrite, marcasite, rhodochrosite, alabandite; organic matter
B. Nonsulfidic ($Ch_2S < 10^{-6}$)	
1. Postoxic	Glauconite and other Fe^{+2}-Fe^{+3} silicates (also siderite, vivianite, rhodochrosite); no sulfide minerals; minor organic matter
2. Methanic	Siderite, vivianite, rhodochrosite; earlier formed sulfide minerals; organic matter

[1]After Berner (1981).

on the other hand, developed a different early diagenetic assemblage (pyrite and chlorite) (Figure 10b).

SUMMARY

The relationship of depositional environment to sediment character, pore-water chemistry, and ultimately to diagenesis is clearly illustrated in Wilcox deltaic sediments, where rapid lateral and vertical changes in depositional environments produced markedly different diagenetic patterns in sand units only a few feet or even inches apart. Controls on early diagenesis included the following factors:

(1) As a result of the combined effects of transport and sorting, different facies have characteristically different detrital mineral assemblages.

(2) Texture influenced diagenesis by controlling fluid throughput. Clay-rich sediments acted as "closed" systems characterized by clay neomorphosis and minor quartz precipitation. Fine-grained sediments acted as "semiclosed" systems with common quartz precipitation and alteration of unstable grains such as micas. Coarse-grained sediments acted as "open" systems characterized by the alteration or dissolution of unstable grains and the development of secondary porosity, as well as the potential for developing late-stage externally derived cements (such as carbonates).

(3) The chemistry of the original pore waters is also reflected in early diagenetic patterns. For example, fluvial/distributary-channel sands flushed by fresh meteoric water are characterized by mica-derived kaolinite. Lakes (and associated splay sands) associated with fresh, but anoxic, water contain abundant siderite. Marine sands flushed by saline pore waters typically exhibit chlorite rims. Marine sediments with reducing or mildly reducing waters contain pyrite or glauconite. Distributary-mouth-bar sands which were rapidly deposited in the mixing zone between fresh and marine waters commonly contain chamosite/berthierine ooids. Shoreline sands commonly contain abundant illite.

Based on our study of the Wilcox, these patterns suggest that lateral and vertical differences in diagenetic patterns, defined on the basis of sediment texture, detrital composition, and early authigenic cements, can be used to complement, and test, traditional sedimentologic methods of interpreting depositional environment.

REFERENCES CITED

Alabach, D. C., 1979, The depositional history of the uppermost Wilcox (lower Eocene) of west-central Beauregard Parish, Louisiana: M.S. thesis, Louisiana State University, Baton Rouge, Louisiana, 98 p.

Bates, R. L., and J. A. Jackson, 1980, Glossary of geology: Falls Church, Virginia, American Geological Institute, 749 p.

Berner, R. A., 1981, A new geochemical classification of sedimentary environments: Journal of Sedimentary Petrology, v. 51, p. 359-365.

Boles, J. R., and S. G. Franks, 1979, Clay mineral diagenesis in Wilcox of south-west Texas: implications of smectite diagenesis in sandstone cementation: Journal of Sedimentary Petrology, v. 49, p. 55-70.

Chan, M., J. B. Hayes, and J. C. Harms, 1981, Petrofacies and depositional environments of Upper Jurassic Naknek Formation, Lower Cook Inlet (Alaska) (abs.): AAPG Bulletin, v. 65, p. 909-910.

Coleman, J. M., and D. B. Prior, 1980, Deltaic sand bodies: AAPG Education Course Note Series 15, 171 p.

Davies, D. K., and F. G. Ethridge, 1975, Sandstone composition and depositional environment: AAPG Bulletin, v. 59, p. 239-264.

Doe, T. W., R. H. Dott, Jr., and I. E. Odom, 1976, Nature of feldspar grain-size relationships in some quartz-rich sandstones: Journal of Sedimentary Petrology, v. 46, p. 862-870.

Edwards, M. B., 1981, Upper Wilcox Rosita delta system of south Texas: growth-faulted shelf-edge deltas: AAPG Bulletin, v. 65, p. 54-73.

Elliott, T., 1986, Deltas, in H. G. Reading, ed., Sedimentary environments and facies: New York, Elsevier, p. 97-142.

Fisher, W. L., 1969, Facies characterization of Gulf Coast Basin delta systems, with some Holocene analogues: Gulf Coast Association of Geological Societies Transactions, v. 19, p. 239-261.

Fisher, W. L., and J. H. McGowen, 1967, Depositional systems in the Wilcox Group of Texas and their relationship to occurrence of oil and gas: Gulf Coast Association of Geological Societies Transactions, v. 17, p. 105-125.

Galloway, W. E., 1968, Depositional systems of the lower Wilcox Group, north-central Gulf Basin: Gulf Coast Association of Geological Societies Transactions, v. 18, p. 275-289.

Galloway, W. E., 1984, Hydrologic regimes of sandstone diagenesis, in D. A. McDonald and R. C. Surdam, eds., Clastic diagenesis: AAPG Memoir 37, p. 3-14.

Hardin, G. C., 1962, Notes on Cenozoic sedimentation in the Gulf Coast geosyncline, U.S.A., in E. H. Rainwater and R. P. Zingula, eds., Geology of the Gulf Coast and Central Texas and Guidebook of Excursions: Houston, Texas, Houston Geological Society, p. 1-15.

Harris, J. R., 1962, Petrology of the Eocene Sabinetown-Carrizo contact, Bastrop County, Texas: Journal of Sedimentary Petrology, v. 32, p. 263-283.

Harrison, W. J., 1989, Modeling fluid/rock interactions in sedimentary basins, in T. A. Cross, ed., Quantitative dynamic stratigraphy: New York, Elsevier, p. 195-232.

Hurst, A., and H. Irwin, 1982, Geological modelling of clay diagenesis in sandstones: Clay Minerals, v. 17, p. 5-22.

Hutcheon, I., A. Oldershaw, and E. D. Ghent, 1980, Diagenesis of Cretaceous sandstones of the Kootenay Formation at Elk Valley (southeastern British Columbia) and Mt. Allen (southwestern Alberta): Geochimica et Cosmochimica Acta, v. 44, p. 1425-1435.

Kantorowicz, J. D., 1985, The petrology and diagenesis of Middle Jurassic clastic sediments, Ravenscar Group, Yorkshire: Sedimentology, v. 32, p. 833-853.

Kastner, M., J. B. Keene, and J. M. Gieskes, 1977, Diagenesis of siliceous oozes, I, Chemical control on the rate of opal-A to opal-CT transformation—an experimental study: Geochimica et Cosmochimica Acta, v. 41, p. 1041-1059.

Loucks, R. G., M. M. Dodge, and W. E. Galloway, 1979, Sandstone consolidation analysis to delineate areas of high-quality reservoirs suitable for production of geopressured geothermal energy along the Texas Gulf Coast: Austin, Texas, Texas Bureau of Economic Geology, report prepared for U.S. Department of Energy, Division of Geothermal Energy, 98 p.

Mackenzie, F. T., R. M. Garrels, O. P. Bricker, and F. Bickley, 1967, Silica in seawater: control by silica minerals: Science, v. 155, p. 1404-1405.

Mann, C. J., and W. A. Thomas, 1968, The ancient Mississippi River: Gulf Coast Association of Geological Societies Transactions, v. 18, p. 187-204.

Murray, G. E., 1961, Geology of the Atlantic and Gulf Coast province of North America: New York, Harper and Brothers, 692 p.

Murray, G. E., and E. P. Thomas, 1945, Midway-Wilcox surface stratigraphy of Sabine uplift, Louisiana and Texas: AAPG Bulletin, v. 29, p. 45-70.

Odin, G. S. and A. Matter, 1981, *De glauconarivm origine:* Sedimentology, v. 28, p. 611-641.

Oliver, R. D., 1980, Depositional environments of the Upper Wilcox Group (Eocene) in Shelby County and portions of Panola and Harrison counties, Texas: M.S. thesis, Stephen F. Austin State University, Nacogdoches, Texas, 98 p.

Siever, R., 1979, Plate-tectonic controls on diagenesis: Journal of Geology, v. 87, p. 127-155.

Stalder, P. J., 1975, Cementation of Pliocene-Quaternary fluviatile clastic deposits in and along the Oman Mountains: Geologie en Mijnbouw, v. 54, p. 148-156.

Stonecipher, S. A., R. D. Winn, Jr., and M. G. Bishop, 1984, Diagenesis of the Frontier Formation, Moxa arch: a function of sandstone geometry, texture and composition, and fluid flux, in D. A. McDonald and R. C. Surdam, eds., Clastic diagenesis: AAPG Memoir 37, p. 289-316.

Thomas, J. B., 1978, Diagenetic sequences in low-permeability argillaceous reservoirs: Quarterly Journal of the Geological Society of London, v. 135, p. 93-100.

Williams, C. E., L. R. Travis, and E. M. Hoover, 1974, Depositional environments interpreted from stratigraphic, seismic, and paleoenvironmental analyses: upper Wilcox Katy field, Texas: Gulf Coast Association of Geological Societies Transactions, v. 24, p. 129-137.

An Overview of Chemical Models and Their Relationship to Porosity Prediction in the Subsurface

Indu D. Meshri
Amoco Production Company
Tulsa, Oklahoma, U.S.A.

> One of the most important problems in petroleum exploration is the capability to predict the type, amount, and location of porosity in potential reservoir rocks ahead of the drill bit. This paper discusses the concepts needed to acquire such information from computerized chemical models, in order to calculate the type and the amount of diagenetic minerals, and hence the porosity modification. The location of porosity may also be predicted when physical mechanisms of mass transfer, such as fluid flow and diffusion, are appropriately coupled with the chemical reactions.
>
> This paper attempts to classify available chemical models into three broad categories—speciation-solubility, reaction-path, and reaction-transport models—to provide a general background for the models referred to in this volume. A brief discussion of theory, inputs and outputs, names of a few (but not all) available codes, and their limitations and applications is presented.
>
> The papers in this volume affirm the increasing reliability of predictions made by using geochemical models. However, the reliability of predictions can be improved if we continue correcting the inaccuracies in thermodynamic and kinetic data and increasing our understanding of mass-transfer mechanisms. In spite of the limitations discussed in this paper, the concepts of chemical modeling based on reversible and irreversible thermodynamics, equilibrium and disequilibrium, and open vs. closed systems, remain of great value in understanding natural rock-water systems. Other advantages of diagenetic simulations using geochemical modeling are: (1) they elucidate the chemical processes responsible for diagenesis, (2) they are much faster than empirical predictive models, which are based on numerous labor-intensive petrographic observations, and (3) they provide for many possible scenarios.
>
> Chemical models are not only appropriate for predicting subsurface porosity but also can be applied toward the solution of problems involving environmental geochemistry and nuclear waste disposal.

INTRODUCTION

Nordstrom et al. (1979) reviewed more than 30 computerized chemical models and described the major differences and similarities between various aqueous-speciation and reaction-path models. Although a few reaction-path calculations were mentioned, the major thrust of their 1979 review centered on aqueous-speciation codes. Reaction-transport models were not available a decade ago.

These investigators sent identical river water and seawater compositions to the modelers and published a comparison of aqueous-speciation and saturation indices (log Q/K) from 13 different models. Calculated activities of major aqueous species were found to be in good agreement, but those of the minor aqueous species showed poor agreement. They concluded that poor agreement for minor species was caused by inconsistent numbers of aqueous complexes in the models and differences in the thermodynamic data bases therein.

The principal thrust of the present review is to divide chemical models into three major types and to document the kinds of information that can be obtained from these models in terms of predictive diagenesis, as well as the type, amount, and timing of diagenetic cements—and therefore the evolution of porosity in space and time (Table 1, Figure 1).

Table 1. Chemical models of diagenesis

No.	Model Type	Theory	Input	Output	Available Computerized Models	Limitations
1.	Speciation-Solubility	Equilibrium thermodynamics. Reversible chemical Eqs. Debye-Hückel Theory of ion assoc. Van't Hoff's extrapolation for heat capacities of minerals, or Temperature dependence of equilibrium constants.	Temperature, Eh, pH of present-day formation water. Water analysis concentration of major and minor ions.	Aqueous spec. (ions complexions, assoc. ions, multivalent ions). Stab. of min. with respect to formation water. Mole ratios of ions.	WATEQ (Truesdell & Jones, 1974, U.S.G.S.) WATEQF (Plummer, et al., 1976, USGS) SOLMINEQ (Kharaka and Barnes, 1973, USGS) EQ3 (Wolery, 1979, 1983, 1986 Lawrence Livermore)	Ion concentration below 35K TDS Temperature below 100°C Ion concentration below 35K TDS Temperatures below 350°C Ion concentration up to 100K TDS Temperatures below 350°C
2.	Reaction-Path	Equilibrium Thermodynamics. Sep. of prod. and reactant into subsystems, thereby indicating a path. Debye-Hückel Theory (ext.) Mair-Kelley power function for heat capacities. Exten. of other thermodyn. functions by Helgeson's program SUPCRT	Composition of rock (e.g. Litharenite, quartzarenite, arkose, subarkose, etc.). Composition of initial water (assumed major and minor ions). Burial history converted to thermal history.	Irreverse rock-water interactions, indicating path and changes in product and reactants in terms of amts. of mineral cements. Qualitative simulations of diagenesis.	PATHI (PATHCALC) (Helgeson, U. Cal Berk.) EQ6 (Wolery, 1979, Lawrence Livermore) PHREEQUE (Parkhurst, 1980)	Very long iteration times (failure to converge) Improved iteration times, but prediction of amounts of cements appears to be inaccurate. Work continues to assess pitfalls.
3.	Reaction-Transport	Nonequilibrium conditions in which chemistry is function of reaction rates. Coupling of chemical reactions with transport via diffusion and fluid flow and texture dependence of permeability.	Reaction coefficients, surface area of reacting minerals. Water analysis (initial composition). Rock analysis (initial composition). Reaction rates. Surface area of reacting mineral. Subsidence profile of formation under consideration.	Nonequilibria rock-water reactions as a function of time. Porosity prediction in space and time.	REACTRAN (Ortoleva et al., 1987b, Geo-Chemical Research) REACTRAN (Lichtner, 1987) GEOKIN (Steefel & Lasaga, 1990)	Limited data base of reaction-kinetics. Appropriate pH and temperature dependence of rate constants lacking for accurate evaluation of porosity evolution in space and time.

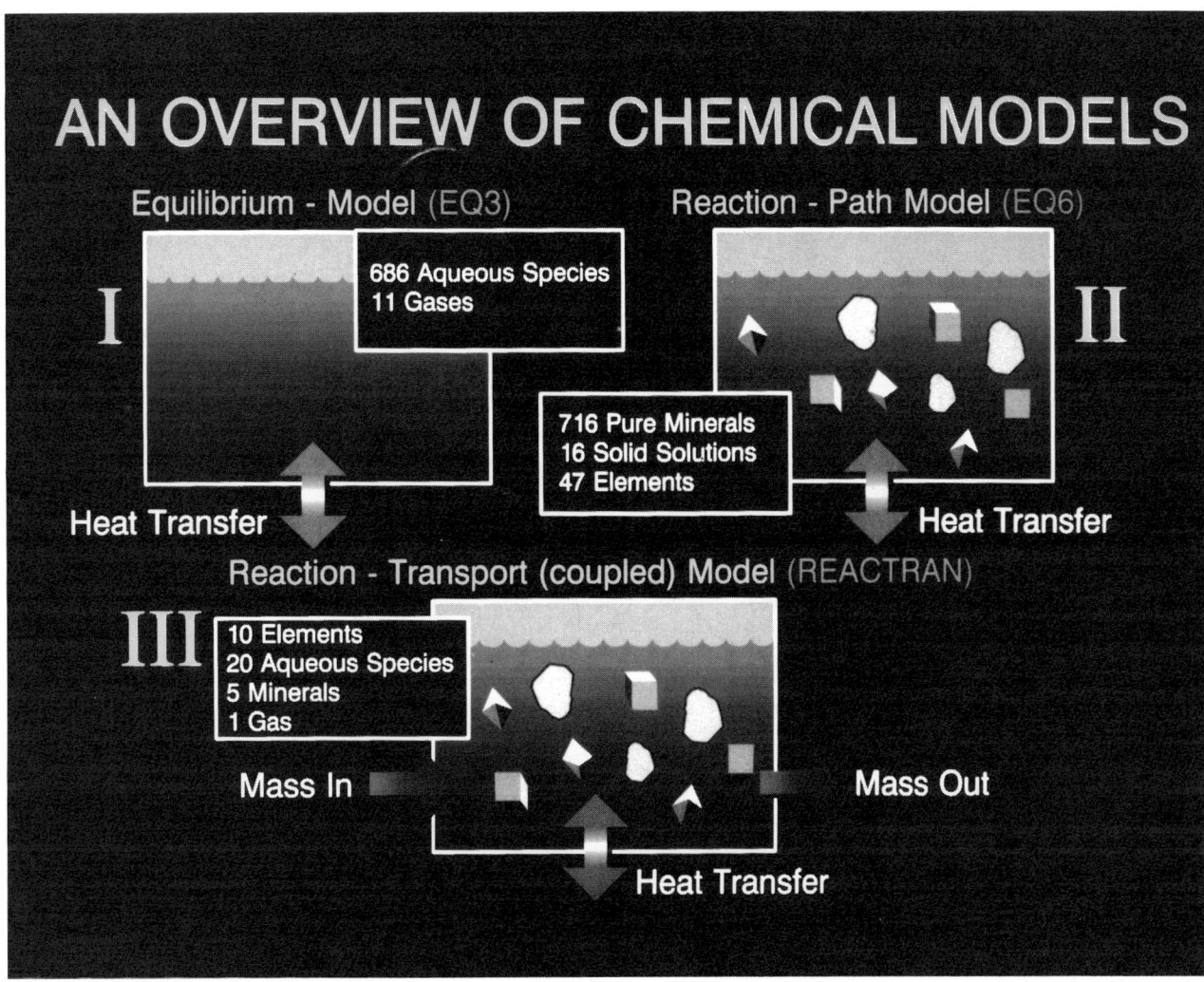

Figure 1. An overview of chemical models

In the past, great emphasis has been placed on single mechanisms for generating secondary porosity; for example, carbon dioxide (Schmidt and McDonald, 1979) and organic acids (Surdam et al., 1984). Although organic acids could keep more calcium and possibly more aluminum in solution, they could also assist in cementation if temperature is hot enough and decarboxylation occurs (Meshri, 1986). Diagenetic events are the result of complex chemical processes that operate through all stages of burial. These complex processes are coupled in chemical models, and thus can provide a better chance of understanding diagenesis if chemical models are used accurately. Porosity modification during diagenesis can be calculated when the volumes of dissolving and precipitating minerals—in other words, diagenetic events—can be quantified in terms of type, amount, and location.

The procedure for developing a conceptual model of any physical or chemical system can be generalized as follows. The first step is to understand the physical or chemical behavior of the system. Cause-effect relationships are determined, and a conceptual model of how the system operates is formulated. The next step is to translate the physics or chemistry into mathematics, that is, to make appropriate simplifying assumptions and develop the governing equations. This constitutes the mathematical model. Once the mathematical model is formulated, the next step is to obtain a solution using one of two general approaches—analytical or numerical. In the analytical approach, one simplifies equations so that solutions may be obtained by analytical methods. In the numerical approach, space and time are discretized and the resulting set of equations is solved on the computer by using an algorithm. The analytical approach is only successful in the very simplest systems.

A conceptual model for porosity prediction can be designed by understanding (a) the relationship between the chemistry of rock-water interactions via equilibrium thermodynamics for fast reactions, and via reaction kinetics for slow reactions and (b) mass transport of dissolved mineral components via processes like diffusion and/or fluid flow.

SPECIATION-SOLUBILITY MODELS

Speciation-solubility models are static models of an aqueous solution, such as a natural formation water. They are commonly used to test whether chemical reactions representing mineral dissolution and precipitation are at or near a state of thermodynamic equilibrium. The corollary follows that they also provide a measure of disequilibrium and hence a measure of thermodynamic drive for the chemical reaction in question. The speciation-solubility models can be used to calculate activities of ions and ion pairs from the analytical concentrations of ions in water. These models generally come complete with the thermodynamic data for a large number of aqueous, mineral dissolution and precipitation, and redox reactions.

Theory

The theory used in these models is that of equilibrium thermodynamics as applied to aqueous systems, which is adequately described in many excellent textbooks (Garrels and Christ, 1965; Drever, 1982). For a more comprehensive treatment of the subject, the reader is referred to physical chemistry texts such as Glasstone (1946) and to special texts on chemical thermodynamics, such as those of Wall (1958) or Lewis and Randall (1961).

Natural waters are aqueous solutions of electrolytes, and all aqueous-speciation models use either a well-known electrochemical theory, propounded by Debye and Hückel (1923), or some extension of it. The electrolyte theories consider the effect, on a single ion, of electrical interactions with the other ions by assuming that oppositely charged ions form a spherical shell around the ion, and that ions are point charges. This assumption is valid only in dilute solutions, but further modifications of the theory allow extensions in concentrated aqueous electrolytes like subsurface brines. A more complete and rigorous theory that can be applied to brines has been proposed by Pitzer (1977), who said of the Debye-Hückel theory, "it gives correct equation for the behavior of electrolytes in the limit of low concentration."

A reversible chemical equilibrium is represented in aqueous-speciation models for a general chemical reaction. A reversible equilibrium simulates running a reaction in a closed vessel, an isolated formation, or a portion of a formation. Under this condition, the thermodynamic system is rock + water + gases, and none of these is allowed to escape. In addition, no mass transfer can be accepted from the surroundings, but energy in the form of heat can be exchanged with the surroundings. In the case of burial, changes in temperature and pressure cause disequilibrium, which is indicated by undersaturated or oversaturated aqueous phases.

Input and Output

A typical input of these models is water-analysis data, including concentration of ions and gases, pH, Eh, and pressure and temperature constraints. A typical output is aqueous speciation at the given temperature and pressure, mole ratios of ions, and the saturation indices for mineral dissolution and precipitation, which are defined below. Speciation results are expressed in molal concentration. Hence, all masses of species and components are relative to 1 kg of water. Mole ratios of ions become a valuable tool in comparison of waters. The affinity (tendency) of mineral dissolution and precipitation is indicated by saturation indices (log Q/K values) of a fluid, with respect to mineral phases. The saturation index can provide a measure of thermodynamic disequilibrium of a particular natural water with the rock composition of the formation.

Available Codes

Many codes are now available, such as SOLMINEQ (Kharaka and Barnes, 1973; Kharaka et al., 1988), WATEQ (Truesdell and Jones, 1974; Plummer et al. 1976), and EQ3/EQ6 (Wolery, 1979, 1983, 1986). The writer has worked with these codes only, and a brief description of each is provided.

Early on (during 1979-1983), we used a version of WATEQ (Truesdell and Jones, 1974). The later revision by Plummer et al. (1976) translated the code from PL-1 to Fortran IV and added several species of manganese to the computer program. Included in this software are thermodynamic data for 193 reactions. This computer program used the Debye-Hückel theory to calculate activity coefficients.

The 1988 version of SOLMINEQ has a database of equilibrium constants precalculated for a temperature range of 25-350°C at intervals of 25°C. This version contains 80 organic and 260 inorganic aqueous species and 220 minerals. Aside from dissolution-precipitation reactions, boiling and mixing of solutions, partition of gases between oil and water phases, and effects of ion-exchange adsorption-desorption can also be studied.

We obtained EQ3.3230 from the National Energy Software Center, Argonne National Laboratory, in 1985. Later, we obtained EQ3NR.3245 from Lawrence Livermore National Laboratory. This software has three options for calculating activities. EQ3NR allows calculation of activities by (1) an extension of the Debye-Hückel equation for relatively dilute waters, (2) Helgeson's B-dot option for concentrated waters, and (3) Pitzer formulations for certain major ions in heavy brines. The temperature range allowed by EQ3 is 0-250°C. The pressure is determined by the steam saturation curve of water at the corresponding temperature. Differential equations can be solved by using the Newton-Raphson iteration method, which provides faster convergence. Detailed discussions of the computer program and input/output constraints are found in Wolery (1979, 1983, 1986).

Applications

In spite of these limitations, we have had reasonable predictions of disequilibrium for commonly occurring diagenetic cements in several formations, one of which, the Upper Almond Sandstone from the Rocky Mountains, is discussed in a companion paper (Meshri and Walker, 1990).

Our tests also show that EQ3 correctly predicts the formation of gypsum crystals from the Great Plains (Oklahoma) waters, which have salinities of about 300 ppt. More tests are needed before a convincing upper limit for the salinity of aqueous solutions can be assessed.

Limitations

In WATEQ the calculation of equilibrium constants at formation temperatures (>25°C but <100°C) was accomplished by using Van't Hoffs' relation, which assumes constant heat capacity. The effect of pressure on the thermodynamic constants was neglected. In principle, the code should be used only for formation waters having a maximum salinity of about 0.1 molal.

Although it gives more accurate estimates of activities in natural formation waters because of many activity correction options, EQ3 is limited in scope for diagenesis because it is an aqueous-speciation model, and all models in this class are static models of rock-water systems. EQ3 is coupled with EQ6, which is a reaction-path model discussed in the next section.

REACTION-PATH MODELS

Aqueous-speciation models discussed in the previous section tend to be static models of an aqueous system. Such models are really inadequate when representing natural systems, which are more likely to be open and to exchange energy and reactants with their surroundings. In contrast with the aqueous-speciation models, the reaction-path models are dynamic, because they indicate a reaction path. Irreversible reactions are modeled in addition to reversible reactions. A typical example of an irreversible reaction is the removal of aluminum in a feldspar dissolution and the capture of aluminum in the precipitation of kaolinite under sedimentary conditions; these reactions indicate a one-way path.

If a certain path is followed, the rock-water system should be modeled as irreversible and/or open.

Theory

Reaction-path models must be initialized by aqueous-speciation models. So the basic theory is that of equilibrium thermodynamics. Reaction-path models can be used to calculate partial equilibrium in aqueous systems. They allow the reaction path of a dynamically reacting system to be computed as a function of a reaction-progress variable but not as a function of true geologic time. These models divide an aqueous system into a reactant subsystem and a product subsystem. The reactant subsystem represents aqueous solution plus minerals in a heterogeneous equilibrium. The product subsystem is another aqueous solution which is not in equilibrium with the other subsystem. If the initial fluid is supersaturated with respect to any mineral phases, the program first "equilibrates" it and calculates the precipitated phases and a new composition of fluid. Reaction then progresses in terms of a response to pressure and temperature changes.

Input and Output

Typical inputs are the mass and mineral composition of the reacting rock and the composition of the reacting water and gas phases. In the burial mode, thermal changes are also needed. Typical output consists of computations that describe the progress of many pertinent chemical reactions in either a flow-through system, or across a thermal gradient, or as a simple titration due to change in concentration of constituents.

Available Codes

The first application of computer techniques to problems of mass transfer in geochemistry began with the work of Helgeson and colleagues at the University of California, Berkeley (Helgeson 1968; Helgeson et al., 1969). The computer program PATHI, also known as PATHCALC, was the first model to attempt mass transfer. It was used to generate models of geochemical processes such as diagenesis and metamorphism. Several difficulties were encountered in using PATHI, such as very long execution times and occasionally abnormal program termination (Wolery, 1979).

Wolery (1979) has written a program similar to PATHI called EQ6, which uses the Newton-Raphson method to solve the system of algebraic equations, instead of their differential counterparts, at each stage of reaction progress. Finite-difference expressions of high order are used to estimate derivatives with respect to reaction progress and then to predict the values of a basis set of unknowns at a subsequent point of reaction progress. The predicted values, which correspond to volume of diagenetic minerals or moles of aqueous species, are then corrected by using the Newton-Raphson iteration method to satisfy the original algebraic equations.

Applications

The major objective of EQ6 was to understand the concentration of radionuclides during disposal of nuclear waste, as documented by many applications discussed in the Proceedings of the Workshop on Geochemical Modeling (Jackson and Bourcier, 1986). The application of interest for porosity prediction is the capability of a reaction-path model to simulate diagenetic sequences. We have demonstrated use of EQ6 in the simulation of diagenetic sequences in the Upper Almond Sandstone (Meshri, 1985; Meshri and Walker, 1990). Bruton (1986) also demonstrated a similar simulation of a generalized diagenetic sequence by using EQ6.

Limitations

In the reaction-path codes the treatment of kinetics is done by using a reaction-progress variable that can generate ambiguities because it does not represent true time. The treatment of solid solutions is limited by availability of thermodynamic data.

Equilibrium constants are corrected for subsurface temperatures but not for subsurface pressures. Although the pressure dependence of equilibrium constants is relatively small for most silicate reactions, this may not be true for carbonate reactions where gas components such as carbon dioxide equilibria vary with pressure.

The approach discussed under the two equilibrium models (aqueous-speciation and reaction-path) does not account for movement of constituents in and out of the system as a function of mass-transfer processes like diffusion and fluid flow. Lichtner (1985) pointed out that reaction-path models for open systems attempt to characterize flowing systems with large Peclet numbers and surface-controlled mineral dissolution rates. However, the distribution of fluid and rock composition in space is not actually determined.

REACTION-TRANSPORT MODELS

Petrographers have provided ample evidence for spatial arrangement of diagenetic minerals in terms of localization of cements. These diagenetic cements, if formed at an appropriate time, can become diagenetic traps for petroleum. Only a few diagenetic traps have been documented thus far (Meshri, 1981, 1989; Wagner and Matthews, 1982). Such traps can be predicted by reaction-transport modeling (Meshri, 1989).

Ortoleva (1979), Harvie et al. (1984), Lichtner (1985), and Ortoleva et al. (1986, 1987a, b) have demonstrated that conditions far from equilibrium may persist in flowing systems over geologic time, and these conditions can be modeled by using reaction-transport equations.

Theory

The general theory behind reaction-transport modeling is based on conservation of mass equations accounting for mass transport and reactions in porous media. These equations are derived from mass-balance bookkeeping, wherein the system is conceptually divided into representative volume elements that must be smaller than the spatial scale to be modeled. Darcy's law is used to relate the fluid velocity to the fluid pressure and mass density. For details of the theory and equations, the reader is referred to Ortoleva et al. (1982, 1987a, b, 1990), Harvie et al. (1984), Lichtner (1985, 1988), and Steefel and Lasaga (1990). An attempt is also made to include mechano-chemical effects, such as pressure solution, in reaction-transport models (Dewers and Ortoleva, 1989a, b, 1990).

Input and Output

The rock-water system is defined as a domain in one- or two-dimensional space where diagenetic changes are to be observed. The diagenesis of interest is driven either by imposed flows or by changes in temperature or pressure during burial. The initial conditions at a time arbitrarily chosen as zero must be defined. The initial conditions include the spatial dimension, data on the spatial gridding to be used, spatial distribution of initial rock composition, and composition and fluid velocity of the imposed fluid. The variables that describe the spatial distribution or rock and fluid composition are evolved in time. The output is generated as a pattern in one- or two-dimensional space for a specified sequence of times. These patterns include spatial variation of ion concentrations in pore fluids, mineral concentrations, porosity and permeability patterns, temperature, and fluid velocity, which is modified through the texture-dependence of permeability.

Available Codes

A reaction-transport code that simulates diagenesis in one-dimensional space and geologic time has been commercially available from Geo-Chemical Research Associates since 1985 (Ortoleva et al., 1987b). Lichtner (1985) also discussed a similar simulation in one-dimensional space and time, but this calculation did not have a complete feedback loop through texture-dependence of permeability (and therefore fluid velocities) and did not allow for user-friendly development of the chemical model. Later, Lichtner (1988) improved these calculations to include a complete feedback loop.

Recent improvements in these models include two-dimensional space and time simulation capability. Steefel and Lasaga (1990) have generated a two-dimensional reaction-transport simulator named GEOKIN. Current versions, however, can only cope with one or two minerals and possibly three to four aqueous reactions. Geo-Chemical Research Associates has recently developed a two-dimensional, general chemistry, fully coupled reaction-transport code that incorporates adaptive gridding and multigridding, and temperature-dependent rate and equilibrium relations.

Applications

Spatial and temporal patterns of diagenesis can be modeled on the reservoir scale (Moore and Ortoleva, 1990). In general, reaction fronts can be studied by the modeling of coupled reaction-transport processes. Application of reaction-transport models can be on the reservoir scale in terms of diagenetic patterns, and on the subgrain scale in the observations of periodic crystal zoning (Pearce and Kolisnik, in press; Reeder et al., in press; Shimizu, in press).

Mechano-chemistry (stress-induced dissolution) is another area where principles of nonequilibrium concepts can be applicable (Dewers and Ortoleva, 1990). Such modeling is useful in understanding compaction, stylolites and repetitively banded alternations of compaction and cementation in the context of preserving porosity and permeability, and the genesis of diagenetic pressure seals at depth. When reaction-transport modeling is coupled with fracturing dynamics, one may also follow the episodic and sometimes periodic release of fluids from sealed domains during burial (Ortoleva et al., 1990).

Limitations

Long execution times are currently a necessary evil when geologically reasonable times and kilometer-scale systems are modeled. Since a typical REACTRAN run covering 70 million years of geologic time can take 2 hours of super computer time, more work is needed on faster numerical techniques. Methods of adaptive gridding and multigridding, such as those used in weather prediction, if employed in reaction-transport modeling, could speed up the simulations. The new adaptive gridding version of REACTRAN (REACTRAN 3.0) and the adaptive gridding, multigridding program GENEX, hold great promise for future work in reaction-transport modeling of diagenesis.

INPUT DATA FOR DIAGENETIC MODELING

Most chemical models come complete with extensive thermodynamic data bases, and reaction-transport codes must also contain reaction-rate data.

Thermodynamic Data

Chemical mechanisms are represented by chemical reactions. A well-known example is the formation of calcite,

$$Ca^{++} + CO_3^{--} \rightleftharpoons CaCO_3.$$

Stability of this transformation is indicated by an equilibrium constant ($K = 10^{-8.6}$) at 25°C and 1 bar. More than 600 reactions are presently accounted for in aqueous-speciation models like EQ3/6.

The thermodynamic constants compiled in the chemical models for each reaction are: (1) heat of reaction, or enthalpy, (2) free energy of reaction, (3) entropy of reaction, (4) equilibrium constant for reaction, and (5) activity coefficients for ions and ion pairs.

The standard entropies of formation for all known chemical substances such as ions and minerals are tabulated in thermodynamic compilations. Using these data, free energies of reaction are calculated and converted to equilibrium constants.

Activity coefficients of ions and ion pairs are functions of formation-water composition and are calculated using either the Debye-Hückel, the extended Debye-Hückel, or the Pitzer theory.

Kinetic Data

In principle, kinetic properties should be predictable from quantum and statistical mechanics and a kinetic theory. However, because of the complexity of time dependent problems, a rigorous and a generally useable theory of chemical kinetics is difficult to develop. In contrast, a practical theory for equilibrium thermodynamics exists and has been used in WATEQ, EQ3, and EQ6.

The kinetics of heterogeneous reactions are extremely sensitive to (1) nature of the surface of the solid phase, (2) chemical reaction at the surface, (3) transport (normally by diffusion) of ions and molecules in solution to and from the surface of the grain, and (4) acidity and temperature of the system.

Identifying the step that determines the rate is often difficult. Approximate theories of chemical kinetics have been developed but are based on average behavior of molecules. Experimental data are also being compiled, but this type of work is tedious and requires a high degree of procedural accuracy (Nagy et al., 1990).

Geologic Data

Geologic data needed to simulate diagenetic sequences in the known area include initial rock composition, initial water composition, and burial and thermal history.

Initial Rock Composition

In an unknown area, a good estimate of the initial rock composition is needed. This may be based on an extrapolation of observations in known (cored) intervals or in outcrop. It is sufficient to know, for example, if the sandstones were quartz arenites, sublitharenites, subarkose, or arkose. However, the more detailed the information, the fewer iterations will have to be made to simulate diagenesis in the known area and then predict diagenesis in an unknown area.

Initial Water Composition

At present, for lack of a better assumption, initial water composition is based on a deduction of the depositional environment; for example, see the paper on the Upper Almond Sandstone (Meshri and Walker, 1990).

Burial and Thermal History

In addition to rock and water data, burial history and thermal history are needed. Such information may already be available in an explored area. For new areas, these data can be obtained by using a heat-flow equation and assuming conductivities for rocks whose stratigraphic characteristics are known.

CONCLUSIONS

1. Chemical models can be classified into three major types—speciation solubility, reaction-path, and reaction-transport models. Table 1 provides a summary of the three types of chemical models and includes a brief description of theory, available software, inputs, outputs, and limitations.
2. Speciation-solubility models provide an understanding of the stability of minerals, aqueous species (ions, multivalent ions, complex ions) and mole ratios of ions up to a temperature of 250°C and salinities of 100 ppt. In other words, qualitative diagenetic inferences about types of cements can be made. WATEQ, SOLMINEQ, and EQ3 are some examples.
3. Reaction-path models can calculate the amounts of cements, provided all related parameters are correct. Amounts of cements dissolved and precipitated may be calculated as a function of reaction progress or in the context of burial history, using known initial mineral and water composition. The amount of cementation may be calculated in an open (flow-through) mode, but more correctly in a closed system. PATHI, EQ3/6, GT, and PHREEQUE are a few examples of reaction-path models.
4. Reaction-transport models can produce spatial and temporal patterns of diagenesis and therefore evolution of porosity in space and time. These models represent the latest and best advance in the science of chemical modeling. More work needs to be done to understand how controls such as temperature, time, and assumed starting compositions affect burial diagenesis. REACTRAN, GEOKIN, and GENEX are a few examples of reaction-transport models.

RECOMMENDATIONS

To build experience in modeling, we must iterate our effort between simulating a particular situation with a current comprehensive model, comparing the simulation to the observed petrographic data, and then revising either the concept, the numerical (chemical and/or geological) data, or the chemical mechanisms, whichever is deemed necessary. The same process should continue until a comfortable degree of confidence is obtained in at least a few examples for each environment, each rock type, and each burial scenario.

ACKNOWLEDGMENTS

I appreciate the careful editing and constructive criticism by Peter Ortoleva, Laurel Babcock, and Jana Walker, which have immensely improved the clarity of this manuscript.

REFERENCES

Bruton, C. S., 1986, Predicting mineral dissolution and precipitation during burial: synthetic diagenetic sequences, in K. J. Jackson and W. L. Bourcier, eds., Proceedings of a workshop on geochemical modeling: Lawrence Livermore National Laboratory, p. 198.

Debye, P., and E. Hückel, 1923, The theory of electrolytes I. Lowering of freezing point and related phenomena: Physik v. 24, p. 185-206.

Dewers, T., and P. Ortoleva, 1989a, Mechano-chemical coupling in stressed rocks: Geochimica et Cosmochimica Acta, v. 53, p. 1243-1258.

Dewers, T., and P. Ortoleva, 1989b, The self-organization of mineralization patterns in metamorphic rocks through mechano-chemical coupling: Journal of Physical Chemistry, v. 93, p. 2842-2848.

Dewers, T., and P. Ortoleva, 1990, Interaction of reaction, mass transport, and rock deformation during diagenesis: mathematical modeling of intergranular pressure solution, stylolites, and differential compaction/solution, this volume.

Drever, J. L., 1982, The geochemistry of natural waters: Englewood Cliffs, New Jersey, Prentice Hall, 388 p.

Garrels, R. M., and C. L. Christ, 1965, Solutions, minerals and equilibria: New York, Harper and Row, 450 p.

Glasstone, S. 1946, A text book of physical chemistry: New York, Van Nostrand, 1320 p.

Harvie, C. E., N. Moller, and J. H. Weare, 1984, The prediction of mineral solubilities in natural waters: the Na-K-Mg-Ca-H-Cl-SO$_4$4-OH-HCO$_3$-CO$_3$-CO$_2$-H$_2$O-system to high ionic strengths at 25°C: Geochimica et Cosmochimica Acta, v. 48, p. 723-751.

Helgeson, H. C., 1968, Evaluation of irreversible reactions in geochemical processes involving minerals and aqueous solutions I. Thermodynamic relations: Geochimica et Cosmochimica Acta, v. 33, p. 853-877.

Jackson K. J., and W. L. Bourcier (eds.), 1986, Proceedings of the workshop on geochemical modeling: Lawrence Livermore National Laboratory, p. 198.

Kharaka, Y. K., and I. Barnes, 1973, SOLMINEQ: A solution mineral equilibrium computation: Springfield, Virginia, National Technical Information Service Report PB 214-897, 82 p.

Kharaka, Y. K., W. D. Gunter, P. K. Aggarwal, E. H. Perkins, and J. D. DeBrall, 1988, SOLMINEQ.88 A computer program for geochemical modeling of water-rock interactions: USGS Water-Resources Investigations Report 88-4227, 420 p.

Lewis, G. N., and M. Randall, 1961, Thermodynamics: New York, McGraw Hill (2nd edition, revised by S. Pitzerk and L. Brewster), p. 723.

Lichtner, P. C., 1985, Continuum model for simultaneous chemical reaction and mass transport in hydrothermal systems: Geochimica et Cosmochimica Acta, v. 49, p. 779-800.

Lichtner, P. C., 1988, The quasi-stationary state approximation to coupled mass transport and fluid-rock interaction in a porous medium: Geochimica et Cosmochimica Acta, v. 52, p. 143-165.

Meshri, I. D., 1981, Deposition and diagenesis of Glauconite sandstone, Berrymore-Lobstick-Bigoray area, south-central Alberta: a study of physical chemistry of cementation: Ph.D. dissertation, University of Tulsa, Tulsa, Oklahoma, 130 p.

Meshri, I. D., 1985, Study of rock-water interactions in the Upper Almond sandstones of the Red Desert and Washakie basins (abs.), in Timing of siliciclastic diagenesis: relationship to hydrocarbon migration: Invited Paper at Sixth Annual Research Conference, Gulf Coast Section SEPM, p. 24.

Meshri, I. D., 1986, On the reactivity of carbonic and organic acids and generation of secondary porosity, in D. L. Gautier, ed., Roles of organic-matter in mineral diagenesis: SEPM Special Volume 8, p. 123-128.

Meshri, I. D., in press, A subtle diagenetic trap in the Glauconite sandstone of southwest Alberta, in P. Ortoleva B. Hallet, A. McBirney, I. Meshri, R. Reeder, and P. Williams, eds., Proceedings of the Workshop on Self-Organization in Geological Systems, Santa Barbara, 1988: Earth Science Reviews (Amsterdam, Elsevier).

Meshri, I. D., and J. M. Walker, 1990, A study of rock-water interaction and simulation of diagenesis in the Upper Almond sandstones of the Red Desert and Washakie basins, this volume.

Moore, C. H., and P. J. Ortoleva, 1990, Effect of fluid and rock compositions on diagenesis: a modeling investigation, this volume.

Nagy, K. L., C. I. Steefel, A. E. Blum, and A. C. Lasaga, 1990, Dissolution and precipitation kinetics of kaolinite: initial results at 80°C with application to porosity evolution of a sandstone, this volume.

Nordstrom, D. K., et al., 1979, A comparison of computerized chemical models for equilibrium calculations in aqueous systems, in E. A. Jenne, ed., Chemical modeling in aqueous systems: American Chemical Society Symposium Series 93, p. 857-892.

Ortoleva, P., 1984, The self-organization of Liesgang bands and other precipitate patterns, in G. Nicolis and F. Baros eds., Chemical instabilities: applications in chemistry, engineering, geology and material science: Dordrecht, Reidel, NATO Advanced Science Institute Series, v. 120, p. 289-297.

Ortoleva, P., E. Merino, and P. Strickholm, 1982, A kinetic theory of metamorphic layering in anisotropically stressed rock: American Journal of Science, v. 282, p. 617-643.

Ortoleva, P., J. Chadam, E. Merino, and A. Sen, 1987a, Geochemical self-organization II: The reactive-infiltration instability: American Journal of Science, v. 287, p. 1007-1040.

Ortoleva, P., E. Merino, C. Moore, and J. Chadam, 1987b, Geochemical self-organization I: reaction-transport feedbacks and modeling approach: American Journal of Science, v. 287, p. 979-1007.

Ortoleva, P., W. Chen, A. Park, and A. Ghaith, 1990, Diagenesis through coupled processes: modeling approach, self-organization, and implications for exploration, this volume.

Ortoleva, P., G. Auchmuty, J. Chadam, E. Merino, C. Moore, and E. Ripley, 1986, Redox front propagation and banding modalities: Amsterdam, Elsevier, Physica 19D, p. 334-354.

Pearce, T. H., and A. M. Kolisnik, in press, Observations of plagioclase zoning using interference imaging, in P. Ortoleva, B. Hallet, A. McBirney, I. Meshri, R. Reeder, and P. Williams, eds., Proceedings of the Workshop on Self-Organization in Geological Systems, Santa Barbara, 1988: Earth Science Reviews (Amsterdam, Elsevier).

Pitzer, K. S., 1977, Electrolyte theory: improvements since Debye and Hückel: Accounts of Chemical Research, v. 10, p. 371-377.

Plummer, L. N., B. F. Jones, and A. H. Truesdell, 1976, A Fortran IV version of WATEQ, a computer program for calculating chemical equilibria in natural waters: USGS Water-Resources Investigations Report 76-13, 61 p.

Reeder, R. J., R. O. Fagioli, and W. J. Meyers, in press, Oscillatory zoning of Mn in solution-grown calcite crystals, in P. Ortoleva, B. Hallet, A. McBirney, I. Meshri, R. Reeder, and P. Williams, eds., Proceedings of the Workshop on Self-Organization in Geological Systems, Santa Barbara, 1988: Earth Science Reviews (Amsterdam, Elsevier).

Schmidt, V., and D. A. McDonald, 1979, The role of secondary porosity in the course of sandstone diagenesis: SEPM Special Publication 26, p. 175-207.

Shimizu, N., in press, The oscillatory trace element zoning of augite phenocrysts, in P. Ortoleva, B. Hallet, A. McBirney, I. Meshri, R. Reeder, and P. Williams, eds., Proceedings of the Workshop on Self-Organization in Geological Systems, Santa Barbara, 1988: Earth Science Reviews (Amsterdam, Elsevier).

Steefel, C. I., and A. C. Lasaga, 1990, Evolution of dissolution patterns, permeability change due to coupled flow and reaction, in D. C. Melchoir and R. L. Bassett, eds., Chemical modeling in aqueous systems II: American Chemical Society Symposium Series 416, p. 212-225.

Surdam, R. C., S. W. Boese, and L. J. Crossey, 1984, The chemistry of secondary porosity: AAPG Memoir 37, p. 127-149.

Truesdell, A. H., and B. F. Jones, 1974, WATEQ, a computer program for calculating chemical equilibria of natural waters: USGS Journal of Research, v. 2, p. 233-274.

Wagner, P. D., and R. K. Matthews, 1982, Porosity preservation in the Upper Smackover (Jurassic) carbonate grainstone, Walker Creek field, Arkansas: response of paleophreatic lenses to burial processes: Journal of Sedimentary Petrology, v. 52, no. 1, p. 3-18.

Wall, F. T., 1958, Chemical thermodynamics: San Francisco, W. H. Freeman, p. 422.

Wolery, T. J., 1979, Calculation of chemical equilibrium between aqueous solutions and minerals: The EQ3/6 software package: Lawrence Livermore National Laboratory, UCRL-52658, 162 p.

Wolery, T. J., 1983, EQ3NR, a computer program for geochemical aqueous specification-solubility calculations: user's guide and documentation: Lawrence Livermore National Laboratory, UCRL-53414, 191 p.

Wolery, T. J., 1986, EQ3/6 status and future development, in K. J. Jackson and W. L. Bourcier, eds., Proceedings of the Workshop on Geochemical Modeling: Lawrence Livermore National Laboratory, p. 198.

A Study of Rock-Water Interaction and Simulation of Diagenesis in the Upper Almond Sandstones of the Red Desert and Washakie Basins, Wyoming

Indu D. Meshri and Jana M. Walker
Amoco Production Company
Tulsa, Oklahoma, U.S.A.

Chemical modeling of the rock-water interactions in the Upper Almond Sandstone within the context of burial (thermal) history of these rocks predicts the petrographically observed sequence of cementation to a fair degree.

The depositional environment of the Upper Almond Sandstone is believed to have been a barrier bar of Late Cretaceous age. Therefore, initial pore-water chemistry is assumed to be a function of seawater composition and, consequently, marks the starting point of diagenesis.

Present-day water compositions along with a knowledge of depositional environment assist in further deduction of early pore-water composition as one-third diluted seawater. Formation waters collected from the Upper Almond Sandstone in the Wamsutter area are distinguished from waters of the underlying Main Almond Sandstone on the basis of total dissolved solids, $\delta^{18}O$, log-derived salinities, and pressure-depth profiles.

The main objective of this study was to test the predictive capability of the existing computerized chemical modeling approach to elucidate diagenesis in the Upper Almond Sandstone. Having noted the chemical processes leading to the simulation of the observed diagenetic sequence, one can follow a similar logic to explore deeper or shallower for similar lithologic units.

INTRODUCTION

This paper attempts to provide the much-needed validation for chemical models by comparing the prediction of diagenetic sequences with those actually observed in the Upper Almond reservoir of southwestern Wyoming. Diagenetic sequences in the Upper Almond are modeled in an initially open system by using the reaction-transport modeling approach (REACTRAN, Ortoleva et al., 1987) and later in a closed system by using the reaction-path approach (EQ3/6, Wolery, 1983) described in this volume.

A wealth of rock and water data available for the Upper Almond in the Wamsutter area has provided an ideal natural laboratory for conducting such rock-water interaction research. A similar study was conducted of the Glauconite Sandstone, a deltaic sandstone of Albian age, in the Alberta basin (Meshri, 1981). Our diagenetic predictions for the Upper Almond are more accurate than those for the Glauconite Sandstone because of additional experience in well-site water collection and analysis procedures, better thermodynamic data, and availability of better chemical models.

Location of Study Area

The study area is located in Sweetwater and Carbon counties in southwestern Wyoming, between lat. 42°30' and 40° 30' N and between long. 108° and

109° W. It includes three basins separated by two cross-basin arches. The east-plunging Wamsutter arch separates the Red Desert basin in the north from the Washakie basin in the center (Figure 1). Farther south, Stateline ridge, also known as the Cherokee ridge, separates the Washakie basin in southwestern Wyoming from the Sand Wash basin in northwestern Colorado. Neither of these cross-basin arches brings Upper Almond rocks to the surface. Outcrops of the Main Almond (Middle and Lower Almond) and the other Upper Almond bar sands bound the basinal area to the east in the Rawlins uplift and to the west in the Rock Springs uplift.

Paleogeography and Regional Geology

Throughout the Cretaceous, the axis of the Western Interior Seaway was the site of deposition of dark, organic-rich marine shales. Depending on their age and location, these shales have been assigned various stratigraphic names, such as Mancos, Lewis, Pierre, and Steel. The shoreline and/or deltaic nature of Cretaceous sandstones and their intertonguing relationship with these shales have been the subject of many studies. Asquith (1972, 1975) has shown that the Lewis shale was one such unit which was deposited on a broad, prograding shelf-slope-basinal facies tract. The distribution of ammonites in these shales is the basis for a detailed biostratigraphic zonation which has been used to map Cretaceous shoreline positions in the Western Interior Seaway (Gill et al., 1970; Gill and Cobban, 1973; Miller, 1977).

General Stratigraphy

In Wyoming, great thicknesses of rocks accumulated along the margins of an epicontinental sea during the Late Cretaceous. Miller (1977) delineated four Almond shorelines by graphic correlation of palynologic microfossils. These shorelines were the site of deposition of Upper Almond sands during the overall westward transgression and accompanying periodic regressions of the epicontinental sea.

The Upper Almond is the uppermost sandstone of the Mesaverde Group, which consists of the Ericson Formation (or its stratigraphic equivalents), the Allen Ridge and Pine Ridge sandstones, and the overlying Almond sandstones. Deposition of the Mesaverde Group must have begun in Late Campanian and ended in early Maestrichtian (Miller, oral communication, 1984).

Depositional Setting as a Starting Point for Diagenesis

The Upper Almond Sandstone was deposited in a nearshore shallow marine environment, which was transitional between a coastal flood plain to the west and a marine environment to the east. It is important to note that the L-sand bar, which is a shoestring bar facies extending along the paleostrike, is encased in marine shales to the east and lagoonal shales to the west, and is underlain by a very tight zone containing oysters, ostracods, and pellets. Known outcrops of Almond in the study area are those of the Main Almond. Because outcrops of the Upper Almond L-sand bar have not been found, late recharge by meteoric waters entering outcrops is unlikely. This is consistent with conclusions based on the examination of present-day water data.

Sample Collection

Core samples were studied from 9 wells (Figure 2). Water samples were collected from 18 producing wells. Eight of these wells were producing from the Upper Almond Sandstone, 7 from the Main Almond, and 3 from the Lewis Sandstone (Figure 2). Cored intervals for each of these wells are shown in Table 1. Well-site water analyses were performed for Eh, pH, HCO_3 (CO_2), alkalinity, NH_3, and SiO_2. Water samples were preserved for later analyses of H_2S, stable isotopes, and major and minor anions and cations. Results of well-site and laboratory analyses are summarized in Table 2 for Upper Almond wells. Procedures for collection, preservation, analysis, correction for condensation, and detection of contaminants are discussed in Meshri (1981).

PRESENT-DAY WATER COMPOSITION

Having determined the depositional environment of the Upper Almond as a barrier bar, as a first scenario, one may assume that early pore waters would have compositions like those of the Cretaceous sea. Although present-day formation waters are more dilute than seawater, they could have been diluted by the condensed water accompanying produced gas. Therefore, present-day waters had to be evaluated carefully. Hence, a dilution correction was calculated.

Dilution Correction

When gas and water are coproduced, some water vapor in the gas may condense and dilute the produced formation water. The amount of dilution is a function of temperature and pressure drop, and the amount of gas and water produced. Such dilution correction is calculated by the McKetta-Wehe (1958) procedure. The variables required for the dilution correction due to condensation are discussed in Meshri (1981).

Although two Upper Almond wells required extensive correction for dilution, all other Upper

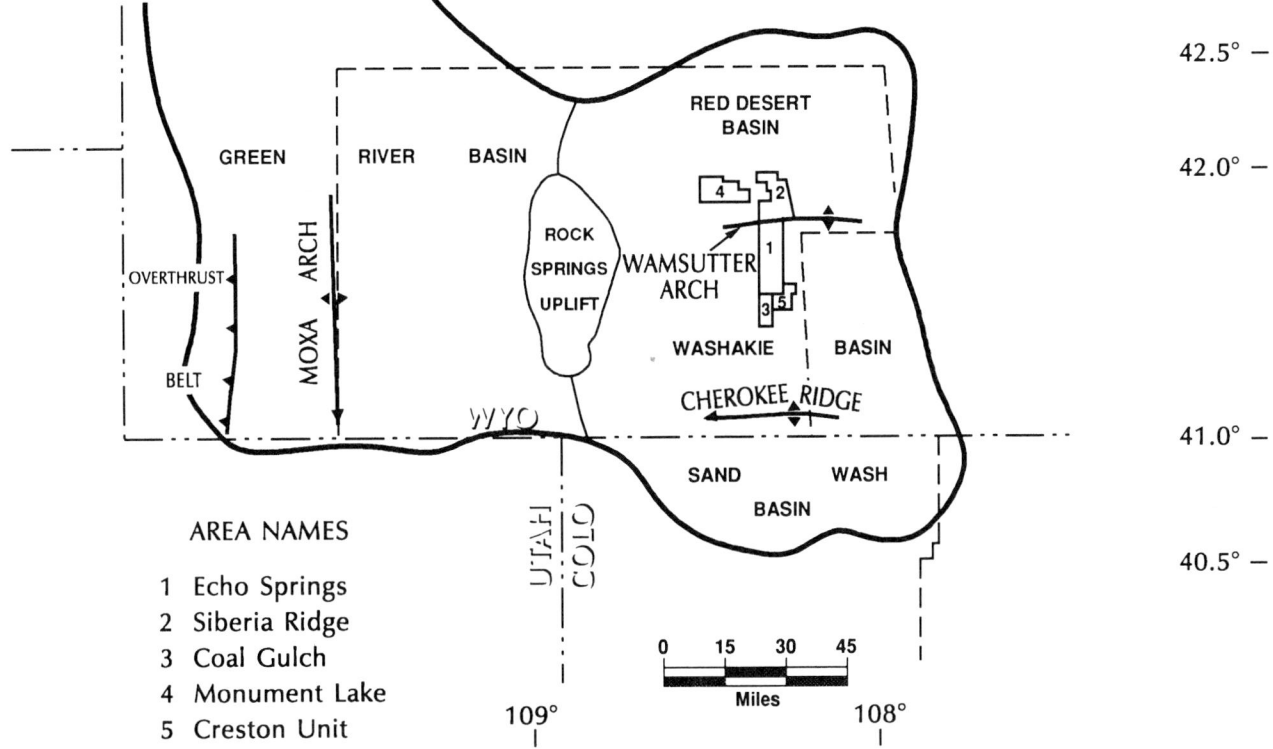

Figure 1. Location of study area, showing three basins and two cross-basin arches.

Almond wells produced essentially 100% formation water (that is, no dilution occurred during production of fluids).

Alkalinity Correction

Organic acid anions can contribute to the alkalinity of formation water. For accurate calculation of mineral equilibria, alkalinity contribution due to bicarbonate must be determined accurately. Discussion of error in HCO_3 alkalinity due to the presence of organic acid anions can be found in Meshri (1981). Analysis of Upper Almond formation waters showed that organic acid anions are present in low concentrations, as would be expected in a gas-producing formation. Thus the contribution to alkalinity by organic acid anions is very small. The alkalinity of Upper Almond formation waters is mainly due to the bicarbonate ion.

Comparison of Log-Calculated and Water-Derived Salinities

Because of the uncertainty in salinity values of two Upper Almond water samples, it became apparent that the reliability of the dilution correction factors should be checked by an independent method. The check consisted of computing R_w (water resistivity) based on corrected total dissolved solids in the formation water at the formation temperatures, and comparing this value with log-calculated R_w for the producing zones. The following conclusions about water resistivities are noteworthy.

1. Upper Almond waters are less saline (TDS range from 8150 to 15,000 mg/L) than Main Almond waters (TDS range from 15,000 to 20,000 mg/L) (Table 2).
2. Values of R_w obtained from analyses of water produced from the Upper Almond range from 0.18 to 0.26 ohm-m. Water produced from the Main Almond interval gives R_w values ranging from 0.12 to 0.15 ohm-m.
3. Log analysis of the Upper Almond intervals indicates R_w values of 0.19 to 0.23 ohm-m for the Upper Almond interval and 0.10 to 0.14 ohm-m for the Main Almond interval. These log-derived values compare well with those obtained from the water analysis data.

Classification of Study Area Waters

The range of salinities and water resistivities discussed previously show that Upper and Main Almond formation waters are different in composition.

Oxygen isotope data and pressure-depth profiles discussed in this section also suggest that Upper and Main Almond are two distinctly separate reservoirs.

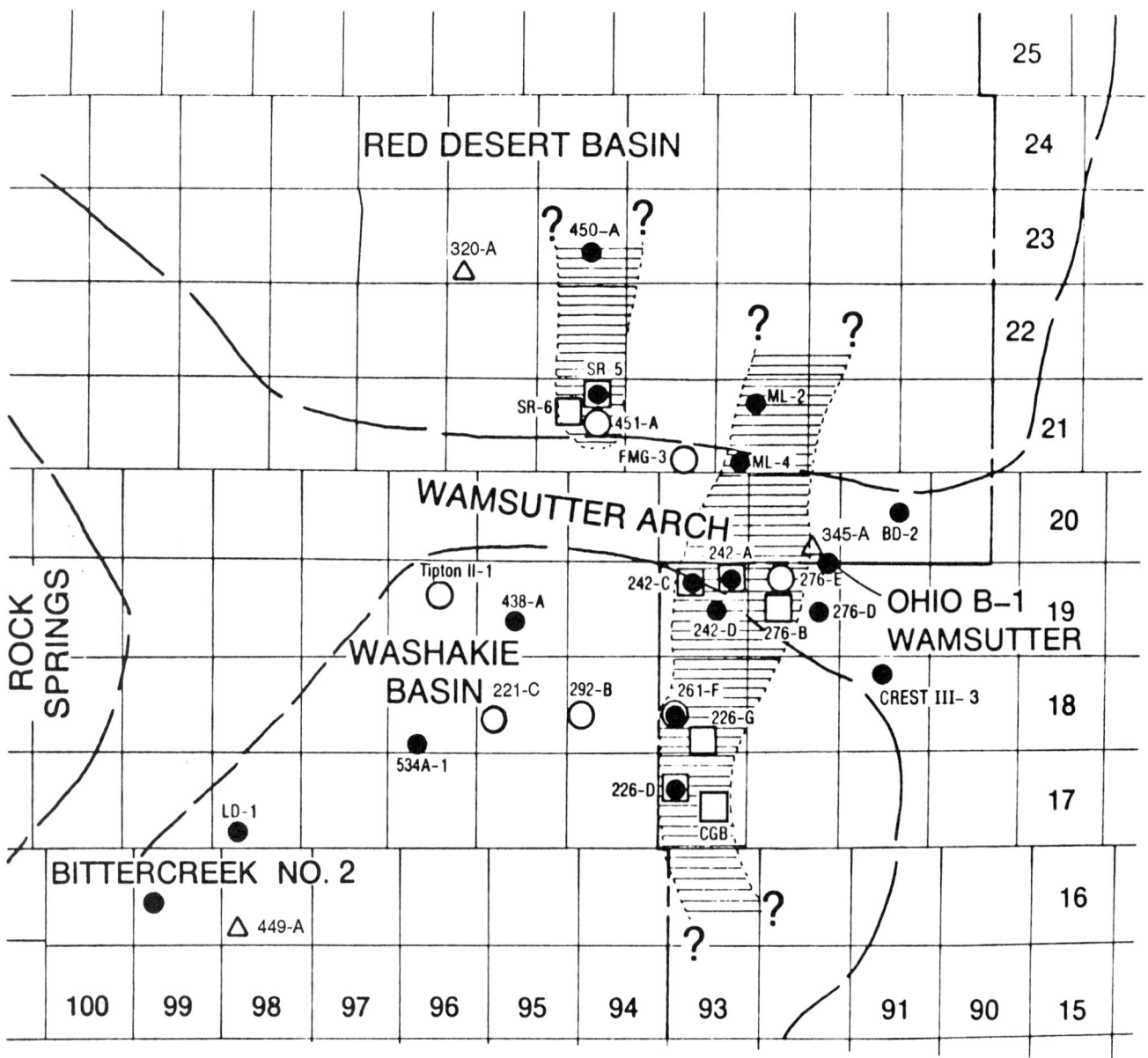

Figure 2. Map of study area showing the locations of wells from which core and water samples were obtained.

$\delta^{18}O$ Vs. TDS Plot

A $\delta^{18}O$ value of about -6‰ separates Upper Almond formation waters (Figure 3) from Main Almond formation waters. For comparison, $\delta^{18}O$ values of Glauconite formation waters (in the Alberta basin), Sussex formation waters (in the Denver basin), and present-day seawater are also plotted vs. TDS on Figure 3. Nearly all waters studied in the Rocky Mountain region seem to have evolved from seawater. This observation is consistent with the interpretation of Upper Almond, Sussex, and

Table 1. Wells from which water samples were obtained

Well Name[1]		Well Location	Depth (ft)	Formation[2]	Cored Interval (ft)
1	Coal Gulch	15-17N-93W	9,068-9,491	U. Almond	No cores
2	Champlin 226-D	5-17N-93W	9,095-9,115	U. Almond	9,060-9,118
3	Champlin 276-B	7-19N-92W	9,298-9,331	U. Almond	No cores
4	Champlin 242-A	1-19N-93W	9,452-9,457	U. Almond	9,475-9,518
5	Champlin 242-C	3-19N-93W	9,470-9,483 9,500-9,532	U. Almond	No cores
6	Siberia Ridge No. 6	9-21N-94W	10,438-10,456	U. Almond	No cores
7	Siberia Ridge No. 5	3-21N-94W	10,586-10,606	U. Almond	10,580-10,636
8	Champlin 226-G[3]	33-18N-93W	8,918-8,936	U. Almond	No cores
9	Champlin 261-F	29-18N-93W	9,055-9,400	U. & M. Almond	9,155-9,195
10	Champlin 276-E	5-19N-92W	9,381-9,391	M. Almond	No cores
11	Champlin FMG No. 3	35-21N-93W	10,664-10,722	M. Almond	10,588-10,866
12	Champlin 292-B	19-18N-94W	10,311-10,458	M. Almond	No cores
13	Champlin 221-C	19-18N-95W	11,088-11,172	M. Almond	11,128-11,237
14	Tipton II No. 1	14-19N-96W	9,390-9,398	M. Almond	No cores
15	Champlin 451A	17-21N-94W	10,384-10,465	M. Almond	No cores
16	Champlin 345A	35-20N-92W	Unknown	Lewis	8,678-8,809
17	Champlin 320A	31-23N-95W	9,825-10,050	Lewis	9,905-10,016
18	Champlin 449A	31-16N-98W	11,753-780	Lewis	11,411-11,450

[1]Well locations shown in Figure 2.
[2]U. Almond indicates Upper Almond; M. Almond indicates Main Almond (Middle and Lower Almond).
[3]Water produced from this well is contaminated.

Glauconite sandstones as normal marine depositional settings.

However, based on total dissolved solids content, the Upper Almond formation waters are more depleted in $\delta^{18}O$ than other formation waters, such as those from the Main Almond. Depletion in $\delta^{18}O$ could be due to dilution of formation water by condensed water. Because the amount of condensation is a function of the amount of gas produced, among other parameters, gas production and therefore condensation is higher for the Upper Almond than for the Main Almond; hence, $\delta^{18}O$ values may be more depleted.

It is possible to correct for dilution of TDS through condensation, but it is not possible to correct for $\delta^{18}O$ depletion. Thus, isotopic values of Upper Almond formation waters cannot be used as the only distinguishing criterion.

Pressure-Depth Profiles

Previous work by D. E. Powley (oral communication, 1984) shows that pressure-depth profiles (Figure 4) of the Upper Almond and Main Almond show almost no variation with depth. Pressures for the Upper Almond cluster around 5000-5200 psi, and those for the Main Almond are in the 6800-7000 psi range. The pressure-depth profile for the Upper Almond is very steep in comparison with the normal pressure-gradient, indicating that the Upper Almond is an overpressured reservoir.

The Main Almond exhibits a pressure-depth profile separate and distinct from the Upper Almond reservoir. Thus, pressure-depth profiles, Rw, and TDS are good distinction criteria for Upper and Main Almond formation waters.

Subclassification of Waters Based on Bromide Content

Water samples from the Upper Almond can be further classified based on their bromide content. Rittenhouse (1967) subdivided oil-field waters into five groups based on their bromide content (Figure 5). In Group I, bromide and TDS content approximate that expected from simple concentration or simple dilution of seawater with water of low TDS and low bromide content. Samples from Coal Gulch and Echo Springs lie close to this ideal dilution line, attesting to the fact that the early pore waters in the Upper Almond barrier bar could have suffered such a dilution.

However, two water samples from wells in Siberia Ridge (N-band bar) lie in Group II, indicating that they were also diluted, with a water of low TDS but somewhat richer in bromide. Bromide and $\delta^{18}O$ values do not differentiate between early coastal meteoric mixing and poststructural recharge.

Deuterium-Oxygen Isotopes and Meteoric Mixing

Craig (1961) was the first to demonstrate that δD and $\delta^{18}O$ values are correlated in meteoric waters, as indicated by a straight line which represents the results of his analyses on several hundred samples of river and lake water and precipitation (Figure 6a):

Table 2. Analyses of Upper Almond waters

Well Number	1	2	3	4	5	6	7
Well-Site Analyses							
Redox (mV)	-120	-140	-142	-145	-140	-155	-155
Dissolved O_2 (mg/L)	0.25	0.15	0.35	0.075	0.07	0.075	0.075
pH (Acidity)	6.65	6.45	7.02	6.45	6.70	6.70	7.3
Density (cm^3)	0.966	0.966	0.966	0.961	0.964	0.970	0.970
Alkalinity (mg/L)	2464	1188	2898	1498	1170	1305	—
$PCO_2(HCO_3)$(mg/L)	2400	1159	2900	1586	1140	1237	1830
NH_4 (mg/L)	10.4	10.8	10.8	7.2	12.9	9.7	5.4
SiO_2 (mg/L)	72	65	100	90	65	70	90
Well Head Temp. (°F)	76	76	80	76	78	90	80
Bot. Hole Temp. (°F)	189	189	187	194	194	210	210
Well Head Pres. (psi)	700	1000	800	800	800	700	780
W Alkalinity	+2.4%	+2.4%	-0.1%	-5.7%	+2.6%	+5.4%	—
Laboratory Analysis (mg/L)							
Ca	140.0	165.0	70.0	44.9	162.0	137.0	157.0	421.9
Mg	34.7	35.0	14.0	8.8	20.9	14.0	14.0	1322.0
Na	3500.0	2747.0	3000.0	2105.0	4190.0	4750.0	4760.0	11019.0
K	245.0	179.0	190.0	145.6	340.0	325.0	320.0	408.0
Cl	4750.0	4400.0	3750.0	2561.0	6400.0	8250.0	8250.0	19805.9
SO_4	21.0	25.3	25.0	14.0	56.0	17.0	17.0	2725.4
HCO_3	2625.0	1260.0	2900.0	2800.0	2140.0	1300.0	1830.0	144.3
SiO_2	72.0	71.0	100.0	160.0	125.0	70.0	90.0	4.38
Fe	11.0	220.0	10.5	350.0	313.0	80.0	80.0	0.0
PO_4	0.0	0.0	0.0	0.0	0.0	0.0	0.0	0.06
Sr	1.7	12.0	0.9	0.7	5.4	13.0	15.0	8.33
F	0.0	0.0	0.0	0.0	0.0	0.0	0.0	1.42
Ba	45.0	15.4	20.0	2.60	3.10	5.50	6.40	0.02
Al	0.20	0.20	0.20	0.20	0.12	0.20	0.20	0.002
NH_4	10.4	10.8	10.8	7.2	12.9	9.72	5.40	0.003
Li								0.18
Mn	0.11	2.50	0.11	3.30	3.50	0.50	0.70	0.0
Br	26.0	22.0	30.0	22.8	51.6	76.0	80.0	68.9

(1) Coal Gulch B
(2) Champlin 226D No. 1
(3) Champlin 276D No. 1
(4) Champlin 242A No. 1
(5) Champlin 242C No. 1
(6) Siberia Ridge No. 6
(7) Siberia Ridge No. 5
(8) Seawater as listed in Nordstorm et al. (1979)

$$\delta D = \delta^{18}O + 10$$

This relationship, though empirical, has been substantiated by the work of other researchers.

Six of seven water samples were analyzed for deuterium (the other water sample was insufficient in quantity). One Siberia Ridge, two Coal Gulch, and three Echo Springs water samples plot very close to the meteoric water line in Figure 6a, indicating that pore waters have been diluted. However, this plot cannot confirm whether dilution was due to early coastal mixing or to late meteoric recharge. Also, there is some doubt about shift in $\delta^{18}O$ and δD values because of possible condensed water dilution.

HCO_3/Cl Enrichment and Meteoric Mixing

Inasmuch as chloride is not involved in diagenesis, it is used as a norm against which other ions are compared. Also, if dilution due to condensation occurs and the ratio of two molar concentrations such as SO_4/Cl and HCO_3/Cl are plotted, any error due to dilution by condensed water cancels. A fairly correlatable straight line is obtained when the HCO_3/Cl ratio is plotted against TDS (Figure 6b), except for sample No. 2, for which there is no explanation. Meteoric waters are rich in HCO_3 and low in chloride and TDS, and upon mixing they enrich formation

ALMOND FORMATION WATERS

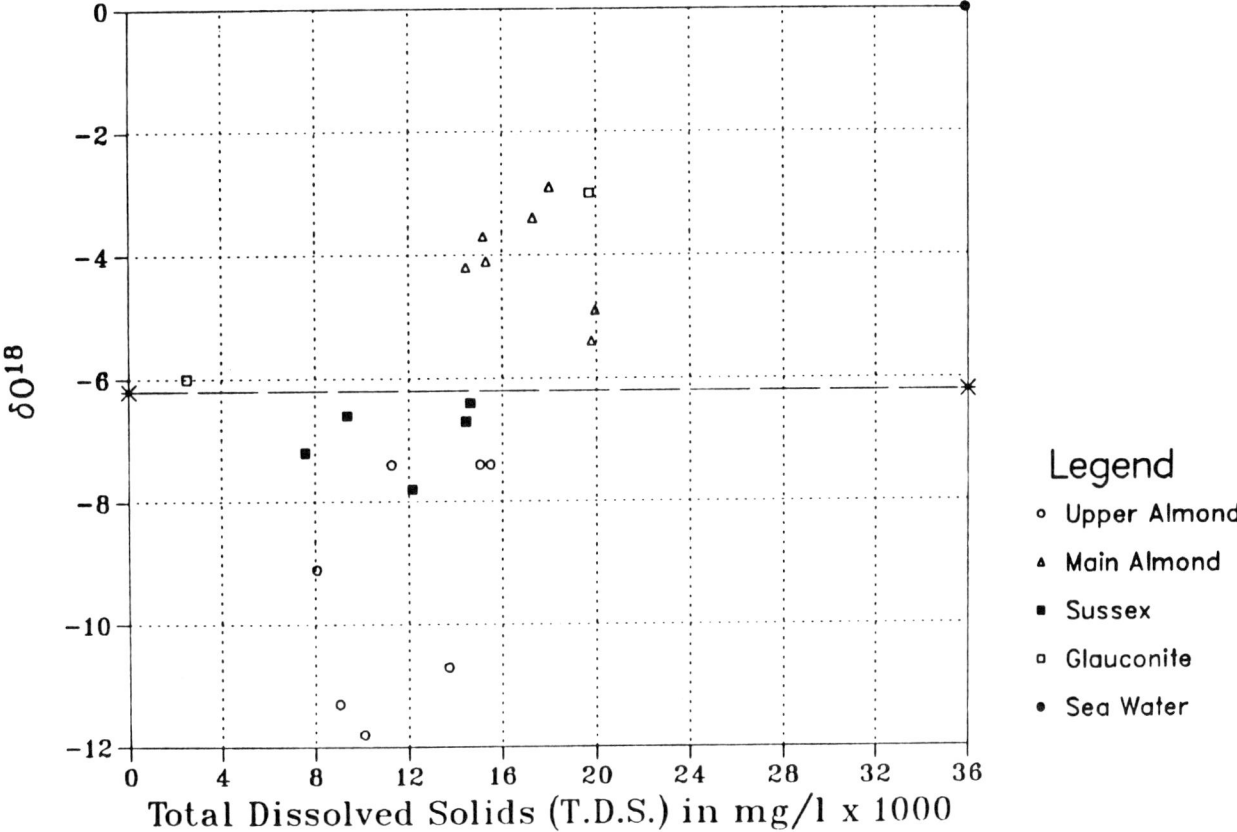

Figure 3. Plot of $\delta^{18}O$ vs. TDS showing distinction between Upper and Main Almond waters and suggesting separation of reservoirs.

waters in HCO_3 and dilute them with respect to TDS. Thus, this trend appears to confirm meteoric dilution, which was also evident on the plots of bromide content and δD vs. $\delta^{18}O$ (Figures 5, 6a). Yet, one cannot distinguish if meteoric mixing was early coastal type or due to late recharge.

Structural Dependency of HCO_3/Cl Ratio and Late Recharge

Data on Glauconite Sandstone waters showed a systematic trend of dilution due to late meteoric recharge from the flanks to the center of the Alberta basin. The HCO_3/Cl ratio was systematically high in structurally high recharge areas and low in the center of the basin (Meshri, 1981).

However, Upper Almond waters show no such correlation (Figure 7) between structurally uplifted areas (Echo Springs—Samples 3, 4, 5; and Coal Gulch—Samples 1, 2) and basin areas (Siberia Ridge—Samples 6, 7). Thus, one may conclude that Upper Almond waters may have been diluted due to early coastal meteoric mixing, or at least some time before the present-day structure formed.

Early Depletion of Sulfate

After early dilution by coastal meteoric mixing, the Upper Almond bar facies underwent burial, and the next probable episode was sulfate depletion. For Almond waters, the SO_4/Cl ratio has dropped from 0.052 for seawater to 0.002 for Almond waters. SO_4 is commonly depleted in most formation waters because of early bacterial reduction and conversion to pyrite and HCO_3 in the following manner:

bacteria $SO_4^{--} + 2CH_2O \rightarrow H_2S + 2HCO_3$

$H_2S + Fe^{++} \rightarrow FeS + 2H^{++}$

The net reaction may be written as

$SO_4^{--} + 2CH_2O + Fe^{++} \rightarrow FeS + 2H_2CO_3$

This appears to be yet another mechanism by which early HCO_3^- enrichment can be explained, in addition to mixing with coastal meteoric waters.

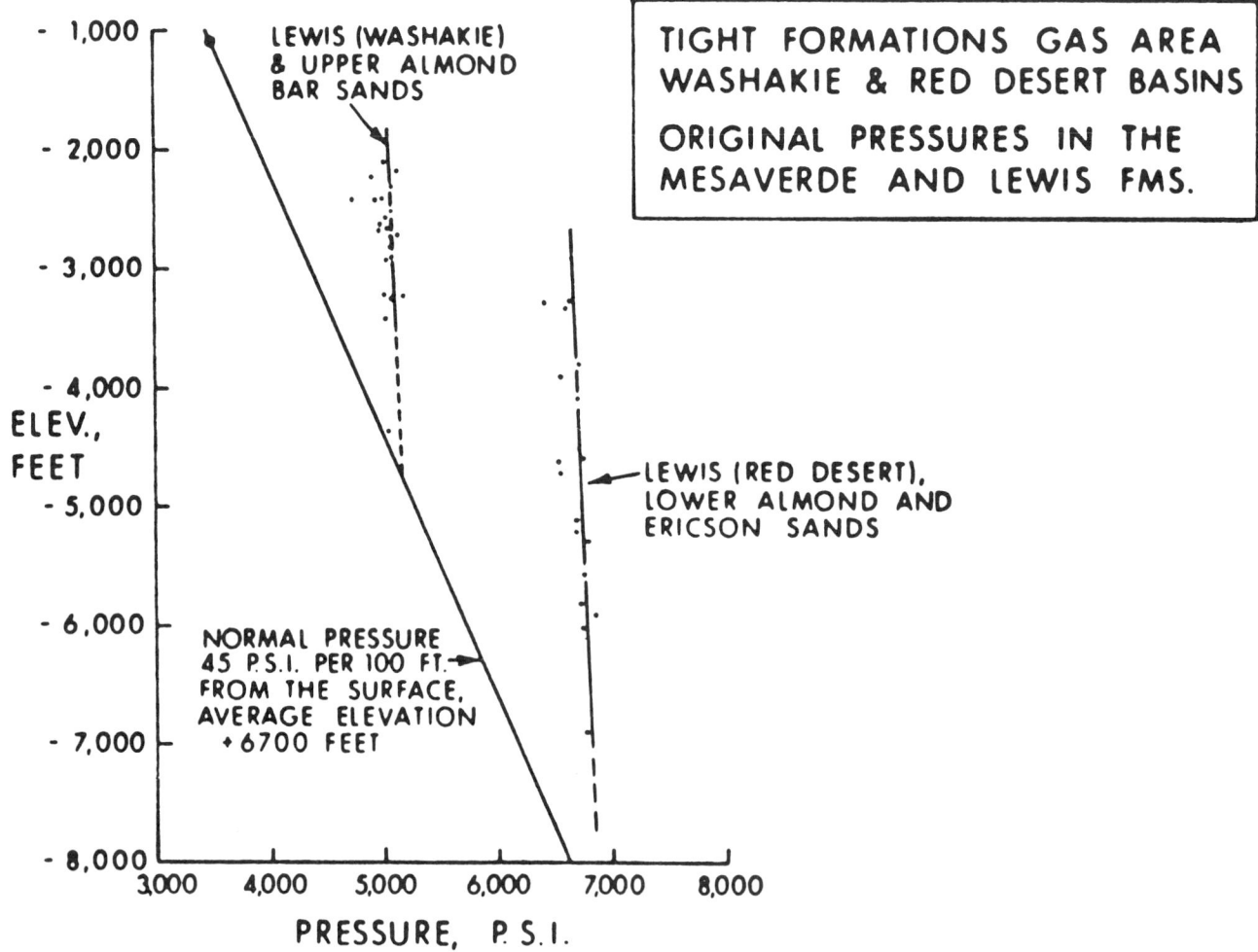

Figure 4. Pressure-depth profile for Upper and Main (lower) Almond bar sand, confirming that Upper and Main Almond are distinctly separate reservoirs (courtesy of D. E. Powley).

CHEMICAL MODELING

We have discussed the three types of chemical models, namely aqueous speciation, reaction-path, and reaction-transport models in the previous paper (Meshri, 1990). Diagenesis in the Upper Almond Sandstone is modeled via reaction-transport modeling (REACTRAN) in an initially open system, until the base of the Upper Almond (because of excessive cementation) becomes closed to the flow of fluid. The rest of the burial is modeled with the reaction-path approach via EQ 3/6.3245 in a closed system.

Modeling Parameters

Input parameters for chemical modeling include initial water composition, initial rock composition, inlet flow velocity, and burial and thermal history.

The initial composition of Upper Almond formation waters was established as one-third diluted seawater that is depleted in sulfate and enriched in bicarbonate, as indicated in the earlier section. Based on Miller (1974), we have generated a burial history profile for the Upper Almond (Figure 8). Although the initial rate of burial for the Upper Almond is almost 6000 ft/6 m.y. (million years), which is about 1000 ft/m.y. (300 m/m.y.), the average burial rate for Upper Almond is about 38 m/m.y. We assume that the fluid is moving up against the rock at the average rate of burial, and therefore we have used an inlet fluid velocity of roughly 40 m/m.y.

The burial history of the Upper Almond is converted to thermal history by using a known geothermal gradient for the area, which is 1.5 F/100 ft (31°C/km).

Initial rock composition is determined petrographically by point counting the grains including some dissolved feldspar grains. We possibly have underestimated the initial amount of feldspar, but the approximate rock composition may have been 90% quartz, 5% potassium feldspar, and 5% plagioclase. In reaction-transport modeling, however, we must include initial porosity as a percentage of total rock

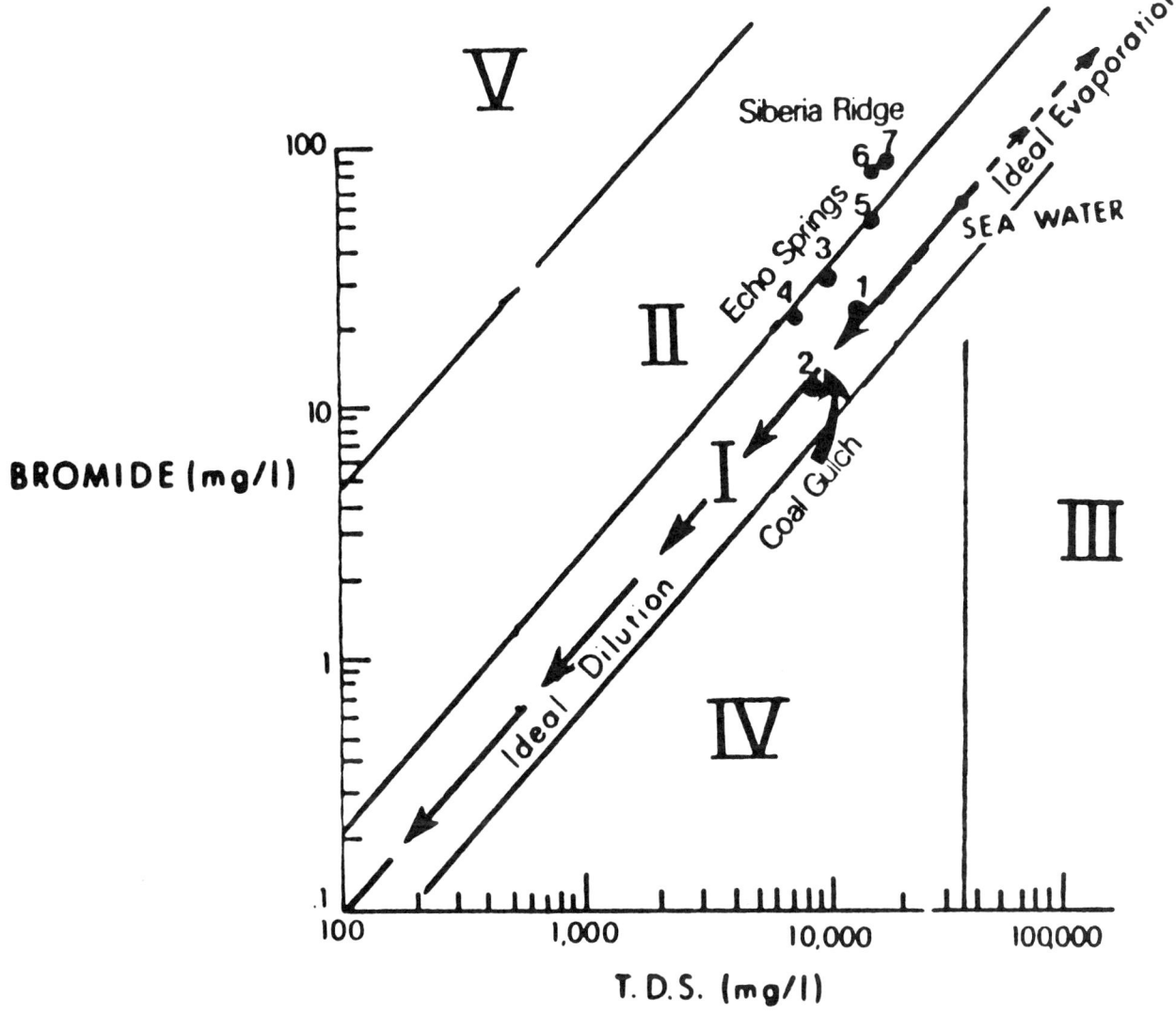

Figure 5. Subdivision of Upper Almond waters from three producing fields, based on bromide content.

volume. So we assume an initial porosity of 25% and recalculate the rock composition normalizing to 75% rock, which is 67.5% quartz, 3.75% feldspar, and 3.75% plagioclase.

Modeling of Upper Almond as an Initially Open System

Although we modeled a 10-m-long rock column, we report changes at the inlet only in Table 3. These simulated episodes of cementation are summarized by shaded bars along the Upper Almond burial profile (Figure 8).

Potassium feldspar dissolves during the first 5 m.y. of burial from initial 3.75% to 0.6%. Between 6 and 10 m.y. it grows back to somewhere between 7.4 and 12.1%, and then it begins to dissolve again between 10 and 20 m.y. These episodes of initial dissolution, feldspar growth, and redissolution are documented in the section on diagenesis.

Plagioclase remains intact up to the first 5 m.y. and then begins to dissolve at 6 m.y. This dissolution episode lasts until 20 m.y., when plagioclase has dissolved from initial 3.75 to 0.7%.

In response to this phenomenon of feldspar dissolution, water begins to become saturated with kaolinite. However, kaolinite does not grow until the first 10 m.y. have passed. Between 10 and 20 m.y., it grows from 8 to 9.5%.

Although dolomite growth occurs from 0.0 to 0.87% in the first few million years, the large quantities of dolomite only begin to form at about 10 m.y. It is also very interesting to note the formation of dolomite up to 12% occurs between 10 and 20 m.y., during the second burial phase. However, as burial

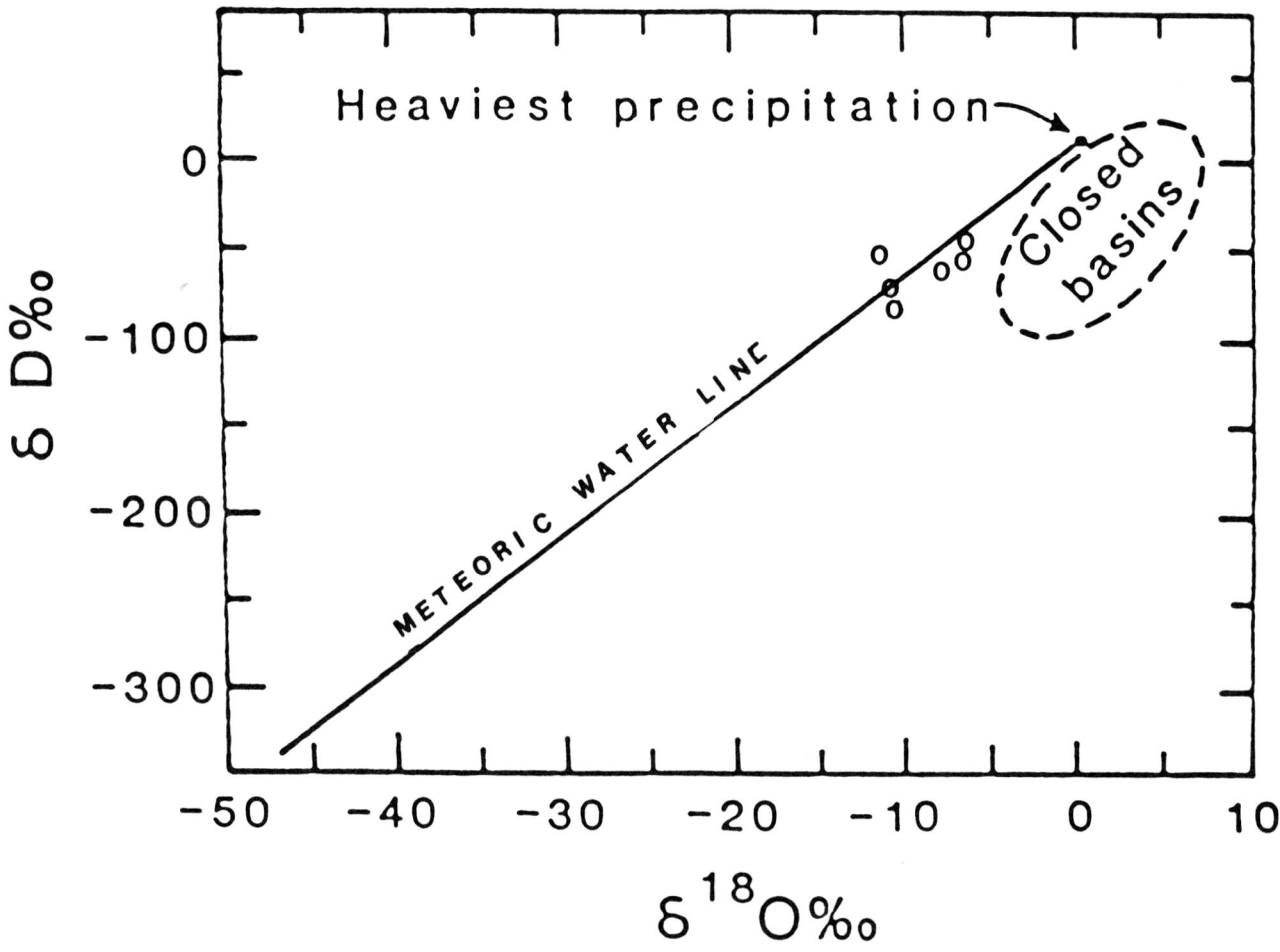

Figure 6a. Plot of δD vs. $\delta^{18}O$ showing dilution of Upper Almond waters by meteoric waters.

proceeds deeper, more dolomite begins to form and initial porosity of 25% drops to 7.6% in 20 m.y. and eventually to 0 at the base of the Upper Almond.

It is important to note that the initial burial (thermal) profile is very steep, and therefore, although not much time has passed, the temperature has increased very rapidly.

We speculate that similar phenomena could take place at the top of the Upper Almond reservoir, where the water would be entering at the boundary from the overlying shales. Anyhow, the reaction transport runs begin to fail, because technically no porosity is left at the inlet. We consider this stage of diagenesis as an indication that the Upper Almond sandstone became a closed system reservoir.

Modeling of Upper Almond as a Closed-System Reservoir

The remaining rock composition is then input into EQ3/6 to proceed with closed-system reaction-path modeling when illite begins to form. So, the simulated sequence of diagenesis should now be compared with the paragenesis actually observed in the Upper Almond Sandstone.

If we have correctly simulated the diagenesis in the Upper Almond, not only should we find a correspondence in the observed cementation sequence, but also we should find present-day waters in near equilibrium with the late-stage cements if, of course, the waters are representative of the formation. Modeling of present-day waters with EQ3, using the pH and temperature measured at the wellsite, gives high oversaturations with respect to two late-stage cements, illite and kaolinite. So we correct the pH for the formation temperature by running one-third diluted seawater, first alone, and then with the subarkose, through the EQ6 reaction-path model (Figure 9a). Figure 9a shows how pH values become acidic for the one-third diluted seawater because of a change in the ionization constant of water and other aqueous equilibria due to increasing temperature. When this one-third diluted seawater is run with a subarkose, the pH values, although more acidic than those at the surface, are alkaline when compared with

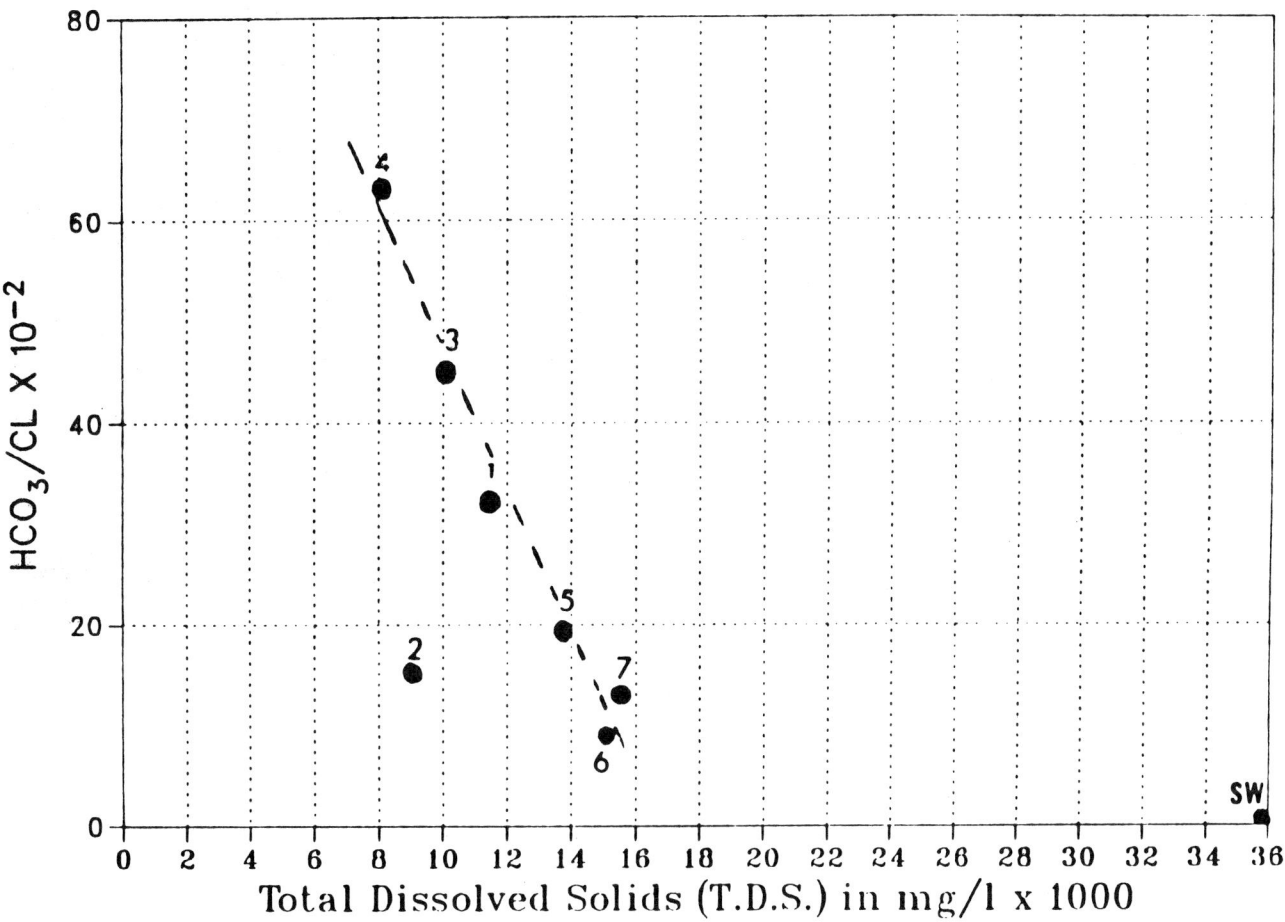

Figure 6b. Plot of HCO_3/Cl vs. TDS showing dilution of Upper Almond waters by meteoric waters.

those for water alone at the appropriate temperatures. We interpret this as a result of hydrolysis of silicates, which buffers the pH values to slightly alkaline. When we calculate silicate equilibria at the correct formation temperature and the corrected pH, we see that illite and kaolinite are only slightly oversaturated to near equilibrium.

In the following section we compare the entire paragenetic sequence.

General Petrographic Attributes of the Upper Almond

Based on the classification system of Folk (1974), the composition of these sandstones falls in the category of chert arenite to subarkose. From the base to the top of the Upper Almond the grain size ranges from 0.06 to 0.19 mm. The porosity ranges from 10 to 13%, and the permeability from 0.2 to 0.6 md.

The best permeabilities are associated with preserved primary intergranular porosity between euhedral quartz overgrowths (Figure 10). Even though some porosity enhancement has occurred because of dissolution of rock fragments (Figure 10a, left-hand side), very little permeability enhancement can be associated with such dissolution. The reduction in porosity and permeability occurs primarily because of pore-filling illite.

Paragenesis

The sequence of diagenesis in the Upper Almond is as follows:

1. Dissolution of plagioclase feldspar and rock fragments.
2. Growth of euhedral quartz.
3. Growth of authigenic feldspar.

Upper Almond Formation Waters

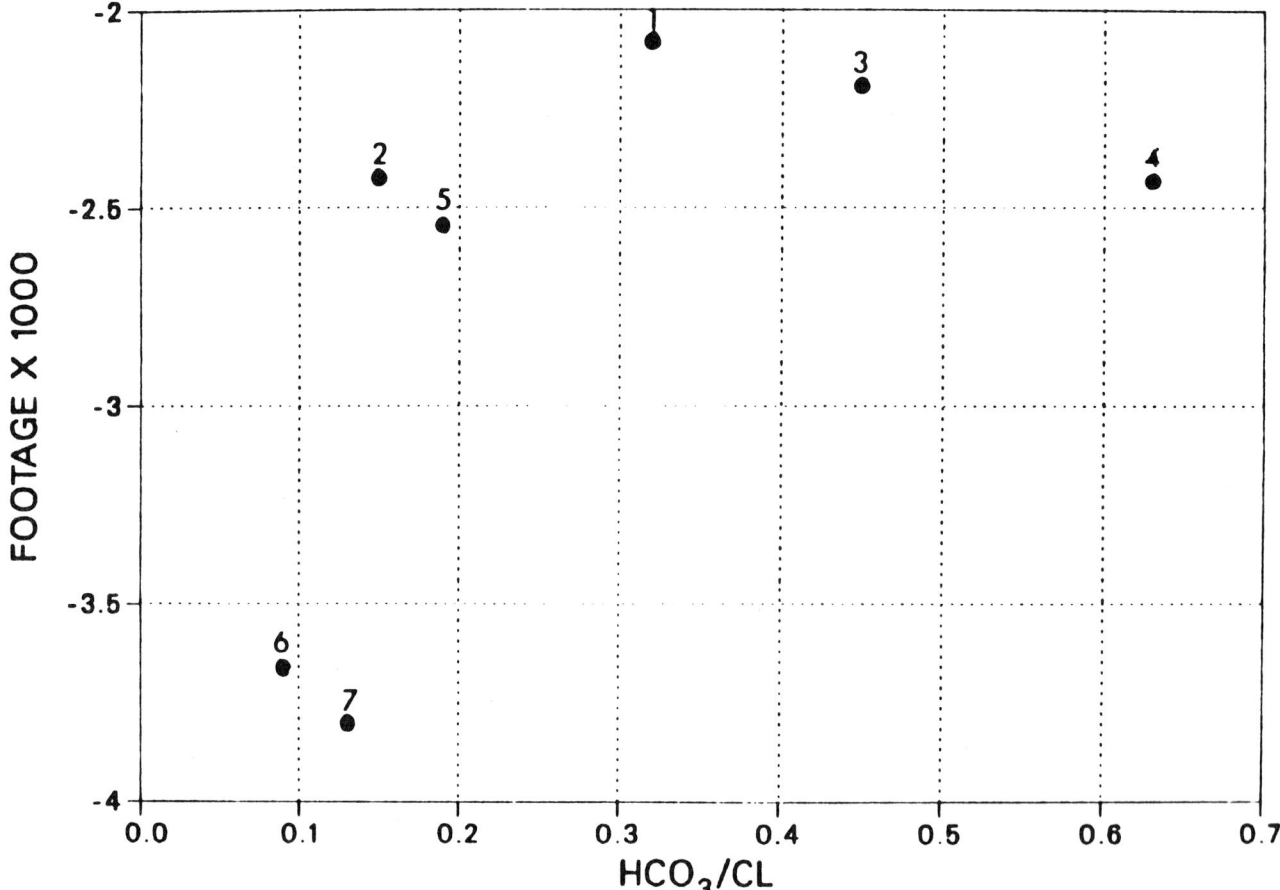

Figure 7. HCO$_3$ enrichment represents no relationship with structural highs. This may mean (a) the meteoric dilution was not a late phenomenon, or (b) if it was late, the HCO$_3$ enrichment in the Almond must be from an additional source of CO$_2$, from reactions not yet understood.

4. Dissolution of authigenic feldspar.
5. Growth of pore-filling kaolinite.
6. Growth of ferroan dolomite (ankerite?).
7. Growth of pore-bridging and pore-filling illite.

In this upper Almond barrier bar (Echo Springs, Wamsutter arch), the first episode of cementation is quartz overgrowths. Figure 11 is a scanning electron micrograph of a sample at 9260 ft from the Champlin 242-D. Euhedral quartz overgrowths formed first, as documented in the thin section photomicrograph (Figure 10).

The simulation in the open system failed to indicate quartz overgrowths. Growth of authigenic feldspar and slight dissolution of this authigenic feldspar can be seen in Figure 11a at point D. Booklets of kaolinite fill the pores in Figure 11a at point B, and booklets of kaolinite can be seen precipitated on the face of authigenic feldspar in Figure 11b at point B.

Overlying the authigenic feldspar is ferroan dolomite cement (Figure 11a, point A). The wisps of illite overlie authigenic carbonate. Some illite grain coatings may have been an alteration product of feldspar.

CONCLUSIONS

1. None of the cross-basin arches, such as Cherokee-ridge or Wamsutter arch, bring Upper Almond rocks to the surface, and the possibility of late meteoric recharge therefore is slim.
2. Log-calculated and compositionally determined values of water resistivity (Rw) for the Upper Almond range from 0.26 to 0.19 ohm-m.
3. Oxygen isotope data, total dissolved solids, Rw, and pressure-depth profiles suggest that the

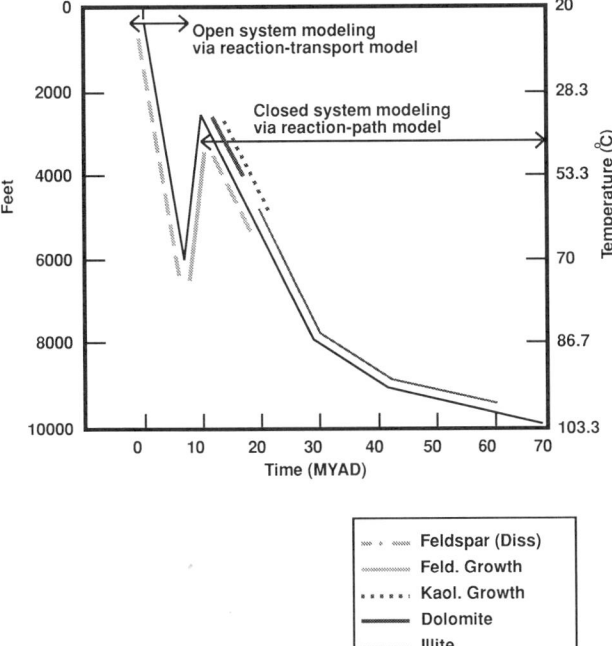

Figure 8. Cementation sequence in Upper Almond Sandstone, as predicted by chemical modeling of subarkose. Sequence was deposited in one-third diluted seawater through the given burial history, which is indicated as MYAD (million years after deposition).

Upper Almond and the Main Almond are two distinctly separate reservoirs.

4. Dilution of early pore waters may have occurred because of coastal meteoric processes, as indicated by deuterium-oxygen isotopes and the HCO_3/Cl ratio.
5. Early enrichment of HCO_3^- is due to (a) coastal meteoric dilution and (b) bacterial reduction of sulfate $SO_4^{--} + 2CH_2O \rightarrow H_2S + 2HCO_3$.
6. Enrichment of Na and K in formation water due to feldspar dissolution continued through burial to a depth of 6000 ft on the Wamsutter arch.
7. Almond waters are currently depleted in Mg, a condition which is consistent with precipitation of late ankerite. Simulations predict dolomite and siderite instead of ankerite, because ankerite is not present in the data base.
8. The simulated sequence is K-feldspar dissolution, growth and dissolution, plagioclase dissolution, dolomite and kaolinite growth, and finally cementation by illite in a closed system, which is consistent with the observations.
9. Exceptions to the observed diagenetic sequence are early-phase quartz overgrowths and observed ankerite cement vs. simulated dolomite cement. We might say there is a general agreement between simulated and observed diagenetic sequence.
10. Present-day Almond waters are in equilibrium to slightly oversaturated with respect to illite and kaolinite, which is consistent with the observed late-stage cementation by illite and kaolinite.

ACKNOWLEDGMENTS

Howard Cotten and Suu Nguyen assisted in well-site analysis. Laboratory analysis of water samples were performed by Suu Nguyen. Analysis of organic acid anions were provided by Bert Fisher. Zane Mcculley provided aluminum and silicon analysis using Graphite Furnace (atomic absorption). Most

Table 3. Volume percent of dissolving and precipitating minerals

Time (ma)	Temp °C	Depth (ft)	Quartz	K-Feldspar	Plagioclase	Albite	Kaolinite	Calcite	Dolomite	Illite	Porosity	Total Rock
0	20.0	0	67.5	3.75	3.75	0	0.00	0	0.00		25	75
1	28.3	1,000	67.5	3.75	3.75	0	0.00	0	0.87		24.1	75.9
2	36.7	2,000	67.1	3.75	3.75	0	0.002	0	0.7		23.6	76.3
3	45.0	3,000	67.0	3.6	3.75	0	0.012	0	0.6		25.1	74.9
4	53.3	4,000	67.2	2.7	3.71	0	0.69	0	0.6		25.1	74.9
5	61.7	5,000	67.0	0.6	3.5	0	0.4	0	0.2		28.3	71.7
6	70.0	6,000	67.0	12.1	1.0	0	3.0	0	0.2		16.4	83.6
10	45.0	3,000	67.0	7.4	0.75	0	8.0	0	10.0		7.3	92.7
20	70.0	6,000	67.0	2.6	0.75	0	9.5	0	12.5		7.6	92.4
30	86.7	8,000										
40	95.0	9,000	Via Reaction-Path Modeling									
50	97.8	9,333								Illite forms		
60	100.6	9,660		K-spar dissolves								
70	103.3	10,000										

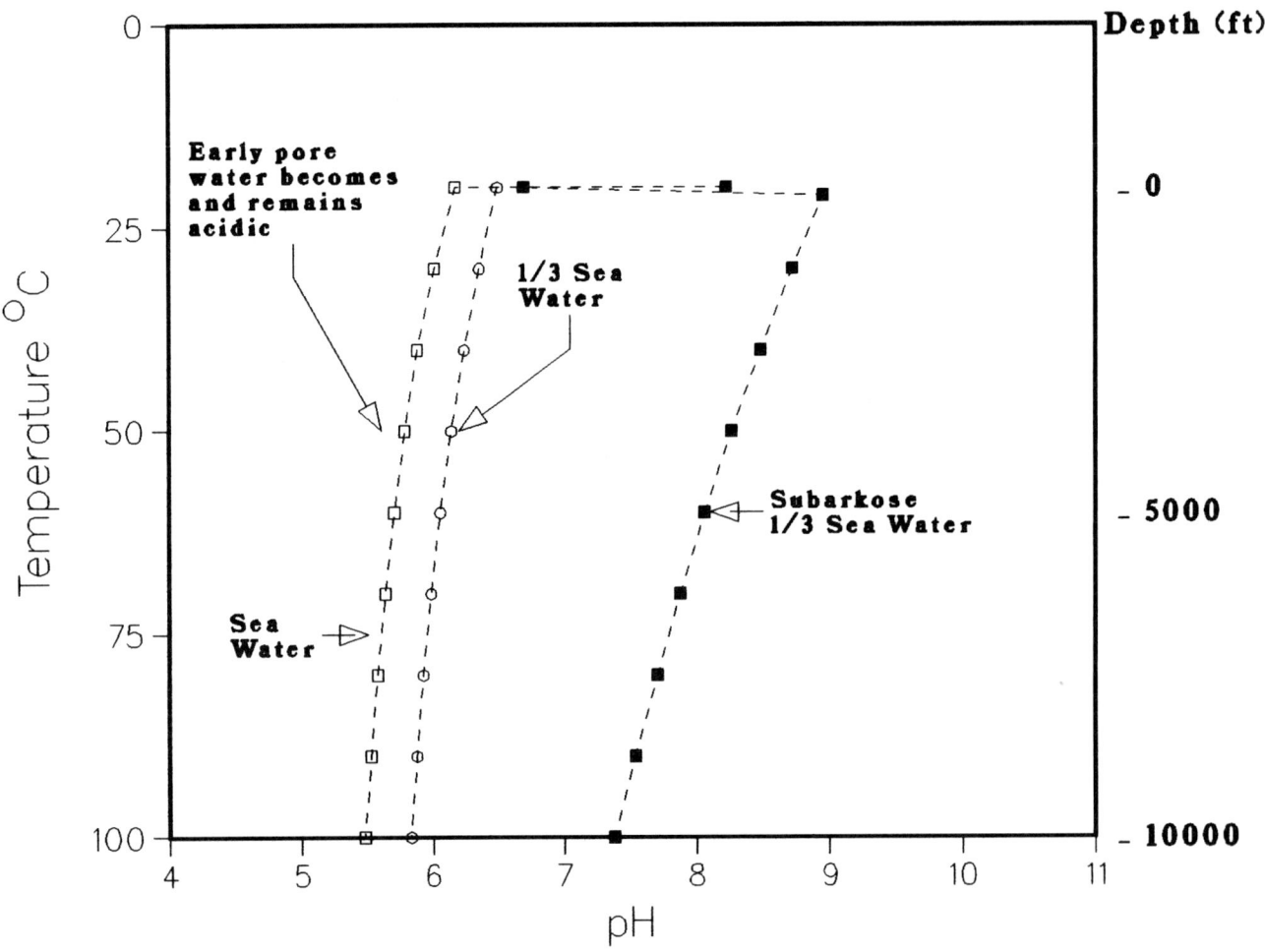

Figure 9a. pH calculated via EQ3/6 for one-third diluted seawater. Increase in acidity with burial is due to changes in ionization constant of water and other aqueous equilibria. When a subarkose is reacted in this system, pH values become more alkaline because of silicate hydrolysis, but are acidic when compared with surface pH.

of the research for this paper was conducted from 1983 to 1984, during which time discussions with geologists in Amoco's Denver Region were of great value. We gratefully acknowledge the support of Amoco Production Company and their permission to publish this work.

REFERENCES CITED

Asquith, D. O., 1972, Depositional topography and major marine environments, Late Cretaceous Wyoming: Ph.D. dissertation, University of California, Los Angeles, California.

Asquith, D. O., 1975, Petroleum potential of deeper Lewis and Mesaverde sandstones in the Red Desert Washakie and Sand Wash basins, Wyoming and Colorado: Rocky Mountain Association of Geologists Field Trip Guidebook.

Craig, H., 1961, Isotopic variations in meteoric waters, Science, v. 133, p. 1702-1703.

Folk, R. L., 1974, Petrology of sedimentary rocks: Austin, Texas, Hemphill Publishing Company, 182 p.

Gill, J. R., and W. A. Cobban, 1973, Stratigraphy and geologic history of Montana Group and equivalent rocks, Montana, Wyoming, North and South Dakota: USGS Professional Paper 776, 37 p.

Gill, J. R., E. A. Merewether, and W. A. Cobban, 1970, Stratigraphy and geologic history of some Upper Cretaceous-Lower Tertiary rocks in south-central Wyoming: USGS Professional Paper 667, 63 p.

McKetta, J. J., and A. H. Wehe, 1958, Chart for the water content of natural gases: Petroleum Refiner, v. 37, no. 8, p. 153-154.

Meshri, I. D., 1981, Deposition and diagenesis of Glauconite Sandstone, Berrymore-Lobstick-Bigoray area, south-central Alberta: a study of physical chemistry of cementation: Ph.D. dissertation, University of Tulsa, Tulsa, Oklahoma, 130 p.

Meshri, I. D., 1990, An overview of chemical models and their relationship to porosity prediction in the subsurface, this volume.

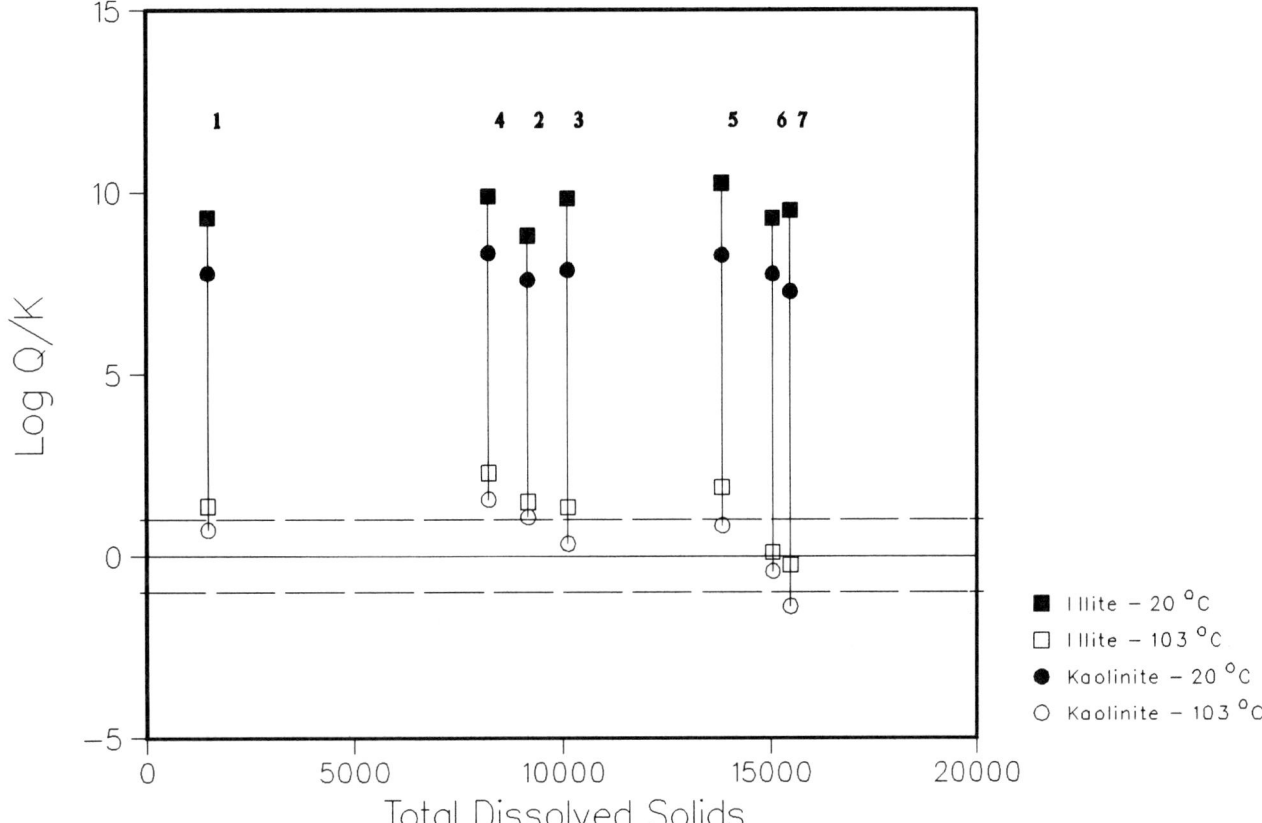

Figure 9b. The Almond waters show high oversaturation if the well-site pH is used to calculate equilibria for kaolinite and illite. If pH is corrected for formation temperature, the present-day waters show near equilibrium to slight oversaturation with respect to late-stage cements like kaolinite and illite.

Miller, F. X., 1977, Biostratigraphic correlation of the Mesaverde Group in southwestern Wyoming and northwestern Colorado, *in* H. K. Veal, ed., Exploration frontiers of the Central and Southern Rockies: Rocky Mountain Association of Geologists, 1977 Symposium, p. 117-138.

Nordstrom, D. K., et al., 1979, A comparison of computerized chemical models for equilibrium calculations in aqueous systems, *in* E. A. Jenne, ed., Chemical modeling in aqueous systems: American Chemical Society Symposium Series 93, p. 857-892.

Ortoleva, P., E. Merino, C. Moore, and J. Chadam, 1987, Geochemical self-organization I: Reaction-transport feedbacks and modeling approach: American Journal of Science, v. 287, p. 979-1007.

Rittenhouse, G., 1967, Bromine in oil-field waters and its use in determining possibilities of origin of these waters: AAPG Bulletin, v. 51, p. 2430-2440.

Wolery, T. J., 1983, EQ3NR: A computer program for geochemical aqueous speciation-solubility calculations: user's guide and documentation: Lawrence Livermore National Laboratory, UCRL-53414, p. 191.

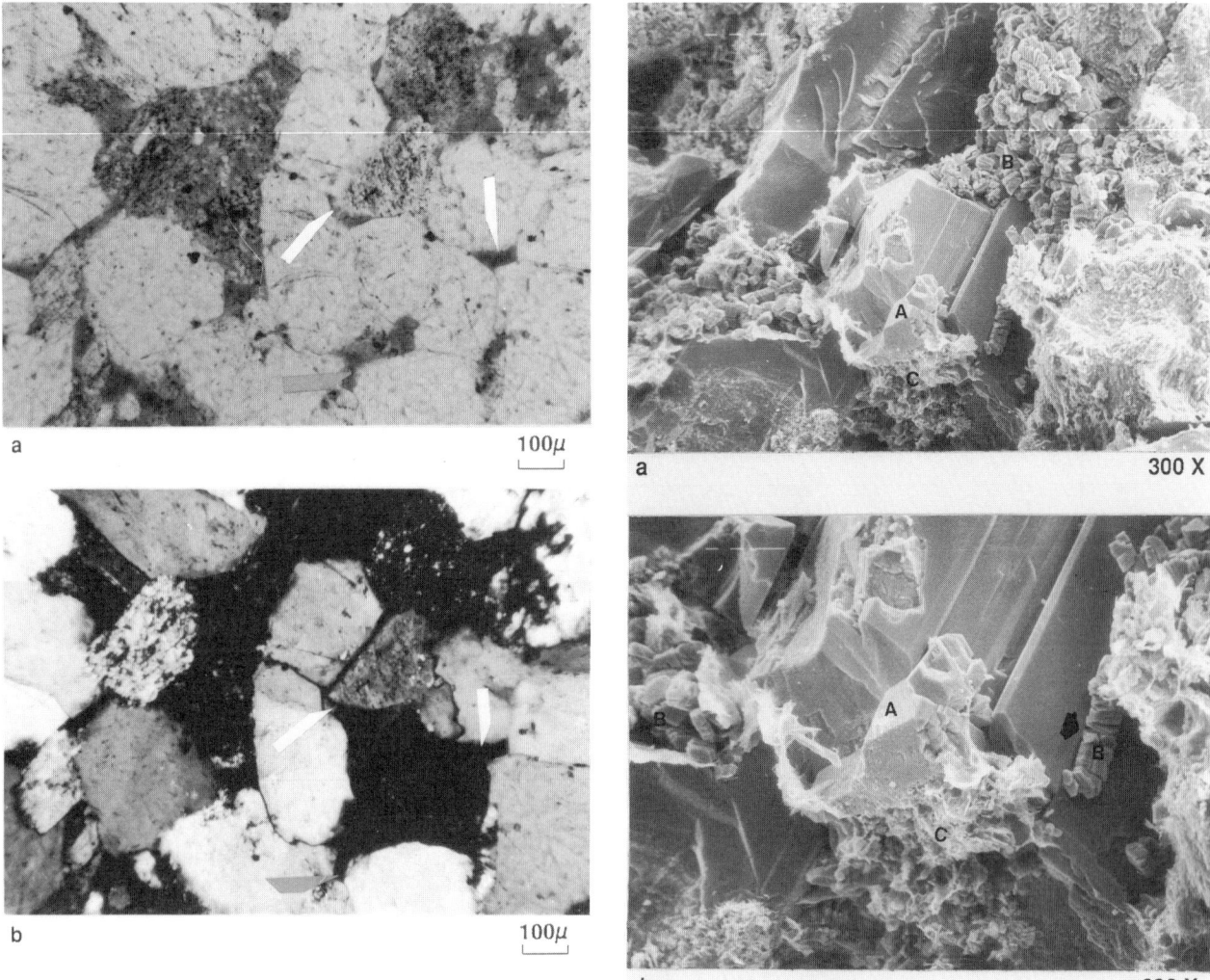

Figure 10. Thin-section photomicrograph showing primary porosity (indicated by arrows) preserved between euhedral quartz overgrowths, Champlin 242-D, 9260 ft, Echo Springs (ϕ = 13.6%, k_{md} = 0.6 md, grain size 0.13 mm).

Figure 11. Scanning-electron micrograph showing continuation of paragenetic sequence from the late stage ankerite at A, to the later kaolinite at B, and illite at C (Champlin 242-D, 9260 ft, Echo Springs).

Bulk Solution Disequilibrium in Aqueous Fluids as Exemplified by Diagenetic Carbonates

Hans-G. Machel
Department of Geology
University of Alberta
Edmonton, Alberta, Canada

Recrystallization, and in some cases precipitation, of minerals in aqueous fluids may occur in thermodynamic disequilibrium with the bulk solution. This compromises applications of the concept of cement stratigraphy and several geochemical models such as those commonly used to identify former subaerial exposure surfaces in carbonate rocks. Bulk solution disequilibrium may be a common phenomenon during recrystallization of carbonates as well as other minerals in aqueous fluids, i.e., in diagenetic, hydrothermal, and metamorphic environments.

Bulk solution disequilibrium generally occurs when dissolution and concomitant reprecipitation take place in thin surface-adsorbed or surface-bonded fluid layers that have compositions different from the solution occupying the bulk of the pore space. The elemental and isotopic compositions of the recrystallized minerals are determined by the compositions of these surface fluid layers, as well as by recrystallization rates. Thereby, paragenetically coeval grains, crystals, or crystal growth zones, may obtain heterogeneous trace element compositions, as exemplified by diagenetic carbonates from a Devonian reef in Western Canada. Modeling of recrystallization in bulk solution disequilibrium yields compositions that do not permit unambiguous interpretation of diagenetic environments such as meteoric, brackish, or marine. These conclusions follow from application of chemical microprofiles of crystal surfaces, diffusion models for dissolving and precipitating crystals, and calculations of trace element distributions.

INTRODUCTION

The pore fluids of many diagenetic environments, such as vadose, meteoric phreatic, marine phreatic, burial diagenetic, etc., commonly have distinctive elemental and isotopic compositions, although there is considerable overlap in some cases (e.g., Veizer, 1983). Diagenetic carbonate minerals (mainly calcite, aragonite, and dolomite) generally are assumed to form in *'bulk solution equilibrium'* with these fluids, i.e., precipitation and recrystallization are assumed to take place in thermodynamic equilibrium with the aqueous solution that occupies the bulk of the pore space. The elemental and isotopic compositions of diagenetic minerals, therefore, are utilized to interpret the diagenetic environment(s) of formation (e.g., James and Choquette, 1984).

Recrystallization, which is common in detrital and diagenetic carbonates and takes place in systems with variable water/rock ratios, has led to qualitative and quantitative models of bulk solution equilibrium (e.g., Morrow and Mayers, 1978; Meyers and Lohmann, 1985; Banner et al., 1988) and *'bulk solution disequilibrium'* (e.g., Brand and Veizer, 1980; Pingitore, 1982). In the case of bulk solution disequilibrium, the compositions of the diagenetic minerals are determined by narrow fluid films that are present as surface-adsorbed or surface-bonded interphase boundary layers with compositions that are different from the composition of the bulk solution. Bulk solution disequilibrium may occur during recrystallization as well as during precipitation. In either case, the elemental and isotopic compositions of any recrystallized mineral are not representative of the diagenetic environment and solution of the bulk aquifer.

Applications of these models have been variably successful in the interpretation and prediction of minor, trace, and rare earth element concentrations, as well as isotopic ratios, of diagenetic carbonates. Often, unambiguous recognition of the diagenetic environment is not possible using geochemical compositions (e.g., Veizer, 1983; James and Choquette, 1984; Machel, 1986). Furthermore, some aspects of diagenetic carbonates have not been sufficiently addressed by any of the above models. For example, paragenetically coeval phases may have very heterogeneous compositions (e.g., Machel, 1986; Martin et al., 1986). Hence, the main objectives of this paper are (a) to advance a hypothesis of bulk solution disequilibrium that explains compositional differences that are too large to be accounted for by bulk solution equilibrium models; and (b) develop a semi-quantitative measure of the extent of bulk solution disequilibrium, using trace elements. Furthermore, systems with relatively large but chemically restricted (semi-closed or closed) pore spaces are compared to those with interphase boundary layers.

Recrystallization of diagenetic calcites is taken as an example, but the hypothesis advanced in this paper is relevant for other minerals, thermal environments, and partly for precipitation from solution. The hypothesis is semi-quantitative and utilizes existing diffusion models and experimental data on crystal dissolution and precipitation kinetics.

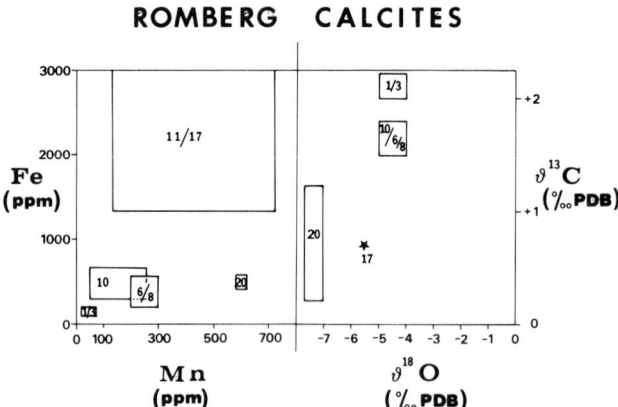

Figure 1. Trace element and isotope data for different types of calcite from the Devonian Romberg Reef Complex in West Germany (Machel, in press). Numbers designate paragenetic events in order of age (1 is the oldest and 20 is the youngest). Box sizes encompass the range of analytical values for each type. Most types can be related to specific diagenetic environments and time intervals. The calcite types were discriminated petrographically (shape and sequence of cements and internal sediments, cross-cutting relationships of fractures and stylolites, etc.) and, in some cases, by cement composition (ferroan vs. nonferroan calcite).

BULK SOLUTION EQUILIBRIUM VERSUS DISEQUILIBRIUM IN CARBONATE DIAGENESIS

Application of bulk solution equilibrium models of diagenetic carbonates generally involves measurements of trace element and/or isotope compositions (less commonly rare earth elements). Cross plots of mineral composition data commonly form clusters of data points that correspond to phases that were identified as paragenetically coeval on the basis of petrography (Fig. 1). Fluid compositions may be back-calculated using elemental distribution coefficients and isotopic fractionation factors. If plotted versus depth, the data may be used to infer diagenetic environments as suggested by several bulk solution equilibrium models (Fig. 2). Such models work, by definition, only where the mineral precipitates are formed in, or at least near, thermodynamic equilibrium with the bulk solution, and if reasonable assumptions can be made regarding burial depth, temperature, and pressure. Thermodynamic bulk solution equilibrium is most likely to be achieved when the fluids are warm or hot, well mixed and/ or replenished, and precipitation is slow. In such cases, chemical gradients within the growing crystals and in the adjacent fluids may be small or absent. If a solution does not change in composition through time, precipitates will be homogeneous and not zoned. Crystal compositions then are indicative of the bulk solution composition, temperature, and pressure.

Many natural diagenetic carbonate crystals are not homogeneous. Inhomogeneities range from concentric to sector and intrasectorial zonation, with or without zonal discontinuities, on any scale between several centimeters and a few micrometers, to rhythmic or irregular zonation and chemical variations on an atomic scale. This is demonstrated particularly well by light and cathodoluminescence microscopy (Machel, 1985), transmission electron microscopy (Wenk et al., 1983), and atomic resolution microscopy (Meike et al., 1988), as well as by a variety of other techniques, such as microsampling across individual crystals and subsequent wet-chemical and isotopic analyses. An inhomogeneous crystal cannot be in thermodynamic equilibrium with the bulk solution because bulk solution equilibrium requires recrystallization to a homogeneous crystal if the solution changes in composition. At best, an active growth surface may form in thermodynamic equilibrium with the solution. Hence, a concentrically zoned crystal generally suggests changes in bulk solution chemistry through time. [This excludes oscillatory zoning on any scale, which is a case of geochemical self-organization that does not necessarily indicate a change in the bulk solution composition:

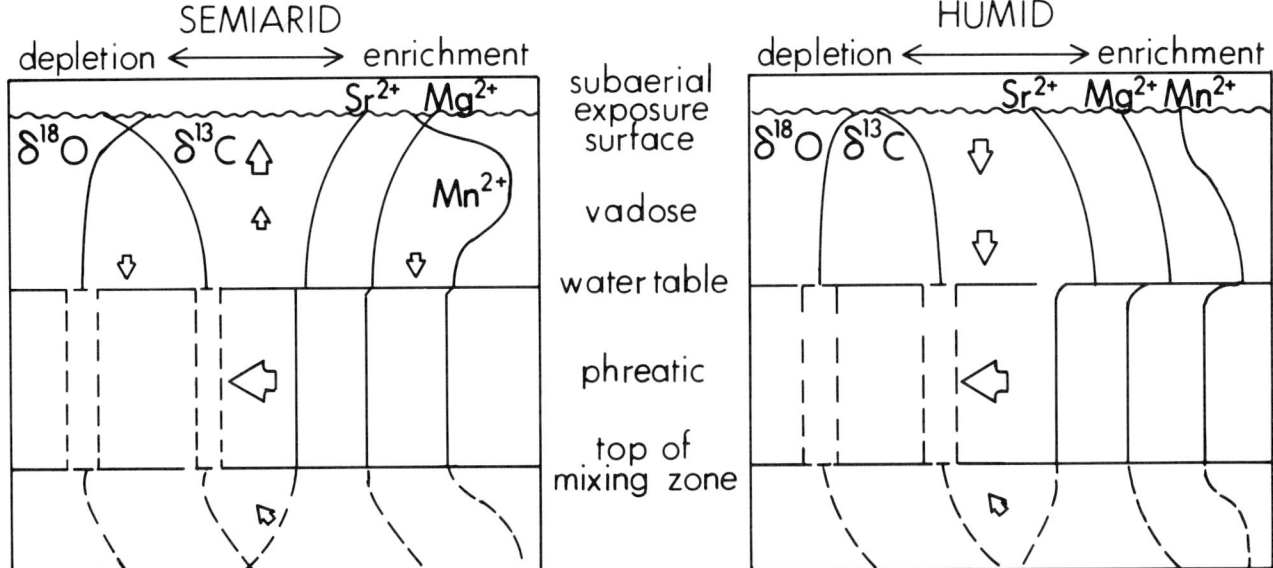

Figure 2. Theoretical geochemical depth profiles of initially aragonite and high-Mg-calcite limestones beneath a subaerial exposure surface in semiarid and humid climates (modified from James and Choquette, 1984). All depth profiles assume bulk solution equilibrium and are based on empirical evidence of known elemental and isotopic variations beneath Cenozoic subaerial exposure surfaces. Open arrows show dominant direction of groundwater movement. Offsets in isotopic ratios and cation concentration occur at the water tables because the vadose and phreatic zones behave, geochemically, as semiclosed systems. Offsets are more pronounced in humid systems. In the semiarid system there are distinctive trends for Sr, Mg, and oxygen isotope ratios in the shallowest vadose zone due to intense evaporation.

e.g., Reeder, 1987.] Non-oscillatory concentric zoning in crystals has been used extensively in carbonate petrology for interpretation of the regional extent and composition of aquifers through burial and time (Meyers, 1978; Grover and Read, 1983; and many others), a concept called 'cement stratigraphy'. Pronounced compositional variations between time-equivalent portions of concentric zones that belong to different growth sectors (Reeder and Grams, 1987) are rarely encountered and generally not utilized in cement stratigraphic interpretations. For the sake of simplicity, sector zonation is disregarded also in this paper, but it is recognized that inter- and intrasectorial compositional differences may be important in some cases.

Cement stratigraphy is a valid and useful concept where time-equivalent grains, crystals, or crystal zones, form compositional clusters such as those depicted in Figure 1. However, there are cases where carbonates that appear to be coeval on the basis of petrography have a wide range of compositions. A striking example are a number of diagenetic calcite types in an Upper Devonian reef from the subsurface of western Canada (Fig. 3). Most data from any given calcite type display large and erratic fluctuations that do not conform to any of the popular bulk solution equilibrium models *even though* petrographic data clearly indicate that (a) this particular reef was subaerially exposed near the top of the section, and (b) a vadose zone probably was present above the depth of −2388 m. [Compare data in Fig. 3 with the predicted trends in Fig. 2; for further details on the petrography, geochemistry, the complete data set, analytical procedures, and discussion of the various applied models, see Machel, 1986]. Furthermore, the observed geochemical variations do not appear to be an artifact produced by the method or scale of sampling. The data plotted in Fig. 3 are from samples that were taken with a dental drill from crystals or crystal zones that are homogeneous under the light and cathodoluminescence microscopes. These considerations lead to three conclusions.

1. The bulk solution equilibrium models depicted in Figure 2 cannot be applied in this case.
2. The chemical compositions of the calcites plotted in Fig. 3 probably represent bulk solution disequilibrium.
3. The concept of cement stratigraphy cannot be applied in this case.

In the following section, a hypothesis is advanced that explains how trace element variations such as those shown in Figure 3 can result during bulk solution disequilibrium recrystallization. This hypothesis is applicable to the recrystallization of other minerals because it is based on general theories of diffusion and kinetics during mineral precipitation and dissolution.

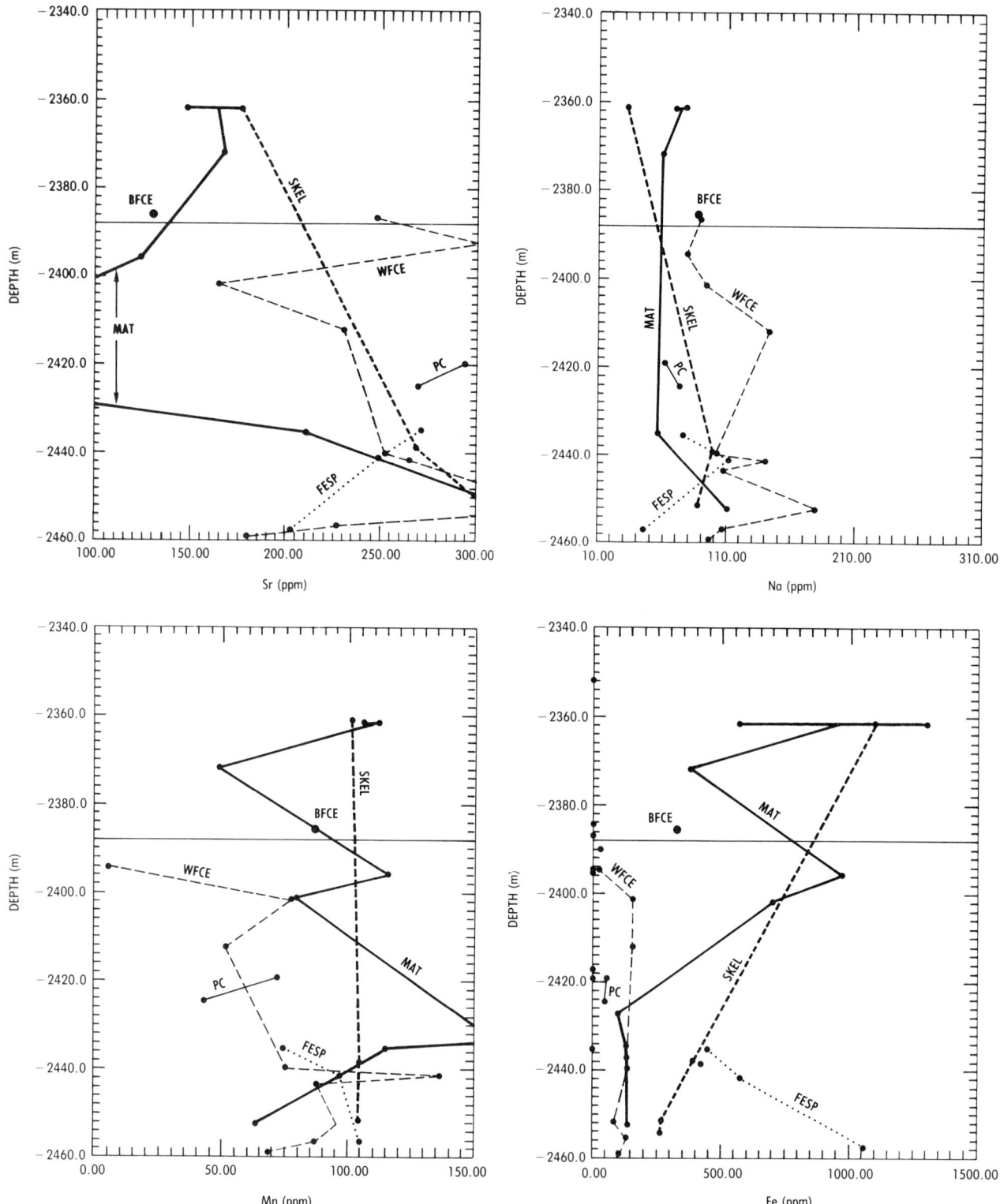

Figure 3. Trace element-depth profiles for diagenetic calcites in a Devonian reef in western Canada (from Machel, 1986). Data points belonging to the same petrographic types are connected (these lines are not calculated regression lines). SKEL = skeletal material; MAT = matrix; PC = prismatic calcite; BFCE = brown fibrous calcite; WFCE = white fibrous calcite; FESP = ferroan spar. Petrographic evidence suggests that all calcites except FESP are recrystallized. Depth −2388 m marks the lowest occurrence of pendant calcite cements (BFCE), and therefore probably indicates the lowest level of former vadose zones. Note that none of the trends shown correspond to any in Figure 2. Samples were taken with a dental drill from optically homogeneous crystals or portions thereof, and then analyzed by ICP. The precision of the data points is better than the width of the symbols used.

RECRYSTALLIZATION

It is generally agreed that diagenetic recrystallization of carbonates is a wet-chemical dissolution-reprecipitation process (Carlson, 1983; Morse, 1983), which is true for other mineral recrystallization in aqueous fluids, e.g., in hydrothermal and in many metamorphic systems. For the sake of simplicity, one may elucidate recrystallization and the accompanying chemical changes in two separate steps, dissolution and precipitation. This leads to a hypothesis on the mechanism of recrystallization as well as the accompanying states of equilibrium or disequilibrium.

Dissolution

Dissolution in aqueous solution can be described by means of ionic concentration profiles adjacent to a dissolving crystal (Berner, 1978). Accordingly, dissolution has been termed to proceed by 'reaction control' or by 'diffusion' / 'transport control' (transport may be by diffusion and/or by advection), or by 'mixed control' (Fig. 4). Implicit assumptions are that the solution/rock ratio is relatively large, and that the rocks (crystals) dissolve congruently. During reaction control, the crystal dissolution rate is slower than the ion transport or diffusion rate adjacent to the crystal. In other words, ions are detached at a rate that is slower than their rate of transport away from the crystal surface. Hence, any detached ion is lost relatively quickly to the bulk solution, and the bulk solution composition abuts directly against the crystal (Fig. 4, top). During diffusion/transport control, the ion detachment rate exceeds diffusion/transport rates away from the crystal. In this case, detached ions accumulate next to the dissolving crystal to form a 'dissolution boundary layer' of a composition that is different from that of the bulk solution (Fig. 4, center). Trends produced by reaction and diffusion control are end members of a spectrum of possibilities. Mixed cases (Fig. 4, bottom) are possible and in nature probably quite common. Such trends are negative for elements whose concentrations in the crystal are lower than in the solution.

One of the best mechanistic models for calcite dissolution (Plummer et al., 1978, 1979; Busenberg and Plummer, 1986) assumes that the dissolution boundary layer is at least partially hydrated and achieves a thickness of about 10-100 Ångstroms. Plummer et al. (1979) and Reddy et al. (1981) further stipulated that the dissolution boundary layer is separated from the bulk solution by a 'hydrodynamic boundary layer', an assumed stagnant volume of solution adjacent to the crystal surface. Depending on experimental conditions, the hydrodynamic boundary layer can vary in thickness from 20 to 150 micrometers (Reddy et al., 1981). The thicknesses of the dissolution boundary and hydrodynamic boundary layers are obviously controlled by factors

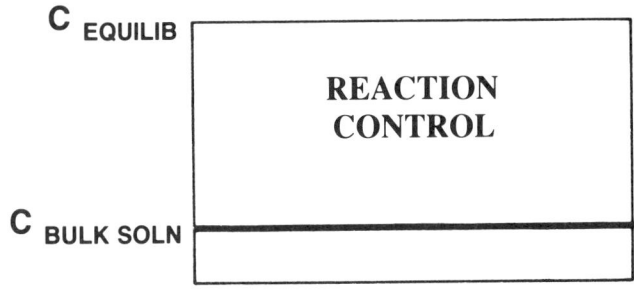

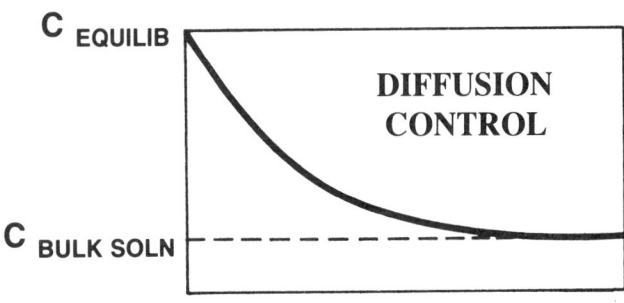

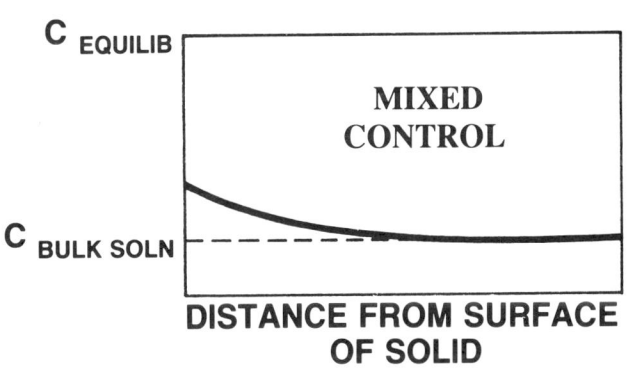

Figure 4. Concentration profiles of ions in solution released during dissolution of a mineral. The mineral surface is the right margin of the diagrams, $C_{EQUILIB}$ is the equilibrium value for the ion(s) in solid solution, and $C_{BULK\ SOLN}$ is the concentration of these ion(s) in the bulk solution. If the rate-controlling step of dissolution is the ion detachment rate, the bulk solution concentration abuts against the mineral surface (top). If the rate-controlling step is diffusion away from the surface, the concentration next to the surface is the equilibrium value (center). A boundary layer with a concentration gradient between $C_{EQUILIB}$ and $C_{BULK\ SOLN}$ and the width of the dashed horizontal line separates the crystal from the bulk solution. Intermediate cases are possible and probably common (bottom). For different ions in the same two-phase system, the gradients between the crystal and bulk solution can also be absent or negative, depending on the initial concentrations of the ion under consideration in both phases.

such as the concentration difference between solid and solution for the ions under consideration, as well as the dissolution, advection, and diffusion rates. Under diffusion/transport control and mixed control, the dissolving crystal surface is not in direct contact with the bulk solution. As a result, the dissolution rate is controlled partially by the compositions and thicknesses of the two boundary layers. Ion exchange with the bulk solution proceeds through these two boundary layers.

Dissolution profiles with temporary reversals away from the crystal surface, as commonly observed in magmatic dissolution and inferred to be caused by 'uphill diffusion' (Watson, 1982; Zhang et al., subm.), are disregarded in the present discussion. Such diffusion profiles cannot develop in the relatively dilute aqueous solutions present in most diagenetic environments, because uphill diffusion can only develop when the fluid is concentrated and viscous enough to develop a structure (which might happen only in the extreme and uncommon cases of brines concentrated beyond halite saturation).

Precipitation

Precipitation of crystals from solution has been described by a number of theories that have been supported to various degrees experimentally (e.g., Sunagawa, 1977; Hartmann, 1982). There are many modes of crystal growth, e.g., dislocation controlled or nucleation controlled, with ion attachment by the spiral mechanism, or by attachment of ion clusters at irregularly spaced kink sites, steps, or terraces on growing surfaces. In the present context it is important to note that, irrespective of the exact mode of growth, the actual crystal growth surfaces in aqueous solution appear to be partially hydrated. This is suggested and substantiated by several lines of evidence. Electrochemical experiments investigating the surface charge of calcite (Somasundaran and Agar, 1967), experimental determination of physisorption and chemisorption isotherms of calcite (Morimoto et al., 1980), reflectance spectroscopy of whole crystals and crystal surfaces of calcite (Gaffey 1985, 1988), and Auger spectroscopy of fresh growth surfaces of calcite (Mucci and Morse 1985; Mucci et al., 1985), as well as the dissolution and precipitation experiments and calculations by Plummer et al. (1978), Reddy et al. (1981), Lahann and Siebert (1982), and Busenberg and Plummer (1986), all indicate partial H_2O and/or OH^--incorporation at the solution-crystal interface.

The data presented in Mucci et al. (1985) may be used to make those points that are of interest in the present context. Mucci et al.'s (1985) Auger spectroscopy data strongly suggest (although they do not unequivocally prove) the following features (as illustrated in Fig. 5):

1. Mg-calcite crystals grown in aqueous solution probably have a partially hydrated surface

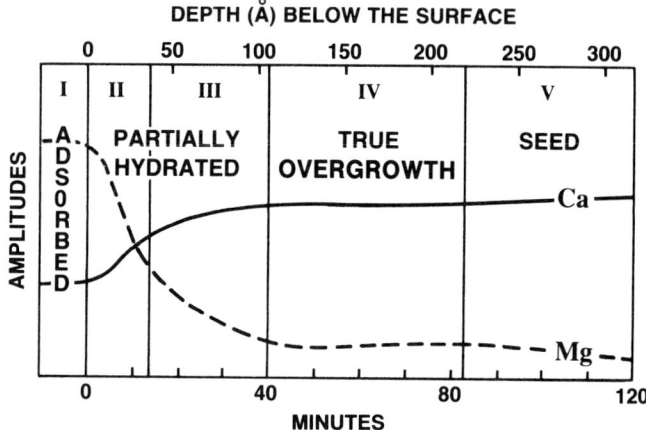

Figure 5. Concentration profile for Mg and Ca in the outer 300+ Ångstroms of a freshly grown calcite overgrowth, as interpreted from Auger spectroscopic data. The Ca and Mg values on the left and on the right are equilibrium values for the bulk solution and for the seed crystal, respectively. The true overgrowth is interpreted to be formed in equilibrium with the bulk solution, and appears to be separated from the bulk solution by a partially hydrated zone with ionic gradients. Modified from Mucci et al. (1985).

adsorption layer. This layer is up to at least 100 Ångstroms in thickness (which is indirectly supported by the data from Somasundaran and Agar, 1967; Reddy et al., 1981; and Davis et al., 1987). The layer separates the so-called 'true' overgrowth from the bulk solution.
2. There are ionic gradients through the partially hydrated surface adsorption layer, as exemplified by magnesium (Fig. 5), and the layer acts as a 'porous membrane' (Mucci et al., 1985).
3. The partially hydrated surface adsorption layer is weakly bonded, as it could be washed off easily. Mucci et al.'s (1985) data suggest that the water was bonded by chemisorption (as H_2O and/or OH^-).

Based on all data from the studies cited above, it is likely that partially hydrated surface adsorption layers with variable thicknesses are common features of growing crystal surfaces in aqueous solution. Such layers probably are attached by physisorption and chemisorption (e.g., Morimoto et al., 1980) [although this distinction is of little significance in the present context]. Furthermore, there probably are many different ionic gradients through these layers that are similar but not identical to the one for Mg shown in Figure 5. Gradients for other ions will be different because Mg does not behave as most other elements that substitute for calcium (e.g., Lorens, 1981; Mucci et al., 1985; Pingitore and Eastman, 1986; Morse and Bender, 1990). Accordingly, each trace element under consideration may display an individual profile through the partially hydrated surface adsorption layers. Furthermore, it is likely that there is a hydrodynamic boundary layer in addition to the partially hydrated surface adsorption layer, similar to the region next to a dissolving grain.

Recrystallization and States of Thermodynamic Equilibrium

During recrystallization of metastable mineral phases, the dissolution boundary and hydrodynamic boundary layers of the dissolving grain may or may not overlap in time and space with the partially hydrated surface adsorption and hydrodynamic boundary layers of the precipitating grain. If a dissolving and a growing crystal are separated spatially by a distance equal to or greater than the sum of the four boundary layers, the mineralogy, trace element and isotope composition of the precipitating crystals are controlled by the composition of the solution in the adjacent pore space. This represents bulk solution equilibrium. Thereby, the compositions of the reprecipitated (recrystallized) crystals are determined by the water/rock ratio. Where the water/rock ratio is large, the bulk solution composition is controlled by extrinsic factors, such as the origin of the diagenetic fluids (meteoric, hydrothermal, etc.) or mineral reactions in aquifers upstream from the diagenetic site under consideration. In the case of a small water-rock ratio, the bulk solution composition is controlled at least partially by the dissolving crystals. In many burial diagenetic environments, for example, bulk solution compositions generally are altered, if not dominated, by rock-water interaction.

Bulk solution disequilibrium occurs if the dissolving and the precipitating crystals are closer to one another than the total sum of all four boundary layers. The distance between the dissolving and the precipitating crystal may become so small that one or the other hydrodynamic boundary layer is no longer present, or that the only water present between the two crystals is a non-Newtonian *interphase boundary layer* that is partially bonded to both surfaces. These situations presumably constitute what has been termed an 'interphase boundary' (Gibbs, 1876), 'messenger film' (Brand and Veizer, 1980), 'reaction zone' (Pingitore, 1982) (as well as other expressions) that is located between the dissolving and reprecipitating parts of a recrystallizing crystal or grain.

As shown experimentally (e.g., Evans et al., 1986, and studies cited therein), recrystallization of minerals (and metals) commonly takes place via interphase boundary migration (also called 'chemically induced grain boundary migration' by Hillert and Purdy, 1978, or 'diffusion-induced grain boundary migration' by Balluffi and Cahn, 1981, and Smith and King, 1981). Interphase boundary layers are driven, at least in part, by chemical free energy reduction (the exact mechanisms are still being debated, see discussion in Evans et al., 1986). Thereby, the composition of the reprecipitating part of a crystal is determined largely by the composition of the adjacent dissolving part of that crystal because (a) there is little or no chemical exchange between the interphase boundary layer and the bulk solution, and (b) the ionic gradients of the dissolving and reprecipitating crystals (Figs. 4 and 5) overlap. Also, the dissolution and reprecipitation rates will be interdependent, which may result in recrystallization rates that vary from crystal to crystal (grain to grain). Because distribution coefficients are rate-dependent, this may result in different 'effective' distribution coefficients for neighboring recrystallizing crystals (grains) (see below).

It is important to emphasize that bulk solution disequilibrium during recrystallization can result in variable compositions (e.g., trace element concentrations) in adjacent crystals or grains even if the bulk solution composition remains constant, which is not possible in systems with bulk solution equilibrium (irrespective of the water/rock ratio). This follows directly from the above considerations, and it has been shown experimentally (e.g., Evans et al., 1986).

The extent of bulk solution disequilibrium, also called 'degree of openness of the system' (e.g., Brand and Veizer, 1980; Veizer, 1983) is a function of (a) the actual spatial distance between the two crystals (which may be as small as a few Ångstroms but at least the thickness of a monomolecular H_2O or OH^- layer, and may vary between grains that are only a few millimeters apart); and (b) the extent of ion exchange between the interphase boundary layer and the bulk solution. The actual spatial distance between the two crystals obviously determines the degree of interference of the ionic profiles with one another, which partially controls the composition of the precipitating phase. Some ionic exchange may take place between the bulk solution and the 'fluid' along the interphase boundary. This was shown experimentally by Evans et al. (1986), and others cited therein, where stable interphase boundaries began to migrate when the bulk solution composition was modified. In addition, where the dissolving crystal or crystal aggregate is porous (as is commonly the case in marine skeletal and non-skeletal carbonates that are subject to recrystallization), ionic exchange is likely to occur also through the primary grain. In this case, exchange with the bulk solution should be much greater than where there is no such porosity, and the precipitating phase may approach bulk solution equilibrium composition.

APPLICATION TO RECRYSTALLIZATION OF DIAGENETIC CARBONATES

As an example, the data shown in Figure 3 can be used to demonstrate the approximate extent of bulk solution disequilibrium during recrystallization of marine and early diagenetic carbonates. These data are part of a comprehensive case study (Machel,

1986). Some additional information from this study is presented below.

Four lines of evidence indicate that the crystals plotted in Figure 3 are recrystallized and represent bulk solution disequilibrium: (a) all crystals consist of low-Mg-calcite (i.e., calcite with less than 3 mole-% Mg substituting for Ca in lattice sites), but some types display textural and compositional evidence of a former high-Mg-calcite mineralogy (calcite with about 3-17 mole-% Mg in Ca-lattice sites); (b) based on textural evidence, these components (cement crystals and skeletal grains) originated as precipitates from normal seawater; (c) there are microtextural features that can only be derived from recrystallization; (d) diagenetically coeval calcites (e.g., petrographically and paragenetically identical calcites) have grossly different trace element contents (Fig. 3). Based on these facts, one may calculate the 'degree of openness' (or water/rock ratio, or bulk solution disequilibrium) using Sr-content changes during recrystallization of marine high-Mg-calcite to diagenetic low-Mg-calcite as an example.

Choice of Starting Composition and Distribution (Partition) Coefficients

For all calculations of trace element concentrations presented below, the Homogeneous Distribution Law (Henderson and Kracek, 1927) was used:

$$D = (m_{TE}/m_{Ca})_s / (m_{TE}/m_{Ca})_l$$

whereby D = partition coefficient (often called distribution coefficient), m = molar concentration, TE = trace element, Ca = calcium, s = solid, and l = liquid. This law permits approximate calculations of thermodynamic equilibrium concentrations of trace elements (molar TE/Ca ratios) in calcite relative to the trace element concentrations (molar TE/Ca ratios) in solution. In the present context, the accuracy of the calculated trace element compositions in natural calcites using this distribution law and the published values of distribution coefficients (all of which were determined experimentally) is adequate. It must be kept in mind, however, that the Homogeneous Distribution Law does not permit calculations of true thermodynamic equilibrium compositions. Firstly, it uses molar concentrations rather than thermodynamic activities; secondly, D is an experimental, nonthermodynamic partition coefficient that is effective only under the specific experimental conditions for which it was determined (e.g., Busenberg and Plummer, 1989; Morse and Bender, 1990). Hence, D has no strict thermodynamic validity, and trace element concentrations calculated using the Homogeneous Distribution Law and published D-values merely approximate thermodynamic equilibrium concentrations. In practice, compositions calculated from thermodynamic models often are very close to those that were determined experimentally (e.g., Busenberg and Plummer, 1989).

A starting Sr-concentration of 1012 ppm is chosen for high-Mg-calcites precipitated in equilibrium with normal marine water of about 25°C. This Sr-concentration is calculated using a distribution coefficient between solid and solution of 0.130 for direct precipitation from solution, which was recommended by Veizer (1983) from the published range of 0.027-0.400 as derived from several studies with different experimental conditions. Pingitore and Eastman (1986) recommended a value of 0.060 for most geological conditions, but they permitted values as large as 0.200 (the Sr-distribution coefficient depends on many factors, among them the absolute concentrations of Sr, Ca, and Mg in the solid and in the fluid, fluid saturation state, temperature, precipitation rate, and type of crystallographic face within the crystal: Lorens, 1981; Mucci and Morse, 1983; Pingitore and Eastman, 1986; Reeder and Grams, 1987; Morse and Bender, 1990). Hence, the choice of a value of 0.130 for the starting high-Mg-calcite is appropriate for the present context. Using the value of 0.060 would merely shift the starting composition of high-Mg-calcite to 467 ppm.

A range of 0.028-0.079 for the Sr-distribution coefficient is used in the calculations discussed below for the *replacement of high-Mg-calcite by low-Mg-calcite*. The midpoint of this range falls near the values of 0.05 and 0.06 that were recommended by Veizer (1983) and Pingitore and Eastman (1986), respectively, for high-Mg-calcite to low-Mg-calcite recrystallization (values higher than 0.079 may be more appropriate if the calcite precipitating during recrystallization contains any Mg: Mucci, pers. comm. 1989). Different distribution coefficients will be effective even in a system of homogeneous bulk solution composition for at least two reasons.

1. Fluids in interphase boundary systems may have compositions different from the bulk solution, and distribution coefficients depend on fluid composition.

2. Interphase boundary fluids of adjacent crystals may have different rates of migration, caused mainly by different fluid compositions, different grain surface areas, and differences in composition of the original grains, all of which combine to yield different rates of dissolution and reprecipitation, i.e., different rates of recrystallization. This, in turn, results in different 'effective' distribution coefficients for most trace (and minor) elements. These distribution coefficients are effective only under the conditions present at the site of recrystallization. Of all the factors influencing trace element partitioning, growth (recrystallization) rate may be the most important one for Sr and most other trace elements. One notable exception is Mg, however, the partitioning of which does not appear to be rate-dependent under

fluid conditions representative of most diagenetic environments (e.g., Mucci and Morse, 1983; Burton and Walter, 1987; Morse and Bender, 1990).

In summary, the concept of interphase migration, as presented here, involves bulk solution disequilibrium, migration rates, and effective distribution coefficients that differ from grain boundary to grain boundary.

Recrystallization of Marine High-Mg-Calcite to Low-Mg-Calcite

The high-Mg-calcite with 1012 ppm Sr may recrystallize to low-Mg-calcite in waters with Sr/Ca-ratios of that of seawater (0.0089) or common 'diagenetic waters' with lower ratios (Table 1; all Sr/Ca-ratios are molar ratios). Such Sr/Ca-ratios may be those of the bulk solution, thus representing different aquifers with diagenetic solutions such as meteoric fresh water, fresh water/seawater mixtures, or normal seawater. Alternatively, these ratios may be present only at the actively growing low-Mg-calcite crystal surface, i.e., within the interphase fluid layers. In this case, the ratios listed in Tab. 1 result from the Sr and Ca derived via diffusion from the bulk solution *and* from the dissolving high-Mg-calcite. For example, the bulk solution may have a Sr/Ca-ratio of 0.0055, which is representative of a typical shallow meteoric groundwater aquifer, and the dissolving high-Mg-calcite has a Sr/Ca-ratio of 0.0011 (equivalent to 1012 ppm). If the diffusion rates from the bulk solution *and* from the dissolving grain are such that the Sr/Ca-ratio at the growing crystal surface is derived in the ratio of 1:1 from both sources, the Sr/Ca-ratio in the 'diagenetic water' at the growing crystal surface is (0.0055+0.0011)/2=0.0033. If a distribution coefficient of 0.050 is taken, the resulting low-Mg-calcite obtains a Sr-content of 144 ppm (line 1, Tab.1). Calcite with 144 ppm Sr, however, may also form in a bulk solution with a Sr/Ca-ratio different from 0.0055, because an interphase 'diagenetic water' of 0.0033 can be present in any bulk solution that has a Sr/Ca-ratio greater than 0.0033 (then proportions different from 1:1 must be contributed from the bulk solution and the dissolving crystal). These cases represent different degrees of bulk solution disequilibrium (or openness) of the system.

Now consider three examples of low-Mg-calcite with 285 ppm Sr, which is the mean Sr-concentration of a common marine cement type in the study area (prismatic calcite in Fig. 3). Calcite with 285 ppm Sr cannot be a primary precipitate from seawater because, depending on the choice of the distribution coefficient, marine calcite (high- or low-Mg-) should have between 467 and 1012 ppm Sr (see above). Calcite with 285 ppm Sr thus may have precipitated from various diagenetic waters with different 'effective' distribution coefficients: (a) the diagenetic water may have had a Sr/Ca-ratio of 0.0041 (perhaps resulting from a bulk solution with a ratio of 0.0055 and dissolving high-Mg-calcite with a ratio of 0.0011 in the proportion of 1.83:1), and the distribution coefficient was 0.079 (line 2, Tab. 1); (b) the diagenetic water may have had a Sr/Ca-ratio of 0.0070 (about midway between values typical for meteoric waters and seawater, respectively), and the distribution coefficient was 0.046 (line 3, Tab. 1); or (c) the diagenetic water may have had a Sr/Ca-ratio like seawater and the distribution coefficient was 0.036 (line 4, Tab. 1). Analogous calculations can be performed for all other marine or synsedimentary calcite types. As a last example, diagenetic low-Mg-calcite with 219 ppm (equivalent to the mean of skeletal calcites in the system) may result from recrystallization in diagenetic water with a marine Sr/Ca-ratio (line 5, Tab. 1).

Table 1. Calculated compositions of diagenetic low-Mg-calcite for various molar Sr/Ca ratios of diagenetic waters (DW) and various distribution coefficients (D_{Sr}). The precursor mineral is high-Mg-calcite with 1012 ppm Sr in all cases.

Interphase Fluid or Bulk Aquifer	$(^mSr/^mCa)_{DW}$	D_{Sr}	ppm
meteoric	0.0033	0.050	144
mixed	0.0041	0.079	285
mixed	0.0070	0.046	285
seawater	0.0089	0.036	285
seawater	0.0089	0.028	219

Implications

These considerations and calculations have several important implications for carbonate diagenesis.

1. Seawater, and diagenetic waters of various Sr/Ca ratios, may lead to identical Sr-concentrations in the resulting diagenetic low-Mg-calcites, as a function of the recrystallization rate and of the degree of bulk solution disequilibrium of the system. The latter is equivalent to different water/rock ratios if the system is 'open', i.e., in bulk solution equilibrium.
2. Point (1) above compromises the diagenetic model proposed by Brand and Veizer (1980) that was designed to assess the degree of openness in *meteoric systems* (Fig. 6). The above considerations imply that Brand and Veizer's (1980) model is valid for *any* bulk solution rather than being restricted to meteoric systems. Therefore, a meteoric diagenetic environment cannot be inferred from plotting data in this model's 'stabilization fields'. The data plotted in Fig. 6 are from the same samples that are represented in Fig. 3 and Tab. 1, and are interpreted to be due to recrystallization in a bulk solution that was chemically modified seawater (see point 3, below, and discussion in Machel, 1986).

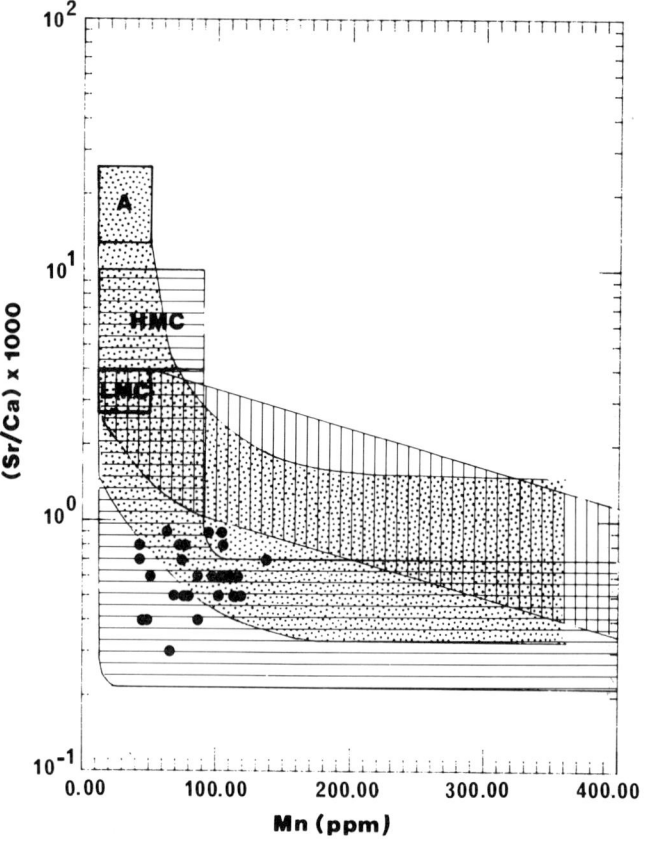

Figure 6. Diagenetic "stabilization fields" for aragonite (A), high-Mg-calcite (HMC), and low-Mg-calcite (LMC) in meteoric waters (after Brand and Veizer, 1980). Ranges inside boxes mark theoretical ranges of inorganic A, HMC, and LMC precipitated in bulk solution equilibrium with modern average seawater. Dotted and horizontally and vertically ruled fields designate progressive diagenetic stabilization (recrystallization), resulting in a loss of Sr and a gain of Mn. The samples from Figure 3, without discrimination into petrographic types, plot in the overlapping stabilization fields of A and HMC.

3. Adjacent grains of originally identical mineralogy and/or composition may obtain grossly different trace element compositions during simultaneous recrystallization. This seriously compromises application of the concept of cement stratigraphy.
4. The composition of the bulk solution *cannot* be back-calculated accurately in systems with pronounced and spatially variable bulk solution disequilibrium. However, one may assume that the composition of any interphase fluid must vary between two end members, that of the dissolving crystal and that of the bulk solution. Hence, there should be a finite range of possible compositions for the interphase fluids. This may permit an *approximate* back-calculation of the bulk solution composition, making reasonable assumptions for the distribution coefficient and the composition of the dissolving solid. This principle was applied in back-calculating the bulk solution composition of the calcites plotted in Fig. 3, which yields bulk fluid compositions between that of typical seawater and certain subsurface waters for the five elements under consideration (Fig. 7). It must be kept in mind, however, that the assumptions made are highly subjective, and that making reasonable assumptions may be impossible in many situations.
5. The compositions of diagenetic low-Mg-calcites may be inconsistent with most popular models for carbonate diagenesis that assume bulk solution equilibrium. This is valid for the trace element compositions (as shown above) as well as for the isotope ratios (however: see point 6 below). Hence, all models summarized in Fig. 2, attempting the recognition of vadose, phreatic, and mixing zones, as well as different climatic settings, cannot, or not necessarily (depending on the degree of bulk solution disequilibrium), be applied in systems with bulk solution disequilibrium.
6. In the case of carbonate diagenesis, the effects of bulk solution disequilibrium recrystallization are much more pronounced for trace elements than for the stable isotopes. This is because the isotopic compositions of most diagenetic fluids and solids do not differ very much, whereas the trace element contents between fluids and rocks commonly differ by several orders of magnitude. Therefore, the effect of bulk solution disequilibrium is greatly attenuated, and perhaps unrecognizable, in the case of the stable isotopes.

The samples plotted in Fig. 3 are a case in point. Although the trace element data clearly indicate bulk solution disequilibrium recrystallization, the oxygen isotope ratios of the same samples represent fairly well the composition and temperature of the formation fluids as being slightly modified marine with temperatures between about 25-35°C (Machel, 1986).
7. The degree of bulk solution disequilibrium may be, and in fact is likely to be, different for several trace elements in the same system ('a single diagenetic site may be relatively open for one cation yet, at the same time, relatively closed for another': Pingitore, 1982, p. 32) [hole, however, that in a strict thermodynamic sense this system is *open*]. This occurs where there is a steep concentration gradient for one but a gentle gradient for another cation, whose concentration in the interphase fluid is similar to that of the bulk solution. In this case, there will be negligible diffusional exchange for the second cation between the interphase and the bulk solution, but appreciable interchange for the first cation.
8. Bulk solution disequilibrium may occur during direct precipitation from solution if continuing precipitation restricts exchange between the bulk solution and the solution adjacent to the growing crystals. For example, the surface fluid layers of two crystals that grow towards one another will

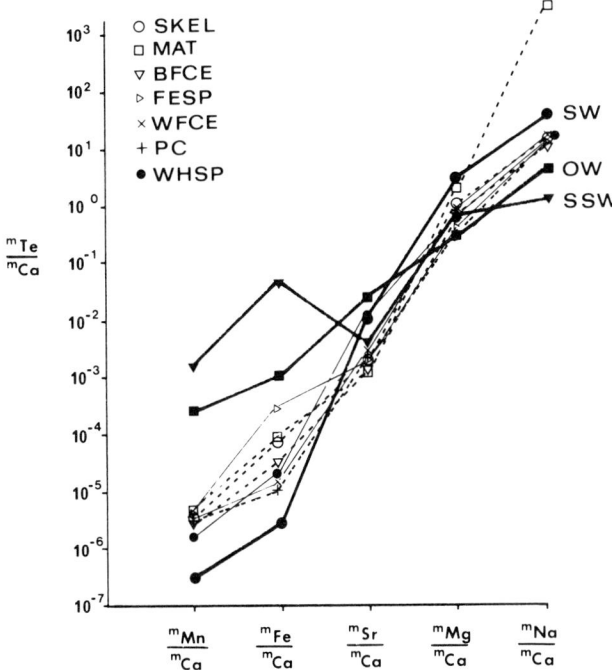

Figure 7. Hypothetical molar concentration ratios of diagenetic waters for the various calcite types plotted in Figures 3, 6 (WHSP = white calcite spar, which is late-diagenetic and not recrystallized calcite; other abbreviations are as in Figure 3; Figure taken from Machel, 1986). Average modern seawater (SW), average modern shallow subsurface (meteoric) water (SSW), and average oil brines (OW) are plotted for comparison (data taken from Veizer, 1983). Te = trace element. Assuming bulk solution equilibrium, the calcite types should have formed from chemically modified seawater that had neither typical SSW nor OW composition.

eventually interfere so that at least the final growth zones of both crystals grow in bulk solution disequilibrium. Another possibility is represented by crystals that grow in relatively large, but restricted pores (see below).

9. Crystal compositions that are not representative of the bulk solution may result from precipitation or recrystallization in relatively large but restricted pore spaces. Such pore spaces, through hydrodynamic and/or diffusional restriction, may not freely communicate and may not be thoroughly mixed with the diagenetic solution in the bulk aquifer. Consequently, (bio-)chemical reactions in the pore and pore walls may alter the composition of the solution within the pore sufficiently to generate a geochemical microenvironment that is compositionally different from the diagenetic solution in the bulk aquifer. Mineral formation in such pore spaces may approach thermodynamic equilibrium with the solution occupying the pore, and the resulting trace element compositions may be significantly different from those predicted by bulk solution equilibrium. Hence, such cases of *'restricted pore solution equilibrium'* may be mistaken for bulk solution disequilibrium, but can be discriminated by means of at least two criteria. Firstly, crystals formed in restricted pore solution equilibrium should be confined spatially to the walls of restricted pores. In contrast, crystals that formed in bulk solution disequilibrium may be distributed throughout an aquifer, irrespective of the porosity and permeability distribution. Secondly, crystals around a pore that were formed in restricted pore solution equilibrium should be compositionally similar, whereas those formed in bulk solution disequilibrium may be compositionally heterogeneous.

CONCLUSIONS

Consideration of recrystallization as a wet-chemical process in two separate steps, dissolution and precipitation, proceeding simultaneously, leads to a hypothesis that involves interphase boundary migration between the metastable and the (more stable) mineral phases. Interphase boundary fluids may be Newtonian fluids with a width of up to several hundred micrometers, or non-Newtonian fluids as narrow as a few Ångstroms. Such fluids commonly have compositions different from that of the bulk solution. Furthermore, interphase fluid compositions may vary widely between adjacent interphase boundaries, and concentration gradients may be present for some but not for other cations. As a consequence, recrystallized grains may form in bulk solution disequilibrium, adjacent paragenetically coeval grains may have grossly different trace element and/or isotope composition, and the composition of the bulk solution cannot be back-calculated accurately. In the case of diagenetic carbonates, the effects of bulk solution disequilibrium commonly are attenuated for the stable isotopes. In addition, compositions similar to those resulting from bulk solution disequilibrium may be obtained by restricted pore solution equilibrium. These two alternatives may be discriminated by means of the spatial restriction of crystals that formed in restricted pore solution equilibrium to the walls of restricted pores, and by means of the compositional homogeneity of such crystals compared to those formed in bulk solution disequilibrium.

The above considerations compromise application of a number of popular geochemical models in carbonate diagenesis. These models can be applied only where bulk solution disequilibrium (and restricted pore solution equilibrium) are absent or negligibly small. Also, some models assume a different meaning, e.g., the model for meteoric diagenesis by Brand and Veizer (1980).

Bulk solution disequilibrium probably is common not only in diagenetic carbonates but also in other minerals that recrystallize in water-dominated systems, i.e., diagenetic, hydrothermal, or metamorphic environments.

ACKNOWLEDGMENTS

The reviews by E. A. Burton, J. B. Fisher, A. Mucci, D. R. Prezbindowski, and R. J. Reeder are greatly appreciated. In addition, several constructive discussions with Liz and Al were particularly influential. This research was supported by the Natural Science and Engineering Research Council of Canada grant No. 55-47939.

REFERENCES

Balluffi, R. W., and J. W. Cahn, 1981, Mechanism for diffusion induced grain boundary migration: Acta Metallurgica, v. 29, p. 493-500.

Banner, J. L., G. N. Hanson, and W. J. Meyers, 1988, Water-rock interaction history of regionally extensive dolomites of the Burlington-Keokuk Formation (Mississippian): isotopic evidence, in V. Shukla and P. A. Baker, eds., Sedimentology and geochemistry of dolostones: SEPM Special Publication 43, p. 97-113.

Berner, R. A., 1978, Rate control of mineral dissolution under Earth surface conditions: American Journal of Science, v. 278, p. 1235-1252.

Brand, U., and J. Veizer, 1980, Chemical diagenesis of a multicomponent carbonate system. I: Trace elements: Journal of Sedimentary Petrology, v. 50, p. 1219-1236.

Burton, L. A., and L. M. Walter, 1987, Relative precipitation rate of aragonite and Mg calcite from seawater: temperature or carbonate ion control?: Geology, v. 15, p. 111-114.

Busenberg, E., and L. N. Plummer, 1986, A comparative study of the dissolution and crystal growth kinetics of calcite and aragonite, in F. A. Mumpton, ed., Studies in diagenesis: USGS Bulletin 1578, p. 139-168.

Busenberg, E., and L. N. Plummer, 1989, Thermodynamics of magnesian calcite-solutions at 25°C and 1 atm total pressure: Geochimica et Cosmochimica Acta, v. 63, p. 1189-1208.

Carlson, W. D., 1983, The polymorphs of $CaCO_3$ and the aragonite-calcite transformation, in R. J. Reeder, ed., Carbonates: mineralogy and chemistry: Reviews in Mineralogy, v. 11, p. 191-226.

Davis, J. A., C. C. Fuller, and A. D. Cook, 1987, A model for trace metal sorption processes at the calcite surface: adsorption of Cd^{2+} and subsequent solid solution formation: Geochimica et Cosmochimica Acta, v. 51, p. 1477-1490.

Evans, B., R. S. Hay, and N. Shimizu, 1986, Diffusion-induced grain boundary migration in calcite: Geology, v. 14, p. 60-63.

Gaffey, S., 1985, Reflectance spectroscopy in the visible and near-infrared (0.35-2.55 microns): applications in carbonate petrology: Geology, v. 13, p. 270-273.

Gaffey, S., 1988, Water in skeletal carbonates: Journal of Sedimentary Petrology, v. 58, p. 397-414.

Gibbs, J. W., 1876, On the equilibrium of heterogeneous substances: Connecticut Academy of Arts and Sciences Transactions, III, p. 108-248.

Grover, G., Jr., and J. F. Read, 1983, Paleoaquifer and deep burial related cements defined by regional cathodoluminescent patterns, Middle Ordovician carbonates, Virginia: AAPG Bulletin, v. 67, p. 1275-1303.

Hartmann, P., 1982, Crystal faces: structure and growth: Geologie en Mijnbouw, v. 61, p. 313-320.

Henderson, L. M., and F. C. Kracek, 1927, The fractional precipitation of barium and radium chromates: American Chemical Society Journal, v. 49, p. 739-749.

Hillert, M., and G. R. Purdy, 1978, Chemically induced grain boundary migration: Acta Metallurgica, v. 26, p. 333-340.

James, N. P., and P. W. Choquette, 1984, Diagenesis 9—Limestones—the meteoric diagenetic environment: Geoscience Canada, v. 11, p. 161-194.

Lahann, R. W., and R. M. Siebert, 1982, A kinetic model for distribution coefficients and application to Mg-calcites: Geochimica et Cosmochimica Acta, v. 46, p. 2229-2237.

Lorens, R. B., 1981, Sr, Cd, Mn, and Co distribution coefficients as a function of calcite precipitation rate: Geochimica et Cosmochimica Acta, v. 45, p. 533-561.

Machel, H. G., 1985, Cathodoluminescence in calcite and dolomite and its chemical interpretation: Geoscience Canada, v. 12, p. 139-147.

Machel, H. G., 1986, Limestone diagenesis of Upper Devonian Nisku carbonates in the subsurface of central Alberta: Canadian Journal of Earth Sciences, v. 23, p. 1804-1822.

Machel, H. G., in press, Submarine Frühdiagenese, Spaltenbildung, und prätektogenetische Spätdiagenese des Briloner Riffs: Geologisches Jahrbuch, Reihe D, Heft 95.

Martin, G. D., B. H. Wilkinson, and K. C. Lohmann, 1986, The role of skeletal porosity in aragonite neomorphism—*Strombus* and *Monastrea* from the Pleistocene Key Largo limestone, Florida: Journal of Sedimentary Petrology, v. 56, p. 194-293.

Meike, A., H. R. Wenk, M. A. O'Keefe, and R. Gronsky, 1988, Atomic resolution microscopy of carbonates, interpretation of contrast: Physics and Chemistry of Minerals, v. 15, p. 427-437.

Meyers, W. J., 1978, Carbonate cements, their regional distribution and interpretation in Mississippian limestones of southwestern New Mexico: Sedimentology, v. 25, p. 371-399.

Meyers, W. J., and K. C. Lohmann, 1985, Isotope geochemistry of regionally extensive calcite cement zones and marine components in Mississippian limestones, New Mexico, in N. Schneidermann and P. M. Harris, eds., Carbonate cements: SEPM Special Publication 36, p. 223-239.

Morimoto, T., J. Kishi, O. Okada, and T. Kadota, 1980, Interaction of water with the surface of calcite: Bulletin of the Chemical Society of Japan, v. 53, p. 1918-1921.

Morrow, D. W., and I. R. Mayers, 1978, Simulation of limestone diagenesis—a model based on strontium depletion: Canadian Journal of Earth Sciences, v. 15, p. 376-396.

Morse, J. W., 1983, The kinetics of calcium carbonate dissolution and precipitation, in R. J. Reeder, ed., Carbonates: mineralogy and chemistry: Reviews in Mineralogy, v. 11, p. 227-264.

Morse, J. W., and M. L. Bender, 1990, Partition coefficients in calcite: Examination of factors influencing the validity of experimental results and their application to natural systems: Chemical Geology, v. 82, p. 265-277.

Mucci, A., and J. W. Morse, 1983, The incorporation of Mg^{2+} and Sr^{2+} into calcite overgrowths: influences of growth rate and solution composition: Geochimica et Cosmochimica Acta, v. 47, p. 217-233.

Mucci, A., and J. W. Morse, 1985, Auger spectroscopy determination of the surface-most adsorbed layer composition on aragonite, calcite, dolomite, and magnesite in synthetic seawater: American Journal of Science, v. 285, p. 306-317.

Mucci, A., J. W. Morse, and M. S. Kaminsky, 1985, Auger spectroscopy analysis of magnesian calcite overgrowths precipitated from seawater and solutions of similar composition: American Journal of Science, v. 285, p. 289-305.

Mullin, J. W., 1972, Crystallization: Cleveland, Ohio, CRC Press (2nd edition), p. 415.

Pingitore, N. E., 1982, The role of diffusion during carbonate diagenesis: Journal of Sedimentary Petrology, v. 52, p. 27-39.

Pingitore, N. E., and M. P. Eastman, 1986, The coprecipitation of Sr^{2+} with calcite at 25°C and 1 atm: Geochimica et Cosmochimica Acta, v. 50, p. 2195-2203.

Plummer, L. N., T. M. L. Wigley, and D. L. Parkhurst, 1978, The kinetics of calcite dissolution in CO_2-water systems at 5°C to 60°C and 0.0 to 1.0 atm CO_2: American Journal of Science, v. 278, p. 179-216.

Plummer, L. N., T. M. L. Wigley, and D. L. Parkhurst, 1979, Critical review of the kinetics of calcite dissolution and precipitation, in E. A. Jenne, ed., Chemical modeling in aqueous systems: American Chemical Society Symposium Series 93, p. 537-573.

Reddy, M. M., L. N. Plummer, and E. Busenberg, 1981, Crystal growth of calcite from calcium bicarbonate solutions at constant $^PCO_2$ and 25°C: a test of a calcite dissolution model: Geochimica et Cosmochimica Acta, v. 45, p. 1281-1289.

Reeder, R. J., 1987, Zoning types and their origins in sedimentary carbonate minerals, *in* Rodrigues-Clemente and Tardy, eds., Geochemistry and mineral formation on the Earth surface: C.S.I.C., Madrid, p. 743-752.

Reeder, R. J., and J. C. Grams, 1987, Sector zoning in calcite cement crystals: implications for trace element distributions in carbonates: Geochimica et Cosmochimica Acta, v. 51, p. 187-194.

Smith, D. A., and A. H. King, 1981, On the mechanism of diffusion-induced boundary migration: Philosophical Magazine, v. 44, p. 333-340.

Somasundaran, P., and G. E. Agar, 1967, The zero point charge of calcite: Journal of Colloid and Interface Science, v. 24, p. 433-440.

Sunagawa, I., 1977, Natural crystallization: Journal of Crystal Growth, v. 42, p. 214-223.

Veizer, J., 1983, Chemical diagenesis of carbonates: theory and application of trace element technique, *in* M. A. Arthur, T. F. Anderson, I. Kaplan, J. Veizer, and L. S. Land, eds., Stable isotopes in sedimentary geology: SEPM Short Course 10: 3-1 - 3-100.

Watson, E. B., 1982, Basalt contamination by continental crust: some experiments and models: Contributions to Mineralogy and Petrology, v. 80, p. 73-87.

Wenk, H. R., D. J. Barber, and R. J. Reeder, 1983, Microstructure in carbonates, *in* R. J. Reeder, ed., Carbonates: mineralogy and chemistry: Reviews in Mineralogy, v. 11, p. 301-367.

Zhang, Y., D. Walker, and C. E. Lesher, in press, Diffusive crystal dissolution: Contributions to Mineralogy and Petrology.

Dissolution and Precipitation Kinetics of Kaolinite: Initial Results at 80°C with Application to Porosity Evolution in a Sandstone

K. L. Nagy, C. I. Steefel, A. E. Blum, and A. C. Lasaga
Department of Geology and Geophysics
Yale University
New Haven, Connecticut, U.S.A.

> Dissolution and precipitation rates of kaolinite were determined at steady-state in 80°C solutions near pH 3 as a function of deviation from equilibrium using flow-through reaction cells. The dependence of dissolution rate (mol/m²/s) on solution saturation state can be described equally well by either of the empirical relations:
>
> $$\text{Rate}_{diss} = -1.20 \times 10^{-12} (1 - \exp(\Delta G/RT))^{1.02}$$
> or
> $$\text{Rate}_{diss} = -1.32 \times 10^{-12} (1 - \exp(0.724 \, \Delta G/RT)).$$
>
> The present data on dissolution rates indicate linearity of the rates near equilibrium. The dissolution and precipitation rates near equilibrium were used to bracket the solubility of the kaolinite, providing a self-consistent thermodynamic reference point for the calculation of solution saturation state.
>
> To investigate the effects of near equilibrium kinetics during diagenesis in a sandstone we have carried out a numerical simulation in which the reactions are driven by fluid flow. Full dissolution and crystal growth laws were obtained from the rates measured for kaolinite and from published dissolution rates far from equilibrium for other minerals, together with the principle of detailed balancing. The results indicate that for a flow velocity of 10 m/yr, the assumption of kaolinite formation under local equilibrium conditions may not be justified. The simulation can account for the timing of diagenetic mineralization inferred from observed textural relationships between quartz, feldspar, and kaolinite. The results emphasize the need for accurate kinetic data on silicate-aqueous reactions and the importance of flow of chemically-reactive fluids in producing disequilibrium mineral assemblages and in controlling the spatial and temporal evolution of porosity in subsurface rocks.

INTRODUCTION

The kinetics of mineral reactions are important to consider when modeling the changes in porosity and permeability of a reservoir rock undergoing subsurface diagenesis. We need to know not only the thermodynamic stabilities of minerals in aqueous solutions, but also the rates at which the minerals dissolve and precipitate. For silicates, these rates are functions, in part, of surface speciation, as shown by the correlation of the pH-dependence of dissolution rates far from equilibrium with H⁺ and OH⁻ adsorption isotherms (Blum and Lasaga, 1988; Carroll-Webb and Walther, 1988). The rates also depend on the degree of disequilibrium of the solutions with respect to the mineral (Lasaga, 1981a, b; Aagaard and Helgeson, 1982; Lasaga, 1984).

Unlike the extensive data set existing for calcite (Morse, 1983), which reacts relatively rapidly, few results have been obtained for silicates. Although data are starting to accumulate on the dissolution rates of silicates far from equilibrium as a function of pH (Grandstaff, 1977, 1986; Chou and Wollast, 1984, 1985; Bales and Morgan, 1985; Knauss and Wolery, 1986, 1988, 1989; Tole et al., 1986; Carroll-Webb and Walther, 1988), little is known about the dissolution and precipitation rates as a function of solution saturation state, especially at diagenetic temperatures. Such rates are necessary for understanding any natural system which may achieve steady-state close to equilibrium or even reach near-equilibrium conditions. For most natural systems, knowledge will be required of rates under conditions different from those in the far-from-equilibrium experiments carried out so far.

Data on the relationship between rates and deviation from equilibrium have been obtained for quartz, cristobalite, and amorphous silica from 18° to 305°C (Rimstidt and Barnes, 1980) and extracted from kinetic and/or phase equilibria experiments to 710°C for a variety of metamorphic minerals (Wood and Walther, 1983; Schramke et al., 1987). Rimstidt and Barnes (1980) observed that both dissolution and precipitation rates of silica phases were linear as a function of deviation from equilibrium as defined by solution composition. Wood and Walther (1983) interpreted the rates of equilibration of metamorphic mineral assemblages as a linear function of deviation from equilibrium as defined by temperature variations about the equilibrium temperature. However, the data of Schramke et al. (1987) suggest nonlinear rate laws for metamorphic reactions (Lasaga, 1986).

Experimentally determined precipitation rates of nonsilicate crystals from solution and theoretical models for precipitation rates controlled by surface reactions show a nonlinear dependence on solution saturation state near equilibrium (Nielsen, 1964). On the other hand, theoretical modeling using Monte Carlo simulations (Blum and Lasaga, 1987) has shown that dissolution rates may be either linear or nonlinear near equilibrium when the rates are complex functions of the free energy of mineral surfaces.

EXPERIMENTAL DETERMINATION OF DISSOLUTION AND PRECIPITATION

Experimental Methods

Materials

Kaolinite from Twiggs County, Georgia (Ward's) was used for all the experiments reported here. The kaolinite was suspended and decanted many times in NaOH solutions (pH ~10). Quartz and iron-oxide concretionary material settled to the bottom in the high-pH solutions and therefore could be separated from the kaolinite, which did not flocculate and settle. The kaolinite suspension was then acidified with HCl to a pH of ~5, which caused the kaolinite to flocculate. The overlying solution was repeatedly decanted and replaced with distilled water in order to remove dissolved species. The kaolinite was then filtered, washed with distilled water, and dried at 45°C. The washed kaolinite had a crystallinity of medium to medium low (Hinckley, 1963) using powder x-ray diffraction. The x-ray diffraction pattern also showed that the washed kaolinite contained a trace of muscovite. Wet chemical analysis indicated a nearly pure composition with trace Fe (1.04% Fe_2O_3), Ti (1.42% TiO_2), and K (0.56% K_2O). SEM photomicrographs showed hexagonal crystals with irregular edges ranging from 0.15 to 4 μm in diameter. Grain thicknesses were approximately 0.1 to 0.3 times the diameter.

Greatly incongruent as well as nonsteady-state results were found to be an artifact of inadequate surface preparation. In order to achieve steady-state and nearly congruent dissolution results, prior equilibration at 80°C in a solution of the same pH was necessary. This equilibration required about 3 to 4 months. The BET surface area of "pre-equilibrated" kaolinite was 17.5 m^2/g ($\pm$ 10%).

Experiments

Steady-state dissolution and precipitation rates of kaolinite were obtained at specific solution saturation states using a flow-through reaction vessel (Figure 1) fully immersed in a constant-temperature water bath. Between 0.25 to 1 g of seed kaolinite was placed in the reaction vessel (~45 ml volume) and allowed to react with a through-flowing fluid of fixed input composition. The flow rate was 0.011 ml/min ($\pm$ 1%) which resulted in a residence time within the cell of about 67 hours. The kaolinite and fluid were kept well mixed by a bottom-mounted stir bar controlled by a stir plate placed directly beneath the bath. Input fluids were concocted at a specific pH, and with specific concentrations of total aluminum and total silicon. Perchloric acid was used to set the pH. Aluminum was fixed using a solution of Al wire dissolved in HCl and diluted to the desired concentration. The source of silicon was dissolved $Na_2SiO_3 \cdot 9H_2O$. Input solutions were pumped through the reaction vessel using an Ismatec brand peristaltic pump. Solutions were filtered on the output side of the vessel with a 0.45 μm Durapore (Millipore) filter. The water bath was maintained at 80° $\pm$ 0.1°C using a Techne temperature-controller. All the experiments were conducted at pH near 3.

Pairs of input and output solutions were analyzed with time for Al, Si, pH, and Na. Total Al and total Si were analyzed colorimetrically with a UV-visible light spectrophotometer, using the catechol violet method (Dougan and Wilson, 1974) and molybdate blue method (Grasshoff, 1976), respectively. No difference in measured Al or Si was observed after filtering the output solution with either a 0.22 μm GS (glass fiber, Millipore) or a 0.05 μm PC (polycarbonate, Nuclepore) filter. Uncertainties in measured Al and Si were $\pm$ 2%. pH was measured

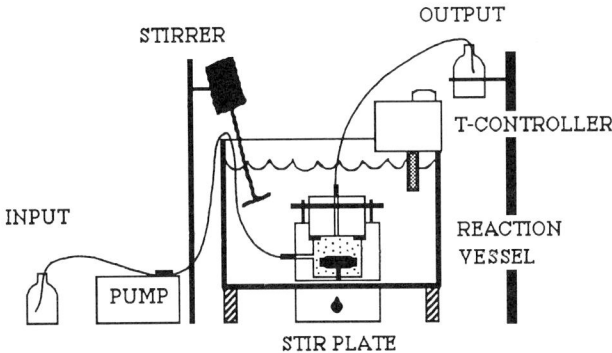

Figure 1. Schematic diagram of the experimental apparatus.

at temperature on an unstirred aliquot of input or output solution using a Ross combination electrode with an uncertainty of ± 0.02 pH units (± 4.5% in concentration of H$^+$). Na was determined by atomic emission spectroscopy and had an analytical uncertainty of 3%. Although Na was monitored to check for possible precipitation, none was observed.

Solids were examined by x-ray diffraction and SEM after the experiments. In most cases, the BET surface area was measured after the experiment. Within the 10% uncertainty of the BET analysis there was no measurable difference between final and starting surface areas.

Calculations

Dissolution and precipitation rates were determined from the measured aluminum and silicon concentrations in the output solutions once a steady-state had been obtained. Steady-state in the flow-through cell is attained when the outlet solution concentrations do not change with time. At steady-state

$$\frac{dN}{dt} = 0 = q_v \cdot \Delta M + 2A_{kaol}R_{kaol}, \quad (1)$$

where N is the number of moles of silicon or aluminum, t is time, q_v is the volume flow rate of the fluid (L/s), ΔM is the molarity of the component (silicon or aluminum) in the output solution minus that in the input solution, A_{kaol} is the total surface area of kaolinite which equals the product of the mass (g) and the surface area (m²/g), and R_{kaol} is the reaction rate (mol/m²/s) of kaolinite in the experimental cell. The factor of 2 in the rate is the stoichiometric coefficient of Si or Al in kaolinite. At all times the composition of the output solution is assumed to be identical to the composition of the solution in the well-stirred reaction cell. Therefore, the measured rate is characteristic of the dissolution (or precipitation) reaction in a solution with a saturation state which equals that of the output solution. The rates were calculated from a rearrangement of the right side of equation 1

$$R_{kaol} = -\frac{\Delta M \cdot q_v}{2 \cdot A_{kaol}}. \quad (2)$$

According to equation 2, the rate is negative for dissolution and positive for precipitation.

The overall reaction is written as

$$Al_2Si_2O_5(OH)_4 + 6H^+ = 2Al^{3+} + 2H_4SiO_4 + H_2O, \quad (3)$$

which, if read from left to right, represents dissolution, and right to left, precipitation.

Equilibrium is represented by the solubility product

$$K_{eq} = \frac{a^2_{Al^{3+}} \, a^2_{H_4SiO_4}}{a^6_{H^+}}, \quad (4)$$

where a_i are the thermodynamic activities of the dissolved species. The saturation state of the solution can be expressed as

$$\Omega = \frac{IAP}{K_{eq}}, \quad (5)$$

the ratio of the ion activity product (IAP) (given by the right side of equation 4) of the output solution at steady-state to the equilibrium ion activity product (K_{eq}).

An equilibrium constant for the Twiggs County kaolinite was determined from the kinetic experiments. Bassett et al. (1979) calculated many different values for $K_{eq,kaol}$ at 25°C using available experimental data and a self-consistent set of thermodynamic hydrolysis constants. They attributed the variation in K_{eq} to such properties as crystallinity, grain size, and composition of the individual kaolinites used in each set of experiments. Our work on kinetics enables us to bracket K_{eq} directly using the solution compositions for the steady-state dissolution and precipitation rates nearest equilibrium, thereby avoiding choosing any particular K_{eq}, and extrapolating to 80°C using published heat capacity functions for kaolinite (Helgeson et al., 1978).

Once K_{eq} was evaluated, the saturation state of the solution was obtained by calculating the distribution of aqueous species using appropriate thermodynamic data. The temperature dependence of the aluminum hydrolysis constants was derived from a second order polynomial fit to the high-temperature data of Bourcier et al. (1987) and the 25°C constants of May et al. (1979). Silica speciation at 80°C was obtained from standard-state thermodynamic values (Naumov et al., 1974), and a temperature dependence for the heat capacity of H_4SiO_4 derived from the temperature dependence of quartz solubility given in Rimstidt and Barnes (1980). The water-dissociation constant was from Busey and Mesmer (1978). Activity coefficients were calculated using the extended Debye Huckel equation with parameters from Wolery (1979) and Truesdell and Jones (1974).

The largest uncertainty in the calculation of saturation state probably arises from the uncertainty in the first hydrolysis constant of aluminum, which we estimate to be 5%. The total uncertainty in an IAP value is 31% (when using equation 4), and includes the additional uncertainties in the measured quantities pH, M_{Al}, and M_{Si}. The uncertainty is also a function of the relatively large powers to which the a_i values are raised in the IAP expression which, in turn, is a result of the stoichiometry of the reaction (equation 3).

The functional dependence of mineral dissolution rates on the deviation from equilibrium has been expressed according to a number of empirical formulas, as well as several theoretical models based on different reaction mechanisms (Ohara and Reid, 1973). In this paper, two possible expressions are considered (Lasaga, 1981b):

$$\text{Rate} = -k \left(1 - e^{\left(\frac{\Delta G}{RT}\right)}\right)^n, \quad (6)$$

and

$$\text{Rate} = -k \left(1 - e^{\left(\frac{n \Delta G}{RT}\right)}\right), \quad (7)$$

where

$$\Delta G = RT \ln(\Omega). \quad (8)$$

ΔG is the Gibbs Free Energy of reaction (3) (where $\Delta G = 0$ represents equilibrium), k (mol/m^2/s) and n are constants, R is the gas constant, and T is temperature (K). These two expressions were chosen to illustrate nonlinear (equation 6, if $n \neq 1$) and linear (equation 7) rate laws near equilibrium.

Equation 6 has been the form used most commonly for representing dissolution and precipitation rates for carbonates as a function of solution saturation state. In general, this form will fit dissolution and growth rates which are controlled by surface defects (Lasaga, 1981a). For small values of ΔG, near equilibrium, equation 6 reduces to

$$\text{Rate} = \pm k \left(\frac{|\Delta G|}{RT}\right)^n \quad (9)$$

Equation 7 is derived from the principle of detailed balancing which applies to an elementary reaction; that is, a reaction which actually takes place at the molecular level as written. The principle of detailed balancing states that at equilibrium the ratio of the forward and reverse rate constants of an elementary reaction is equal to the equilibrium constant for that reaction. For an overall reaction (as in equation 3) made up of a series of elementary steps, however, the ratio of the forward and reverse rate constants of the overall reaction does not necessarily equal the equilibrium constant. This is because the rate-limiting mechanisms for each of the forward and reverse reactions may be different (Lasaga, 1981a). At constant pH and temperature and for small values of ΔG, equation 7 reduces to

$$\text{Rate} = k \, n \left(\frac{\Delta G}{RT}\right) \quad (10)$$

which states that the rate is linearly proportional to $\Delta G/RT$ near equilibrium for some constant values of k and n. This will be true for any reaction and for a sufficiently small deviation from equilibrium. Farther from equilibrium, equation 7 results in nonlinearity in the rate versus ΔG curve.

Experimental Results and Discussion

We present the results from five dissolution and three precipitation experiments, and discuss three aspects of the data (Table 1)—congruency of the steady-state dissolution rates, evaluation of the steady-state precipitation results, and determination of equilibrium between kaolinite and solution at 80°C.

Typically, and as long as the pre-equilibrated kaolinite was used, steady-state was reached in the dissolution experiments after 200 to 300 hours. In all cases, dissolution of the kaolinite at steady-state was nearly congruent, and is exemplified by the results of experiments KP6 and KP6' (Figure 2). Because of the nature of the flow-through apparatus, the input solution can be changed without dismantling the experiment. In this manner, dissolution rates at more than one saturation state can be obtained. After 1000 hours, the input solution of KP6 was changed to a composition closer to equilibrium, yielding the results of KP6' (Figure 2). Note that the dissolution rate is slower for KP6'. Also, the pH for KP6 and KP6' rose in the output solution over that of the input solution (Figure 2c) as would be expected for the dissolution of kaolinite (see equation 3).

The data on precipitation, at this point in the overall experimental study, are not as conclusive as the dissolution results. We emphasize that the precipitation rates reported here are preliminary. In order to set up the precipitation experiments it is necessary to supersaturate the input solutions with respect to kaolinite but to undersaturate them with respect to other potential phases. Because of the relatively high solubility of the Twiggs County kaolinite and the position of the equilibrium boundary with respect to the stability fields of other phases (see below), the silicon has to be fairly concentrated and the aluminum relatively dilute in order to satisfy this requirement for the input solution. Results of experiment KP15 demonstrate the uncertainties in the interpretation of congruency (Figure 3). The measured steady-state decrease in aluminum in the output solution of KP15 (Figure 3a) was only 20 μmolar. Correspondingly, the congruent decrease in silicon would have been only

Table 1. Dissolution (negative values) and precipitation (positive values) rates of kaolinite at 80°C and pH near 3 as a function of the ion activity product of the solution at steady state. Aluminum and silicon concentrations and pH correspond to the output solution at steady state. Δ_{Al} and Δ_{Si} are the output concentration minus the input concentration of aluminum and silicon, respectively. All rates were calculated using Δ_{Si}, except for two precipitation experiments, where Δ_{Si} was indeterminable.

Experiment Number	Rate ($mol_{kaol}/m^2/s$)	$\ln(IAP/K_{eq})$	Al ($\mu mol/L$)	Si ($\mu mol/L$)	pH	Δ_{Al} ($\mu mol/L/g_{kaol}$)	Δ_{Si} ($\mu mol/L/g_{kaol}$)
KD3	-1.32×10^{-12}	-5.686	59	63	3.15	227	242
KP6	-8.00×10^{-13}	-1.453	205	313	3.04	140	144
KP11	-6.12×10^{-13}	-0.380	129	528	3.12	87	113
KP10	-3.34×10^{-13}	-0.228	110	570	3.14	30	61
KP6'	-3.17×10^{-13}	-0.693	119	566	3.10	42	58
KP13	1.30×10^{-13}	0.205	200	546	3.10	-14	-25
KP12	2.04×10^{-13}	0.748	299	559	3.08	-37	(1)
KP15	2.34×10^{-13}	0.514	160	1,961	2.97	-52	(1)

[1]Could not be determined within experimental uncertainty (see text).

1% of the total Si concentration. It was impossible to observe this decrease in silicon because it was within the uncertainty of the analytical procedure (Figure 3b).

If solution compositions lay within the stability fields of other minerals, it would be theoretically possible to precipitate these minerals provided the degree of supersaturation were high enough to overcome any nucleation barriers. The other potential phases for the experimental solution compositions dealt with here are quartz, cristobalite, amorphous silica, gibbsite, and boehmite (Figure 4). All solutions were highly undersaturated with respect to albite and paragonite, which therefore are not included in Figure 4.

Ideally, the kaolinite would have had a part of its stability field below saturation with respect to both an aluminous and a siliceous phase; that is, the equilibrium boundary shown for kaolinite in Figure 4 would have been shifted to lower $a_{H_4SiO_4}$ and $(a_{Al^{3+}}/a^3_{H^+})$ values. Thermodynamic calculations using data from Robie et al. (1978) yield such a boundary. However, the rate data presented in this paper permit the bracketing of equilibrium for the Twiggs County kaolinite, and this boundary lies wholly within the stability fields of either an aluminous and/or a siliceous phase (Figure 4). Using the kinetics experiments to bracket the equilibrium constant (equation 4), we arrived at a value of $\log(K_{eq}) = 3.88$.

A number of the dissolution experiments and two of the three precipitation experiments lie within the stability field of gibbsite, based on the solubility data of Wesolowski et al. (1988) (Figure 4). Because our knowledge of the solubility of boehmite is uncertain (Apps et al., 1989), we have chosen not to place its stability boundary on the phase diagram. It is possible that boehmite is more stable than gibbsite at 80°C, which would result in all the experimental rates falling within the field of an aluminous phase. At this point we interpret the minor observed incongruencies (i.e., Al < Si) measured in some of the dissolution experiments as either preferential enrichment in aluminum of the kaolinite surface or precipitation of small amounts of an aluminous phase on the kaolinite surface.

Some of the output solutions for the precipitation experiments were supersaturated with respect to quartz as well as to gibbsite. Because of the suspected difficulty of nucleating either phase at its calculated supersaturation, we interpret the decreases in both aluminum and silicon (observed in experiment KP 13, only) as an indication of kaolinite formation on the kaolinite seed material. However, if other phases had formed, the small mass of quartz, gibbsite, or boehmite that could have precipitated in any experiment precludes identification by x-ray diffraction.

Using our derived value of K_{eq} and equations 4, 5, and 8, the experimental rates were plotted against ($\Delta G/RT$) in order to see the form of the variation of the rates with deviation from equilibrium (Figure 5). The dissolution rate data define a curve that can be fit well within experimental error by either equation 6 or equation 7 (Figure 5). When fit to the form given by equation 6,

$$\text{Rate}_{diss} = -1.20 \times 10^{-12} (1 - \exp(\Delta G/RT))^{1.02}. \quad (11)$$

If fit to equation 7,

$$\text{Rate}_{diss} = -1.32 \times 10^{-12} (1 - \exp(0.724 \, \Delta G/RT)). \quad (12)$$

In general, the dissolution rates are relatively constant far from equilibrium and then decrease at an increasing rate as equilibrium is approached. Either fit to the dissolution rate data yields an essentially linear rate law near equilibrium, where, according to equation 9

$$\text{Rate}_{diss} = -1.20 \times 10^{-12} (|\Delta G|/RT)^{1.02}, \quad (13)$$

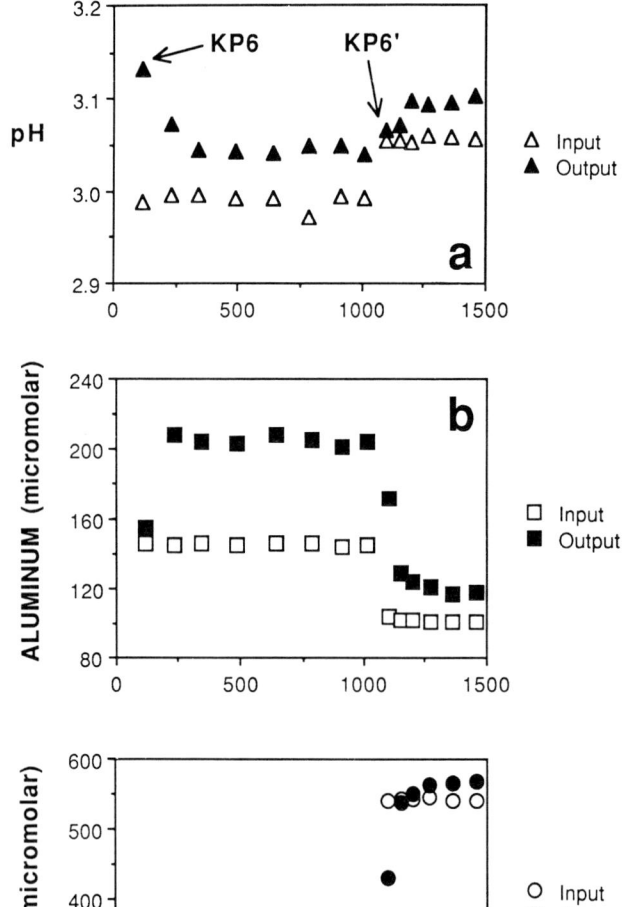

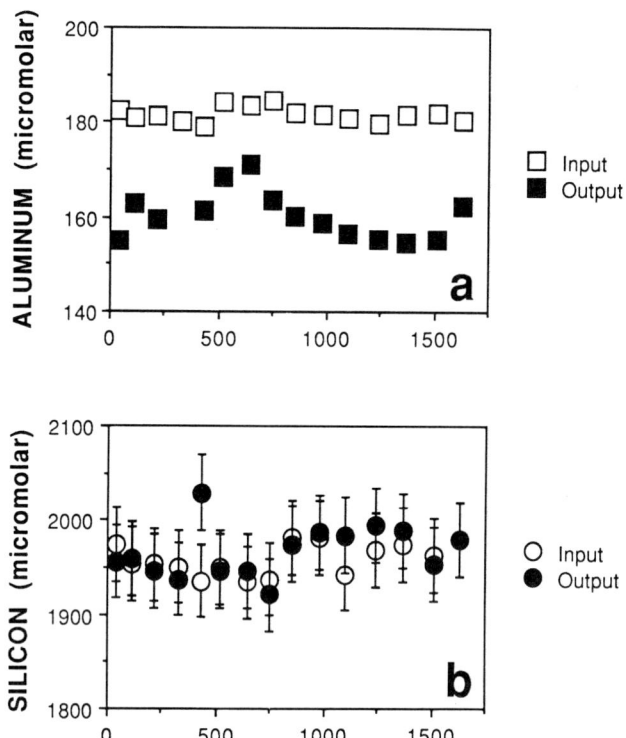

Figure 2. Analytical results of dissolution experiments KP6 and KP6'. (a) pH, (b) total aluminum, (c) total silicon. Arrows point to beginning of the series of KP6 and KP6' input and output solutions.

Figure 3. Analytical results of precipitation experiment KP15. (a) total aluminum, (b) total silicon.

and, according to equation 10

$$\text{Rate}_{diss} = 0.96 \times 10^{-12}\, (\Delta G/RT). \tag{14}$$

Near equilibrium, the two empirical relations yield the same rate ± 17%. This is well within the scatter of the data.

The three preliminary precipitation rates are similar in value and in their corresponding solution saturation states. The precipitation data are still too preliminary to extract the functional dependence of precipitation rates on deviation from equilibrium. Specifically, we can make no statements on the applicability of the principle of detailed balancing to kaolinite dissolution and precipitation rates near equilibrium.

Interestingly, the shape of the curve that represents dissolution rates vs. ΔG mimics a theoretically calculated curve for dissolution of an imperfect crystal; that is, one with dislocations (Blum and Lasaga, 1987) (Figure 6a). The measured precipitation rates are confined to a narrow range of saturation states, making it difficult to determine if the rates are in a nonlinear or linear regime for crystal growth (Figure 6b). Clearly, more rates over a broader range of supersaturation and especially far from equilibrium are needed to define the dependence of the growth law on solution saturation state. Such experiments are currently in progress.

OCCURRENCE OF KAOLINITE IN SANDSTONE RESERVOIRS

Before applying the experimental data to simulations of diagenesis in a sandstone, it is useful to summarize some of the geological observations. We will review both generalities and specifics of authigenic kaolinite formation and concomitant mineral dissolution and precipitation. Examples are

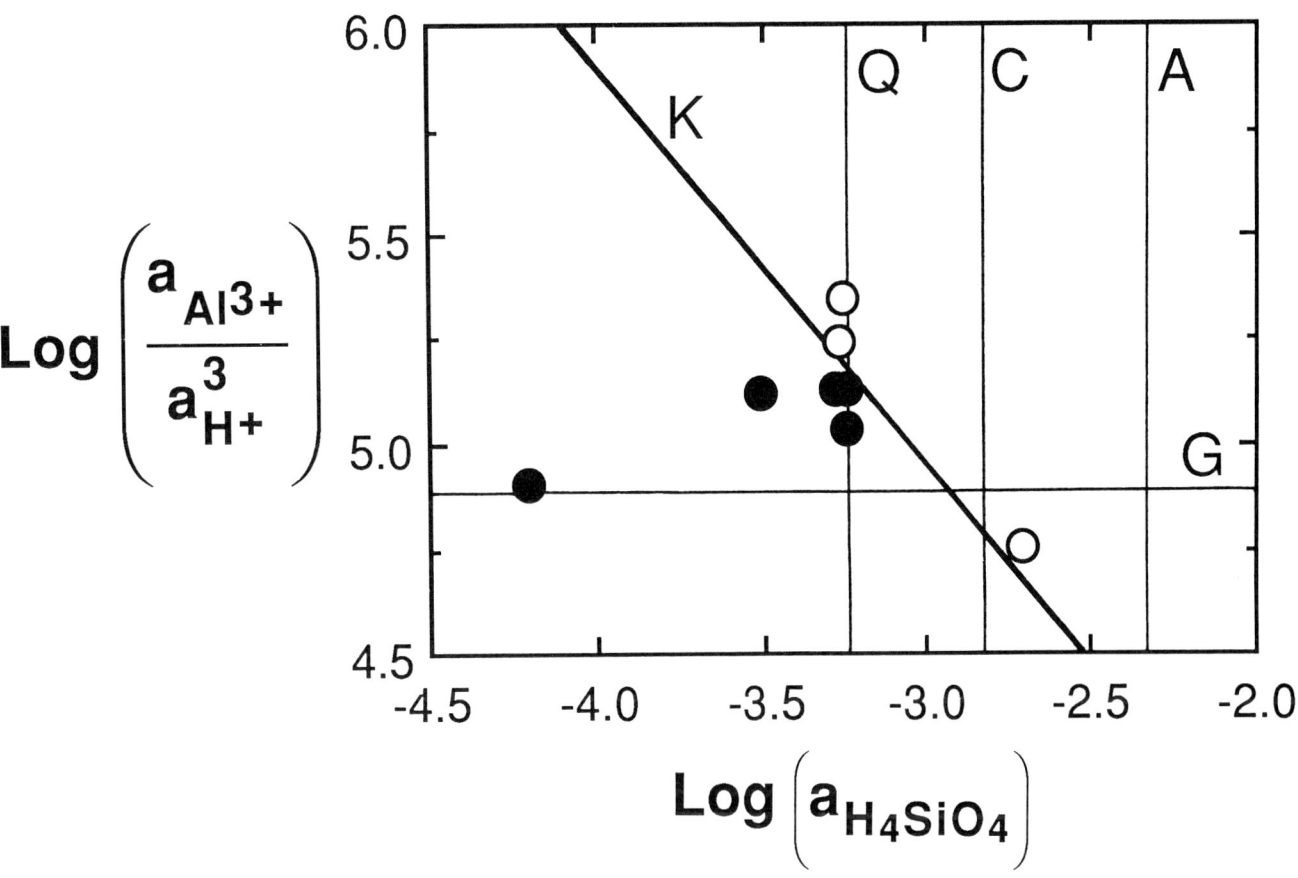

Figure 4. Mineral stability diagram at 80°C for the possible phases that could form in the experimental solutions: K = kaolinite, Q = quartz, C = cristobalite, A = amorphous silica, and G = gibbsite. The boundary for kaolinite was determined in this study; those for the silica phases are from Rimstidt and Barnes (1980); and that for gibbsite is from Wesolowski et al. (1988). The solution compositions at steady-state for the dissolution rates (closed circles) and precipitation rates (open circles) obtained in this study are shown.

taken from Jurassic sandstone reservoirs in the North Sea basin and the Gulf Coast Tertiary section.

Kaolinite is one of the dominant clay minerals in clastic reservoir rocks, occurring both as a detrital and an authigenic phase. It can participate as both a reactant and a product in a variety of natural diagenetic reactions (Table 2). Kaolinite can also react to form secondary minerals during enhanced recovery involving steam injection or in-situ combustion (Hutcheon, 1984).

Experimental studies of the temperature and pressure stability of kaolinite indicate that it will break down in the presence of quartz at a temperature of ~260 to 275°C over the $P_{H_2O} = P_{Total}$ range of ~0.05 to 2 kb (Hemley et al., 1980). If P_{Total} < ~0.05 kb, the dehydration reaction will occur at lower temperatures. Therefore, kaolinite should be stable throughout most of a sedimentary basin, assuming normal temperature and pressure gradients. Whether or not kaolinite will dissolve or precipitate then depends on its reaction kinetics and the solution composition.

The compositional dependence of the stability of kaolinite in relation to other solid phases and a fluid phase in arkosic sandstones can be represented by diagrams showing the activities of aqueous silica, aluminum, hydrogen ion and other cations (Bowers et al., 1984). For example, at high enough values of the activity ratios K^+/H^+ or Na^+/H^+, a feldspar, mica, or even zeolite will be the equilibrium aluminosilicate phase rather than kaolinite. What cannot be ascertained from such diagrams is how fast these equilibrium assemblages will be achieved if a nonequilibrium solution infiltrates the sandstone.

North Sea Jurassic Sandstones

The Jurassic Brent Sand Formation in the North Sea basin exhibits a characteristic diagenetic

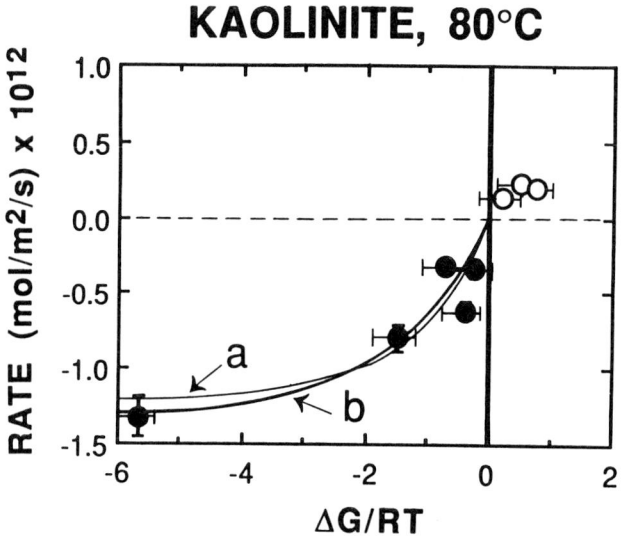

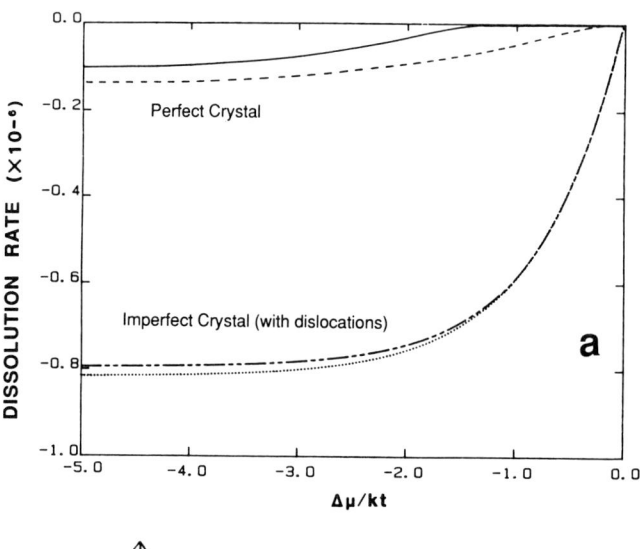

Figure 5. Plot of dissolution (closed circles) and precipitation (open circles) rates vs. solution saturation state expressed as $\Delta G/RT$ (see equation 8 in text). Increasing negative values of $\Delta G/RT$ indicate increasing undersaturation; increasing positive values represent increasing supersaturation. Dissolution rates are negative and precipitation rates are positive. Curve (a) is the fit to equation 6 of the text. Curve (b) is the fit to equation 7. The solid vertical line represents equilibrium of the solution with kaolinite as determined in this study. The horizontal dashed line represents no net dissolution or precipitation.

sequence, recognized petrographically, in which kaolinite is interpreted to have formed from K-feldspar and/or mica (biotite and muscovite) (Blanche and Whitaker, 1978; Hancock and Taylor, 1978; Hawkins, 1978; Sommer, 1978; Bjørlykke, 1984). Based on these studies the diagenetic sequence consists of early quartz cementation, followed by kaolinite ± quartz cementation, and then by illite formation. Kaolinite precipitation is hypothesized to be related to meteoric water flow resulting from tectonic uplift in the Late Jurassic to Early Cretaceous.

From observations of thin sections and SEM photomicrographs, the above workers noted the following textural relationships between kaolinite and other silicates. The formation of kaolinite from feldspar is inferred from the partial replacement of K-feldspar grains by kaolinite books, and from the existence of patches of dense kaolinite books having the outline of the shape of a cleaved feldspar grain. Also, in general, K-feldspar grains show evidence of dissolution, especially along cleavage surfaces. The formation of kaolinite from mica is inferred in instances where kaolinite grains are intergrown with mica grains. In such cases, the kaolinite books occur as a splayed-out array at the end of a mica grain (that is, perpendicular to the c-axis of the mica). For the most part, illite is interpreted to have formed

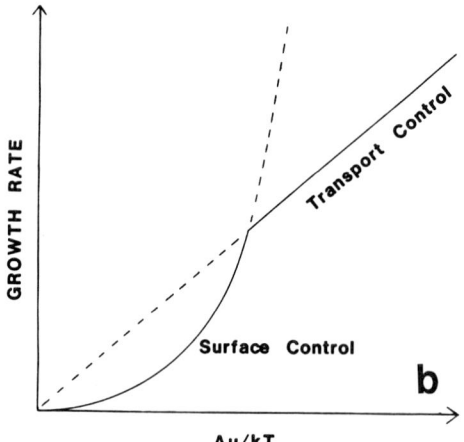

Figure 6. (a) Predicted dissolution rate for an imperfect crystal as a function of the saturation state of the solution from Monte Carlo simulations of dissolution. $\Delta\mu/kT = 0$ is equilibrium, and the solution becomes increasingly undersaturated as $\Delta\mu/kt$ becomes more negative. Note the similarity in shape of the theoretical curve with the experimental dissolution data for kaolinite shown in Figure 5. (b) Theoretical shape of crystal growth rate laws demonstrating the change in mechanism between surface-reaction-controlled and transport-controlled growth. Surface-reaction-controlled growth is nonlinear with respect to solution saturation state, whereas transport-controlled growth is linear. Both figures are from Blum and Lasaga (1987).

from the kaolinite, both by pseudomorphic replacement of kaolinite and by growth outward into pore spaces from kaolinite grains.

The occurrence of authigenic kaolinite in the Brent Sand Formation is also characteristic of a particular facies. Well-crystallized kaolinite (kaolinite books) is found more extensively in the more porous sandstone channel facies. In contrast, less abundant, poorly crystalline kaolinite (vermicular kaolinite) is found in the micaceous shales.

Table 2. Examples of proposed diagenetic mineral/solution reactions involving dissolution and precipitation of kaolinite

Reaction	Reference
$CaAl_2Si_2O_8 + 3H_2O + 2CO_2 = Al_2Si_2O_5(OH)_4 + Ca^{2+} + 2HCO_3^-$ (anorthite) (kaolinite)	Land (1984)
$2KAlSi_3O_8 + 2CO_2 + 11H_2O = 2K^+ + Al_2Si_2O_5(OH)_4 + 4H_4SiO_4 + 2HCO_3^-$ (K-feldspar) (kaolinite)	Land (1984)
$2NaAlSi_3O_8 \cdot CaAl_2Si_2O_8 + 4H^+ + 2H_2O = 2Al_2Si_2O_5(OH)_4 + 4SiO_2 + 2Na^+ + Ca^{2+}$ (plagioclase) (kaolinite) (quartz)	Siebert et al. (1984)
$CaAl_2Si_4O_{12} \cdot 4H_2O + 2H^+ = Al_2Si_2O_5(OH)_4 + 2SiO_2 + 3H_2O + Ca^{2+}$ (laumontite) (kaolinite) (quartz)	Crossey et al. (1984)
$7Al_2Si_2O_5(OH)_4 + 8SiO_2 + 2Na^+ = 2NaAl_7Si_{11}O_{30}(OH)_6 + 7H_2O + 2H^+$ (kaolinite) (quartz) (smectite)	Hutcheon (1984)

Gulf Coast Tertiary Sandstones

Two Tertiary sandstone units from the Gulf Coast sedimentary basin show diagenetic sequences in which kaolinite is interpreted to have formed at different stages of the burial history. One is the Oligocene Frio Formation and the other is the Eocene Wilcox Group (Fisher and Land, 1986).

In the simplified diagenetic sequence based on $\delta^{18}O$ values in the Frio Formation in Brazoria County, Texas (Milliken et al., 1981), authigenic mineral formation is correlated with temperature of precipitation assuming isotopic exchange in an open system; that is, in the presence of fluid flow. The authors concluded that quartz formed at 65° and 80°C; kaolinite formed near 100°C; and albite precipitated most rapidly near 150°C. Carbonate minerals formed over the range of temperatures represented by the burial depths of the cores that were examined. Kaolinite was observed within pore spaces and as feldspar replacements.

In the Wilcox sandstones, Fisher and Land (1986) place the precipitation of kaolinite between that of quartz and calcite cements, but the range of occurrence of each mineral considerably overlaps in time. Texturally, kaolinite appears to have grown after quartz, although intergrown kaolinite and quartz can be found. Fisher and Land (1986) state that kaolinite did not form in multiple events and formed prior to calcite cementation. They noticed no difference in morphology (crystallinity) or abundance of kaolinite over the depth of the studied cores (1528 to 4446 m) in rocks ranging from fine-grained feldspathic litharenites to subarkoses. On the basis of this evidence, Fisher and Land (1986) concluded that kaolinite formed relatively early during the diagenetic history of the sandstone, and they estimate a temperature range of 80° to 110°C based on oxygen isotopic compositions of all the diagenetic minerals.

SIMULATIONS OF DIAGENESIS

Many diagenetic sequences, such as those described in the previous section, may form in open systems where the reactions are driven by the influx of basinal fluids which are out of equilibrium with mineral assemblages in the rock. This could occur, for example, where fluids flow up or down a temperature gradient, or where fluids enter a rock from a chemically different lithology. Where diagenetic reactions occur as a result of fluid flow, the rates of the reactions will exert an important control on both the temporal and the spatial sequence of minerals. The following simulation illustrates the importance of mineral reaction rates by considering the formation of kaolinite from feldspar.

The problem we consider is the infiltration of slightly acid, near equilibrium subsurface water at 80°C into a subarkose containing a small amount of calcite cement. The initial porosity and grain size distribution are representative of a channel sand facies.

Numerical Methods

The reaction-transport simulations are carried out using the code GEOKIN which treats multicomponent, finite-rate reactions and transport in one dimension (Steefel and Lasaga, 1990). The code is based on numerical solution of the "reaction-transport" equation, which can be written as

$$\frac{\partial(\Phi C_i)}{\partial t} = -\nabla \cdot J_i - R_i, \quad (i = 1, 2, \ldots, N_c) \quad (15)$$

where C_i (mol$_i$/m$^3_{fluid}$) refers to the concentration of species i in solution, N_c is the number of chemical components in the system, Φ is the porosity of the medium, J are the fluxes (advection, dispersion, and diffusion), and R_i is the net rate (mol$_i$/m$^3_{rock}$/s) for all the chemical reactions involving i that result from mineral-solution interaction (Lasaga, 1984). The simulation is carried out using the form given by equation 7, with n = 1. For the multicomponent case, R_i is written more explicitly as (Lasaga, 1984)

$$R_i = -\sum_{j=1}^{N_r} \nu_{ij} A_j k_j \left(1 - \frac{IAP_j}{K_j}\right) (i = 1, 2, \ldots, N_c), \quad (16)$$

where N_r is the total number of minerals in the system, and ν_{ij} are the stoichiometric coefficients for species i in mineral j. A_j refers to the surface area per unit volume of porous medium for each of the minerals, k_j is the rate constant for the dissolution reaction (mol/m²/s), and IAP_j and K_j are the ion activity product and equilibrium constant, respectively, for mineral j.

Initial and Boundary Conditions

The porosity changes caused by the mineral-solution reactions will result in changes in the permeability of the medium. One empirical relationship between the porosity (ϕ) and permeability (k_p) which is often cited (Lerman, 1979) is

$$k_p \propto \phi^3. \quad (17)$$

The permeability is linearly related to the Darcian flow velocity, q (m³_fluid/m²_rock/s), in Darcy's Law (Bear, 1979),

$$q = -\frac{k_p}{\mu}(\nabla P - \rho g) \quad (18)$$

where μ is the fluid viscosity, ∇P is the fluid pressure gradient, ρ is the fluid density, and g is the vector of gravity acceleration. In one dimension, the velocity will evolve according to

$$\frac{q(t)}{q_o} = \frac{k_p(t)}{k_{po}} = \left(\frac{\phi_{min}(t)}{\phi_o}\right)^3, \quad (19)$$

where q_o is the initial flow velocity of 10 m/yr, ϕ_o is the initial connected porosity of 20%, and ϕ_{min} is the minimum porosity over the flow path. The conservation of fluid mass requires that the flow rate be determined by the minimum porosity along the flow path. This model, therefore, does not take into account the full nonlinear effects which appear when changes in permeability cause flow to be diverted in two dimensions (Ortoleva et al., 1987; Steefel and Lasaga, 1990). A flow velocity of 10 m/yr is chosen as a reasonable upper limit of flow rates expected in a permeable unit where flow is gravity driven (Garven and Freeze, 1984). The chosen porosity of 20% is common for a subsurface sandstone which already has undergone some chemical diagenesis. The diffusion-dispersion coefficient is fixed at 10^{-3} m²/yr for the course of the simulation. The simulation results are examined over a distance of 25 m from the inlet boundary for a flow period of 20,000 years.

The model sandstone is composed of quartz, K-feldspar, albite, calcite, and trace kaolinite and muscovite (Table 3) in proportions that classify it as a subarkose (Folk, 1968). Initial grain sizes and surface areas of the minerals also are listed in Table 3. The minerals gibbsite and boehmite were included in the simulation, although they are not initially present in the rock.

Equilibrium constants and dissolution rate constants are listed for all the minerals in Table 4. The equilibrium constant for kaolinite was set at a value an order of magnitude lower than that for the Twiggs County, Georgia kaolinite used in the experimental study. This was done to represent more accurately the solubility of a subsurface kaolinite which, having formed at higher temperature, would probably be more well-crystalline than the kaolinite we used (Hinckley, 1963) and therefore would have a lower solubility (Bassett et al., 1979). The dissolution rate constant for kaolinite determined in this study at pH 3 was extrapolated to pH 5, by assuming that k decreased by a factor of 5 over this pH range (Carroll-Webb and Walther, 1988). We assumed that there is no pH dependence to the dissolution rate constant between pH 5 and 8, based on the data of Carroll-Webb and Walther (1988) for kaolinite dissolution at 60°C. We used the equilibrium constant for muscovite as a substitute for that of illite (Table 4). Also, the dissolution rate of muscovite was set equal to that of kaolinite, in the absence of any rate data at 80°C. Data of Knauss and Wolery (1989) on muscovite dissolution at 70°C as a function of pH, published since this calculation was performed, indicate that this is a reasonable approximation.

Eighteen aqueous species are considered in the simulation (Table 5), eight of which are independent basis components; that is, they totally describe the composition of the fluid. The concentrations of the independent aqueous species are fixed at the inlet (Table 6). The inlet fluids are slightly supersaturated with respect to kaolinite, muscovite, and quartz; in equilibrium with albite and gibbsite; and slightly undersaturated with respect to K-feldspar, boehmite, and calcite. Total dissolved solids (TDS) were 6100 ppm and ionic strength was 0.1 Molar. This would be a typical composition for a porewater with a large meteoric fluid component. Inlet pH was fixed at 6.30 and evolved to a maximum of 7.65 over the length of the flow path. Over this near neutral pH range, dissolution rates of silicates far from equilibrium have been shown to be approximately pH-independent, justifying the exclusion of a pH-dependent term in the calculation of reaction rates (equation 16).

Modeling Results and Discussion

During the evolution of the system with time, the volume fractions of minerals in the sandstone (expressed as percentages in Figure 7) vary as a result of precipitation and dissolution reactions driven by the influx of reactive fluids. When the reaction-transport equations are integrated over geologic time periods, the reaction fronts move slowly downstream because of the gradual change in the volume fractions

Table 3. Initial volume percentage, total surface area, and grain size of minerals which compose the sandstone treated in the diagenetic model

Mineral	Volume Percentage	Total Surface Area [m^2/m^3_{rock}]	Grain Radius [mm]
Kaolinite	0.033	10.0	0.10
Muscovite	0.0	0.0	
Quartz	70.0	933.0	2.25
Albite	2.5	33.3	2.25
Gibbsite	0.0	0.0	
K-feldspar	2.5	33.3	2.25
Boehmite	0.0	0.0	
Calcite	5.0	300.0	0.50

Table 4. Equilibrium constants at 80°C and dissolution rate constants at near-neutral pH and 80°C for the minerals composing the sandstone treated in the diagenetic model

Mineral	Reaction	Log(K_{eq})	Ref.	log(k) [k in mol/m^2/s]	Ref.
Kaolinite	$Al_2Si_2O_5(OH)_4 + 6H^+ = 2Al^{3+} + 2SiO_2(aq) + H_2O$	2.88	1	-12.50	1,5
Muscovite	$KAl_2(AlSi_3O_{10})(OH)_2 + 10H^+ = K^+ + 3Al^{3+} + 3SiO_2(aq) + 6H_2O$	8.23	2	-12.50	6
Quartz	$SiO_2 = SiO_2(aq)$	-3.28	2	-11.32	7
Albite	$NaAlSi_3O_8 + 4H^+ = Na^+ + Al^{3+} + 3SiO_2(aq) + 2H_2O$	1.45	2	-10.86	8
Gibbsite	$Al(OH)_3 + 3H^+ = Al^{3+} + 3H_2O$	4.88	3	-8.03	9,10
K-feldspar	$KAlSi_3O_8 + 4H^+ = K^+ + Al^{3+} + 3SiO_2(aq) + 2H_2O$	-0.81	2	-10.48	11
Boehmite	$AlO(OH) + 3H^+ = Al^{3+} + 2H_2O$	6.42	2	-9.61	10
Calcite	$CaCO_3 = Ca^{2+} + CO_3^{2-}$	-8.99	4	-4.47	12

References:
1. This study
2. Bowers et al. (1984)
3. Wesolowski et al. (1988)
4. Plummer and Busenberg (1982)
5. Nagy et al. (1988)
6. Chosen to equal that of kaolinite
7. Rimstidt and Barnes (1980)
8. Knauss and Wolery (1986)
9. Bloom and Erich (1987)
10. Scotford and Glastonbury (1972)
11. Helgeson et al. (1984)
12. Sjöberg (1978)

of the reacting minerals. This is seen most clearly in the sharp front for the volume fraction of calcite. Upstream of this front, all the calcite has been dissolved from the rock; downstream, the calcite is only slightly affected. The volume fractions of the silicates change more slowly, although the important effect of the calcite front which acts to buffer the solution pH is clear from the change in slope of many of the profiles (for example, kaolinite and muscovite form in significant quantities only upstream of the calcite front). K-feldspar decreases with time throughout the rock, although the actual decrease in its volume percentage varies spatially. Albite decreases near the inlet and increases slightly downstream of the calcite front. Quartz precipitates throughout the rock during the simulation.

The evolution of porosity in the subarkose when infiltrated by a reactive fluid is also a function of both time and space (Figure 8). The net porosity change is the sum of all the changes in the volume fractions of the various minerals in the system, that is

$$\Phi(t) = 1 - \sum_{j=1}^{N_r} \Phi_j(t), \quad (20)$$

Table 5. Primary and secondary solution species used in the numerical simulation

Primary Species	Secondary Species
H^+	OH^-
Na^+	$Al(OH)^{2+}$
K^+	$Al(OH)_2^+$
Al^{3+}	$Al(OH)_3^0$
SiO_2	$Al(OH)_4^-$
Ca^{2+}	HCO_3^-
CO_3^{2-}	H_2CO_3
Cl^-	$CaHCO_3^-$
	$CaCO_3^0$
	$H_3SiO_4^-$

Table 6. Inlet and initial fluid composition used in numerical simulation of diagenesis in subarkosic sandstone at 80°C

Composition			Saturation	State
Element	Molarity	ppm	Mineral	$Log(\Omega)$
Na	0.100	2300	Kaolinite	1.16
K	3.16×10^{-4}	12	Muscovite	0.50
Al	2.00×10^{-6}	0.054	Quartz	0.42
Si	1.26×10^{-3}	35	Albite	0.03
C	3.16×10^{-4}	3.8	Gibbsite	0.00
Ca	2.00×10^{-3}	80	K-feldspar	-0.22
Cl^1	0.105	3700	Boehmite	-1.54
			Calcite	-1.83

pH = 6.30   I = 0.1 M   TDS = 6150 ppm

[1]Obtained by charge balance.

where Φ is the porosity of the sandstone and the Φ_j are the volume fractions of the minerals. Understanding the overall porosity change requires an understanding of all the mineral-solution reactions as a function of space and time in the flow field.

Figure 9 shows the instantaneous precipitation rate of quartz and the dissolution rate of calcite (in units of mol/m$^3_{rock}$/s) after 20,000 years. At this point in time, the calcite front has migrated to a position about 14.5 m from the inlet to the system and is marked by a sharp increase in the dissolution rate. Immediately downstream of the front, minor precipitation of calcite occurs (Figure 10). Quartz, which is supersaturated at the inlet, precipitates throughout the aquifer, but its rate of precipitation gradually decreases downstream as the fluids approach its equilibrium solubility. The contrasting behavior of calcite and quartz illustrates an important concept about the formation of reaction fronts. The shape of the precipitation curve (sharp versus gradual), and therefore the morphology of the reaction front as expressed by the mineral volume fractions depends on the ratio of the reaction rate to the flow velocity (Lichtner, 1988; Steefel and Lasaga, 1990). The rate constant for calcite, for example, is large enough for equilibrium between the solution and calcite to be achieved quickly, before the fluid has a chance to move a significant distance downstream. This accounts for both the spike in the dissolution rate for calcite (Figure 9) and the steplike nature of its volume fraction (Figure 7). In contrast, quartz, which has a reaction rate constant 7 orders of magnitude smaller than calcite (Table 4), precipitates slowly enough for equilibrium to be achieved only over a significant distance along the flow path. In the case of quartz and the other silicates considered, their slow rates of reaction relative to the flow velocity means that their reaction fronts will be diffuse (Figures 9, 10).

The slow rates of reaction of the silicate minerals do not allow buffering of the solution pH as does the calcite reaction rate (Figure 11). Equilibration of the solution with respect to calcite takes place essentially instantaneously when fluid comes in contact with calcite. Downstream of the calcite front the pH is effectively constant.

Changes in speciation in solution, which affect the saturation state of the fluid with respect to the minerals, will affect the precipitation and dissolution rate profiles. This effect can be seen in the discontinuities in the dissolution rates of the feldspars (Figure 10) and the small decrease in the quartz precipitation rate (Figure 9) at 14.5 m, which is caused by the abrupt change in solution pH at the calcite reaction front (Figure 11). The decrease in the rates of dissolution of both albite and K-feldspar immediately on the upstream side of the calcite front is due to a decrease in the degree of undersaturation of the porewater with respect to the feldspars. This results from a gradual increase in pH and total aluminum in solution caused by dissolution of the feldspars. When the fluid reaches the calcite front, however, the pH increases dramatically as calcite dissolves (Figure 11). The sudden increases in the dissolution rates of the feldspars (which appear as discontinuities over a single grid point) are due to the greater degree of undersaturation caused by changes in the speciation of aluminum and silicon. The fact that these speciation changes which are caused by the abrupt rise in pH override the direct effect of the pH rise on feldspar saturation (which is to make the solution closer to equilibrium) illustrates the importance of considering the full range of possible species. The change in silicon speciation also causes the supersaturation of the solution with respect to quartz to decrease slightly, resulting in a slower precipitation rate. With further flow downstream, the undersaturation of the solutions with respect to the feldspars decreases: the K-feldspar dissolves less rapidly and albite eventually precipitates.

The change in porosity (Figure 8) in the sandstone is therefore primarily the result of the combined

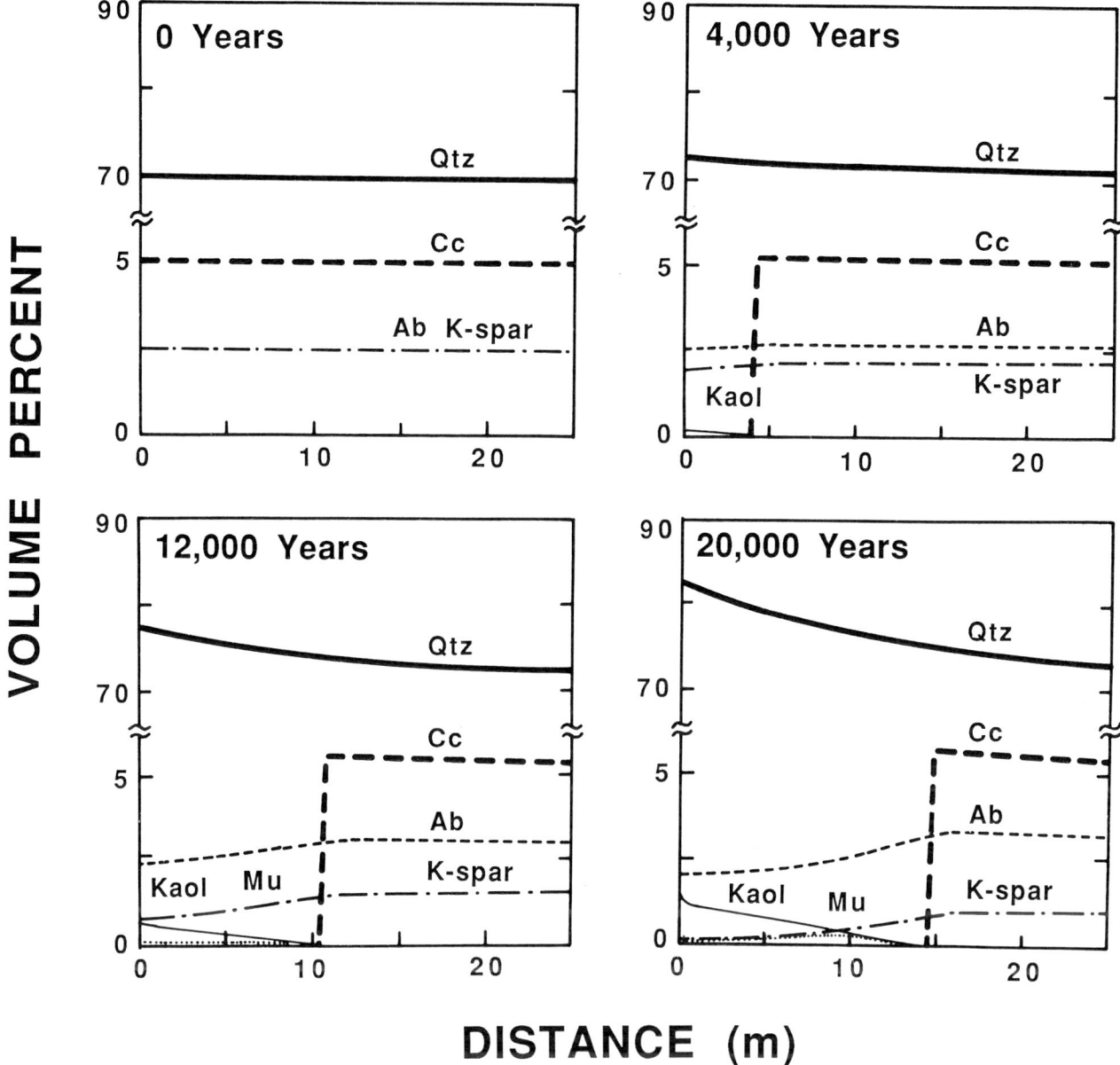

Figure 7. Temporal and spatial evolution of the volume percentages of the minerals in the subarkose modeled in this study: Kaol = kaolinite, Mu = muscovite, Cc = calcite, Ab = albite, K-spar = K-feldspar, Q = quartz. Note the contrast between the sharp front for calcite, which reacts rapidly, and the more diffuse profiles for the silicates, which react more slowly. Flow is from left to right.

effects of the precipitation of quartz and the dissolution of calcite. The other reactions in the system have a much less pronounced effect because their net molar volume change is relatively small. The initial increase in porosity near the inlet is due to the relatively rapid dissolution of calcite (which makes up 5% of the rock at the beginning of the simulation). Again, the location of the calcite front is clearly marked by the sharp decrease in porosity which moves downstream with time. The gradual decrease in porosity near the inlet is primarily due to quartz precipitation, inasmuch as SiO_2 is the chief component added to the rock. Although the quartz precipitation rates are relatively slower, quartz is the dominant mineral comprising the sandstone, and the inlet solution is fairly supersaturated with respect to quartz. Kaolinite, although more supersaturated than quartz at the inlet, is volumetrically a minor mineral in this simulation, because it forms from the dissolution of feldspar which makes up only 5% of the rock. In a more feldspathic sandstone, kaolinite precipitation would be more important in affecting changes in porosity.

One important control on the evolution of the porosity is the rate of fluid flow, which provides the driving force for the diagenetic reactions. As the total

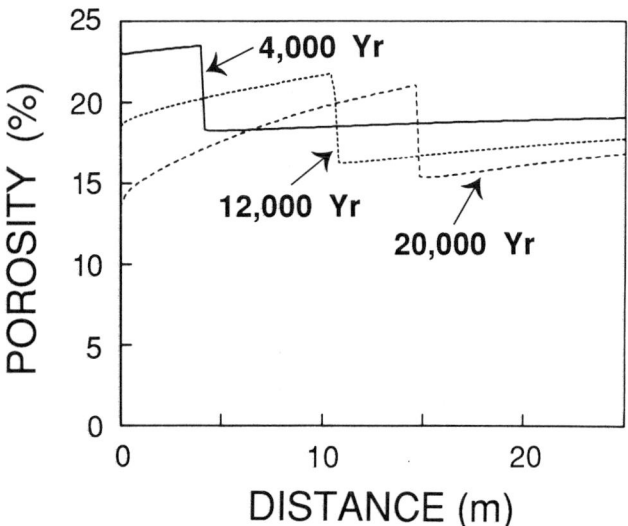

Figure 8. Temporal and spatial evolution of porosity in the subarkose modeled in this study. Flow is from left to right.

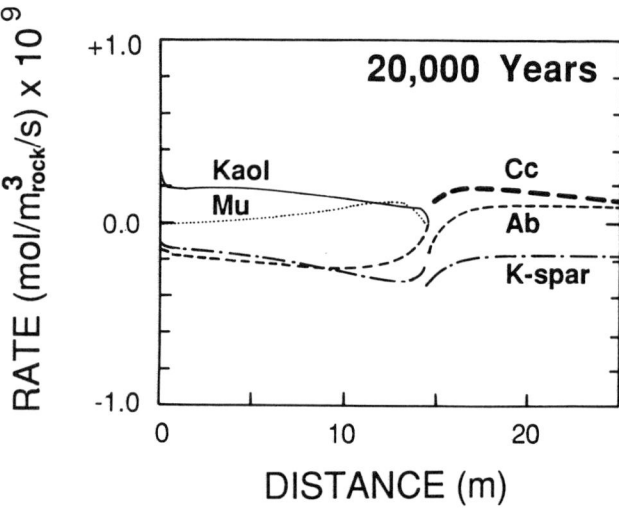

Figure 10. Precipitation (positive) and dissolution (negative) rates of kaolinite, muscovite, albite, and K-feldspar after 20,000 years. For calcite only the precipitation rate is shown, and it is located downstream of the calcite dissolution profile seen in Figure 9. Flow is from left to right.

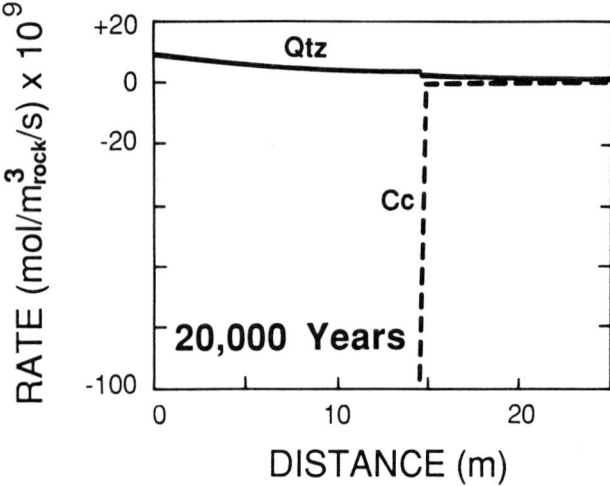

Figure 9. Precipitation rate of quartz and dissolution rate of calcite after 20,000 years. The precipitation rate is positive and the dissolution rate is negative. Notice the sharpness of the reaction rate profile of calcite relative to the more diffuse profile of quartz. Flow is from left to right.

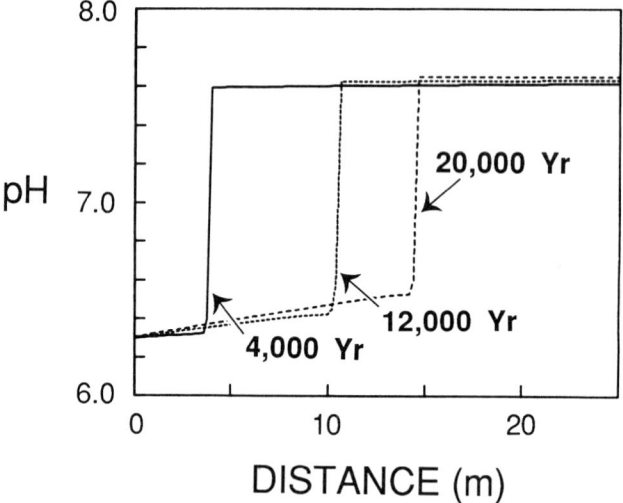

Figure 11. Temporal and spatial evolution of pH in the solution. The sharp fronts in pH correspond to the sharp reaction rate profiles of calcite. Flow is from left to right.

porosity of the sandstone decreases because of mineral reactions, the permeability of the medium must change as well. Because a reduction in porosity (and implicitly, the permeability) must result in a decrease in Darcian flow velocity (equation 18), there is a corresponding gradual decrease in the propagation rate of the reaction fronts (compare the locations of the calcite reaction front in Figure 8 at 4,000, 12,000, and 20,000 years).

This isothermal simulation of diagenesis in a subarkosic sandstone, although limited in applica-bility because of the assumptions involved, nonetheless reproduces many of the reactions that have been proposed based on textural evidence. Upstream of the calcite reaction front, albite, K-feldspar, and calcite break down to form kaolinite, muscovite, and quartz. These numerical results support the proposed reactions resulting in kaolinite authigenesis in both the Jurassic North Sea and Tertiary Gulf Coast sandstones. In both these cases, quartz and kaolinite were interpreted to have precipitated simultaneously based on textural evidence. The onset of muscovite precipitation takes place slightly downstream from the inlet and could represent the formation of an illitic phase from the feldspars. In this particular simulation, we see no evidence for the direct formation of illite (muscovite) from kaolinite as Bjørlykke (1984) has proposed for the Brent Sand Formation in the North Sea.

It is important to point out that the simulation presented here does not necessarily provide a unique description of subsurface sandstone diagenesis. Although the code GEOKIN can incorporate temperature variations, we chose to neglect them in this simulation. Over the 25 m of the flow path, the assumption of isothermal conditions is reasonable. Over a longer flow path, however, temperature effects would have to be considered for a realistic simulation, because the equilibrium constants and reaction rates are both functions of temperature. Fluids moving up or down a temperature gradient will experience continually changing saturation states.

It should also be pointed out that the results presented here strongly depend on the fluid flow rate, for two reasons. Because the amount of any mineral that can be dissolved or precipitated depends on the rate at which reactive fluids can be brought into the rock, the velocity of the reaction fronts will be proportional to the flow rate. In addition, the geometry of the reaction zones, as discussed above, depends on the ratio of the mineral reaction rates to the flow rate and not simply on the reaction rates themselves. If the flow rate is slow enough, the local equilibrium assumption may be valid even for the silicate minerals.

Another important factor that would change the results of the simulation is the form of dissolution and the precipitation rate law for each mineral. For the simulation presented here, we assumed that the rates over the entire range of solution saturation states could be defined using the principle of detailed balancing (Lasaga, 1981b; Aagaard and Helgeson, 1982). This strictly applies only to rates near equilibrium and for elementary reactions. If the reaction mechanism varies as a function of solution saturation state and/or is different for precipitation than for dissolution (which is probable), then the principle of detailed balancing does not apply. The determination of the proper form of the rate laws as a function of saturation state is the objective of our ongoing experimental work.

CONCLUSIONS

Experimental data on mineral/solution reaction kinetics and thermodynamics, numerical modeling which combines fluid flow and reaction kinetics, and observational data must all be synthesized in order to be able to make reasonable statements about porosity and permeability distributions in reservoir sandstones. In particular:

1. Reaction rate laws for silicate mineral dissolution and growth are needed, both as a function of deviation from equilibrium and pH, in order to accurately model the temporal and spatial evolution of porosity and permeability.
2. These rate laws for silicates can be obtained experimentally, using flow-through reaction cells. Both dissolution and precipitation rates at steady-state can be measured for specific solution compositions and saturation states.
3. The acquisition of kinetic data through a range of saturation states at constant temperature using the flow-through reaction cell permits the determination of the equilibrium solubility of the mineral. This amounts to bracketing the equilibrium constant by performing both dissolution and precipitation experiments close to equilibrium. The solubility of the kaolinite used in this study was determined in this manner.
4. Rate data obtained thus far for kaolinite at 80°C and pH near 3 indicate that a linear relationship between dissolution rates near equilibrium and solution saturation state may be valid. More dissolution rates near equilibrium, as well as precipitation rates over a wider range of saturation states, are needed to verify the applicability of the principle of detailed balancing.
5. Numerical simulations of diagenesis which involve the coupling of silicate rates with a reasonably high fluid velocity indicate that the relatively slow kaolinite rates do not permit buffering of the solution composition along the flow path. The simulations indicate that even if fluids eventually achieve equilibrium with respect to diagenetic minerals such as kaolinite downstream along a flow path, the actual formation of diagenetic reaction fronts may be a nonequilibrium feature. Disequilibrium mineral assemblages can persist over meter-length scales and can vary along a constant temperature flow path.
6. The numerical coupling of one-dimensional fluid flow and mineral dissolution and precipitation rates at constant temperature can yield reasonable representations of observed diagenetic mineral assemblages and their textural relationships. Thus, mineral textures could be interpreted as results of flow-driven reaction fronts instead of burial metamorphism.

ACKNOWLEDGMENTS

The experimental results presented in this paper were carried out under contract to Lawrence Livermore National Laboratory, in support of the Yucca Mountain Project. The Yucca Mountain Project is administered by the Office of Civilian Radioactive Waste Management of the Department of Energy. We would also like to acknowledge Amoco Foundation for a fellowship to C. I. Steefel and for their continuing support of our experimental work. I. Meshri, T. Park, P. Schroeder, and P. Van Cappellen provided useful comments on earlier versions of the manuscript.

REFERENCES CITED

Aagaard, P., and H. C. Helgeson, 1982, Thermodynamic and kinetic constraints on reaction rates among minerals and aqueous solutions. I. theoretical considerations: American Journal of Science, v. 282, p. 237-285.

Apps, J. A., J. M. Neil, and C.-H. Jun, 1989, Thermochemical properties of gibbsite, bayerite, boehmite, diaspore, and the aluminate ion between 0 and 350°C: U.S. Nuclear Regulatory Commission Technical Report NUREG/CR-5271,LBL-21482, 321 p.

Bales, R. C., and J. J. Morgan, 1985, Dissolution kinetics of chrysotile at pH 7 to 10: Geochimica et Cosmochimica Acta, v. 49, p. 2281-2288.

Bassett, R. L., Y. K. Kharaka, and D. Langmuir, 1979, Critical review of the equilibrium constants for kaolinite and sepiolite, in E. Jenne, ed., Chemical modeling in aqueous systems: American Chemical Society Symposium Series 93, p. 389-400.

Bear, J., 1979, Hydraulics of groundwaters: New York, McGraw-Hill, 569 p.

Bjørlykke, K., 1984, Formation of secondary porosity: how important is it?, in D. A. McDonald and R. C. Surdam, eds., Clastic diagenesis: AAPG Memoir 37, p. 277-286.

Blanche, J. B., and J. H. McD. Whitaker, 1978, Diagenesis of part of the Brent Sand Formation (Middle Jurassic) of the northern North Sea Basin: Journal of the Geological Society of London, v. 135, p. 73-82.

Bloom, P. R., and M. S. Erich, 1987, Effect of solution composition on the rate and mechanism of gibbsite dissolution in acid solutions: Soil Science Society of America Journal, v. 51, p. 1131-1136.

Blum, A. E., and A. C. Lasaga, 1987, Monte Carlo simulations of surface reaction rate laws, in W. Stumm, ed., Aquatic surface chemistry: chemical processes at the particle-water interface: New York, J. Wiley & Sons, Inc., p. 255-292.

Blum, A. E., and A. C. Lasaga, 1988, Role of surface speciation in the low-temperature dissolution of minerals: Nature, v. 331, p. 431-433.

Bourcier, W. L., K. G. Knauss, and K. J. Jackson, 1987, Aluminum hydrolysis constants to 250°C determined from boehmite solubility measurements (abs.): GSA Abstracts with Programs, v. 19, p. 596.

Bowers, T. S., K. J. Jackson, and H. C. Helgeson, 1984, Equilibrium activity diagrams for coexisting minerals and aqueous solutions at pressures and temperatures to 5 kb and 600°C: New York, Springer-Verlag, 397 p.

Burton, W. K., N. Cabrera, and F. C. Frank, 1951, The growth of crystals and the equilibrium structure of their surfaces: Philosophical Transactions of the Royal Society of London, v. A 243, p. 299-358.

Busey, R. H., and R. E. Mesmer, 1978, Thermodynamic quantities for the ionization of water in sodium chloride media to 300°C: Journal of Chemical Engineering Data, v. 23, p. 175-176.

Carroll-Webb, S. A., and J. V. Walther, 1988, A surface complex reaction model for the pH-dependence of corundum and kaolinite dissolution rates: Geochimica et Cosmochimica Acta, v. 52, p. 2609-2623.

Chou, L., and R. Wollast, 1984, Study of the weathering of albite at room temperature and pressure with a fluidized bed reactor: Geochimica et Cosmochimica Acta, v. 48, p. 2205-2217.

Chou, L., and R. Wollast, 1985, Steady-state kinetics and dissolution mechanisms of albite: American Journal of Science, v. 285, p. 963-993.

Crossey, L. J., B. R. Frost, and R. C. Surdam, 1984, Secondary porosity in laumontite-bearing sandstones, in D. A. McDonald and R. C. Surdam, Clastic diagenesis: AAPG Memoir 37, p. 225-237.

Dougan, W. K., and A. L. Wilson, 1974, The absorptiometric determination of aluminium in water. A comparison of some chromogenic reagents and the development of an improved method: Analyst, v. 99, p. 413-430.

Fisher, R. S., and L. S. Land, 1986, Diagenetic history of Eocene Wilcox sandstones, south-central Texas: Geochimica et Cosmochimica Acta, v. 50, p. 551-561.

Folk, R. L., 1968, Petrology of sedimentary rocks: Austin, Texas, Hemphill Publishing Company, 182 p.

Garven, G., and R. A. Freeze, 1984, Theoretical analysis of the role of groundwater flow in the genesis of stratabound ore deposits: American Journal of Science, v. 284, p. 1085-1124.

Grandstaff, D. E., 1977, Some kinetics of bronzite orthopyroxene dissolution: Geochimica et Cosmochimica Acta, v. 41, p. 1097-1103.

Grandstaff, D. E., 1986, The dissolution rate of forsteritic olivine from Hawaiian beach sand, in S. M. Colman and D. P. Dethier, eds., Rates of chemical weathering of rocks and minerals: New York, Academic Press, Inc., p. 41-59.

Grasshoff, K., 1976, Methods of seawater analysis: New York, Springer-Verlag, 315 p.

Hancock, N. J., and A. M. Taylor, 1978, Clay mineral diagenesis and oil migration in the Middle Jurassic Brent Sand Formation: Journal of the Geological Society of London, v. 135, p. 69-72.

Hawkins, P. J., 1978, Relationship between diagenesis, porosity reduction, and oil emplacement in late Carboniferous sandstone reservoirs, Bothamsall oilfield, E. Midlands: Journal of the Geological Society of London, v. 135, p. 7-24.

Helgeson, H. C., J. M. Delany, H. W. Nesbitt, and D. K. Bird, 1978, Summary and critique of the thermodynamic properties of rock-forming minerals: American Journal of Science, v. 278, p. 1-229.

Helgeson, H. C., W. M. Murphy, and P. Aagaard, 1984, Thermodynamic and kinetic constraints on reaction rates among minerals and aqueous solution. II. Rate constants, effective surface area, and the hydrolysis of feldspar: Geochimica et Cosmochimica Acta, v. 48, p. 2405-2432.

Hemley, J. J., J. W. Montoya, J. W. Marinenko, and R. W. Luce, 1980, Equilibria in the system Al_2O_3-SiO_2-H_2O and some general implications for alteration/mineralization processes: Economic Geology, v. 75, p. 210-228.

Hinckley, D. N., 1963, Variability in "crystallinity" values among the kaolin deposits of the Coastal Plain of Georgia and South Carolina: Clays and Clay Minerals, v. 11, p. 229-235.

Hutcheon, I., 1984, A review of artificial diagenesis during thermally enhanced recovery, in D. A. McDonald and R. C. Surdam, eds., Clastic diagenesis: AAPG Memoir 37, p. 413-429.

Knauss, K. G., and T. J. Wolery, 1986, Dependence of albite dissolution kinetics on pH and time at 25°C and 70°C: Geochimica et Cosmochimica Acta, v. 50, p. 2481-2497.

Knauss, K. G., and T. J. Wolery, 1988, The dissolution kinetics of quartz as a function of pH and time at 70°C: Geochimica et Cosmochimica Acta, v. 52, p. 43-53.

Knauss, K. G., and T. J. Wolery, 1989, Muscovite dissolution kinetics as a function of pH and time at 70°C: Geochimica et Cosmochimica Acta, v. 53, p. 1493-1502.

Land, L. S., 1984, Frio Sandstone diagenesis, Texas Gulf Coast: a regional isotopic study, in D. A. McDonald and R. C. Surdam, eds., Clastic diagenesis: AAPG Memoir 37, p. 47-62.

Lasaga, A. C., 1981a, Rate laws in chemical reactions, in A. C. Lasaga and R. J. Kirkpatrick, eds., Kinetics of geochemical processes: Mineralogical Society of America, Reviews in Mineralogy, v. 8, p. 1-68.

Lasaga, A. C., 1981b, Transition state theory, in A. C. Lasaga and R. J. Kirkpatrick, eds., Kinetics of geochemical processes: Mineralogical Society of America, Reviews in Mineralogy, v. 8, p. 135-169.

Lasaga, A. C., 1984, Chemical kinetics of water-rock interactions: Journal of Geophysical Research, v. 89, p. 4009-4025.

Lasaga, A. C., 1986, Metamorphic reaction rate laws and development of isograds: Mineralogical Magazine, v. 50, p. 359-373.

Lerman, A., 1979, Geochemical processes: water and sediment environments: New York, Wiley-Interscience, 481 p.

Lichtner, P. C., 1988, The quasi-stationary state approximation to coupled mass transport and fluid-rock interaction in a porous medium: Geochimica et Cosmochimica Acta, v. 52, p. 143-165.

May, H. M., P. A. Helmke, and M. L. Jackson, 1979, Gibbsite solubility and thermodynamic properties of hydroxy-aluminum ions in aqueous solution at 25°C: Geochimica et Cosmochimica Acta, v. 43, p. 861-868.

Milliken, K. L., L. S. Land, and R. G. Loucks, 1981, History of burial diagenesis determined from isotopic geochemistry, Frio Formation, Brazoria County, Texas: AAPG Bulletin, v. 65, p. 1397-1413.

Morse, J. W., 1983, The kinetics of calcium carbonate dissolution and precipitation, *in* R. J. Reeder, ed., Carbonates: mineralogy and chemistry: Mineralogical Society of America, Reviews in Mineralogy, v. 11, p. 227-264.

Nagy, K. L., A. E. Blum, and A. C. Lasaga, 1988, Precipitation kinetics and solubility of kaolinite in dilute aqueous solutions at 80°C (abs.): GSA Abstracts with Programs, v. 20, p. A42.

Naumov, G. B., B. N. Ryzhenko, and I. L. Khodakovsky, 1974, Handbook of thermodynamic data (translation of Russian report): USGS Report WRD 74-001, 328 p.

Nielsen, A. E., 1964, Kinetics of precipitation: New York, Pergamon Press, 151 p.

Ohara, M., and R. C. Reid, 1973, Modeling crystal growth rates from solution: Englewood Cliffs, New Jersey, Prentice-Hall, 272 p.

Ortoleva, P., J. Chaddam, E. Merino, and A. Sen, 1987, Geochemical self-organization II. The reactive infiltration instability: American Journal of Science, v. 287, p. 1008-1040.

Plummer, L. N., and E. Busenberg, 1982, The solubilities of calcite, aragonite, and vaterite in CO_2-H_2O solutions between 0 and 90°C, and an evaluation of the aqueous model for the system $CaCO_3$-CO_2-H_2O: Geochimica et Cosmochimica Acta, v. 46, p. 1011-1040.

Press, W. H., B. P. Flannery, S. A. Teukolsky, and W. T. Vetterling, 1986, Numerical recipes: the art of scientific computing: Cambridge, England, Cambridge University Press, 818 p.

Robie, R. A., B. S. Hemingway, and J. R. Fisher, 1978, Thermodynamic properties of minerals and related substances at 298.15 K and 1 Bar (10^5 Pascals) pressure and at higher temperatures: USGS Bulletin 1452, 456 p.

Rimstidt, J. D., and H. L. Barnes, 1980, The kinetics of silica-water reactions: Geochimica et Cosmochimica Acta, v. 44, p. 1683-1700.

Schramke, J. A., D. M. Kerrick, and A. C. Lasaga, 1987, The reaction muscovite + quartz = andalusite + K-feldspar + water. Part 1. Growth kinetics and mechanism: American Journal of Science, v. 287, p. 517-559.

Scotford, R. F., and J. R. Glastonbury, 1972, The effect of concentration on the rates of dissolution of gibbsite and boehmite: Canadian Journal of Chemical Engineering, v. 50, p. 754-758.

Siebert, R. M., G. K. Moncure, and R. W. Lahann, 1984, A theory of framework grain dissolution in sandstones, *in* D. A. McDonald and R. C. Surdam, eds., Clastic diagenesis: AAPG Memoir 37, p. 163-175.

Sjöberg, E. L., 1978, Kinetics and mechanism of calcite dissolution in aqueous solutions at low temperatures: Stockholm Contributions in Geology, v. 32, p. 1-92.

Sommer, F., 1978, Diagenesis of Jurassic sandstones in the Viking Graben: Journal of the Geological Society of London, v. 135, p. 63-67.

Steefel, C. I., and A. C. Lasaga, 1990, Evolution of dissolution patterns: permeability change due to coupled flow and reaction, *in* D. C. Melchior and R. L. Bassett, eds., Chemical modeling in aqueous systems II: American Chemical Society Symposium Series 416, p. 212-225.

Tole, M. P., A. C. Lasaga, C. Pantano, and W. B. White, 1986, The kinetics of dissolution of nepheline ($NaAlSiO_4$): Geochimica et Cosmochimica Acta, v. 50, p. 379-392.

Truesdell, A. H., and B. F. Jones, 1974, WATEQ, a computer program for calculating chemical equilibria of natural waters: USGS Journal of Research, v. 2, p. 233-248.

Wesolowski, D. J., D. A. Palmer, and S. E. Drummond, 1988, Solubility of gibbsite and speciation of aluminum in H-Na-K-Cl-OH-acetate brines in the range 6-125°C and 0-5 molal ionic strength (abs.): GSA Abstracts with Programs, v. 20, p. A42.

Wolery, T. J., 1979, Calculation of chemical equilibrium between aqueous solution and minerals: the EQ3/6 software package: Lawrence Livermore Laboratory URCL-52658, 41 p.

Wood, B. J., and J. V. Walther, 1983, Rates of hydrothermal reactions: Science, v. 222, p. 413-415.

Diagenesis Through Coupled Processes: Modeling Approach, Self-Organization, and Implications for Exploration

W. Chen
A. Ghaith
Department of Chemistry
Indiana University
Bloomington, Indiana, U.S.A.

A. Park
Department of Geology
Indiana University
Bloomington, Indiana, U.S.A.

P. Ortoleva
Department of Chemistry
Indiana University
Bloomington, Indiana, U.S.A.

The coupling of diagenetic reaction-transport processes can lead to patterns of petroleum and mineral distributions that are not a trivial reflection of imposed basin features such as sedimentary beds, folds, and faults. In this article we review examples of diagenetically differentiated features of this type. They include oscillatory intracrystalline zoning, banded cements (observed to serve as diagenetic petroleum traps), reaction-front fingering (which can lead to local inhomogeneities of porosity and permeability), oscillatory ejection of fluids from overpressurized, sealed compartments, and patterns of buoyancy-driven convection cells that can arise from mineral or kerogen reactions or the geothermal gradient. Mathematical reaction-transport modeling provides a method for determining conditions under which these phenomena may exist and for assessing their importance in petroleum genesis, migration, and trapping.

Both qualitative and quantitative arguments are used to demonstrate the feedback mechanisms and orders of magnitude of the properties of these diagenetic structures. As illustrative examples, we focus on patterns of the migration of methane driven by its own buoyancy, flow self-focusing at reaction fronts in carbonate rocks, and autonomous oscillatory fluid release from overpressurized compartments. The development of banded pressure seals through a mechano-chemical feedback destabilizing the state of uniform compaction is studied in a companion chapter (Dewers and Ortoleva, 1990). With these examples, we illustrate how reaction-transport coupling can generate spatial patterns of petroleum and minerals that can play a central role in determining reservoir quality and the diagenetic influences on petroleum distribution within the basin.

Because experiments on geological length and time scales are often unfeasible, the development of quantitative models and the implementation of computer codes to simulate them are becoming a third branch of geological investigation that is distinct from classical, experimental, and theoretical geoscience. In the present context, then, this "computational geochemistry" allows us to identify and characterize these diagenetic phenomena born of the strong coupling between reaction, transport, and mechanical processes.

DIAGENETIC REACTION-TRANSPORT FEEDBACK AND SELF-ORGANIZATION

A variety of spatial and temporal distributions of petroleum or mineralization in a sedimentary basin have no direct origin in pre-existing features such as original bedding, folding, or faulting. These phenomena have apparently arisen as a result of reaction, transport, and mechanical processes. In that they are not directly related to pre-existing features ("templates"), they are said to have "self-organized" (Nicolis and Prigogine, 1977; Nicolis and Nicolis, 1987; Ortoleva et al., 1987a; Ortoleva, 1990; Ortoleva et al., 1990). Identifying such phenomena and delineating conditions for their formation are challenging new problems in geochemistry. Because some of them play a key role in the migration and trapping of petroleum, they are of practical importance in that predicting their existence is not in the realm of classical stratigraphic or tectonic analysis (Dewers and Ortoleva, 1988).

Although we tend to think of a single process, it often happens that a variety of processes are coupled so strongly that qualitatively new effects and system behaviors arise because of this coupling. Such strong coupling phenomena are not understandable from the perspective of decoupled processes. In this article we attempt to demonstrate that the coupled network of diagenetic reaction, transport, and mechanical processes can support the feedback needed to generate a variety of diagenetic patterning phenomena. We use mathematical reaction-transport models to show how these patterns naturally emerge without the imposition of a template.

Perhaps the most familiar example of a self-organizing system is a layer of fluid heated from below in a gravitational field. As suggested in Figure 1, this system can develop an array of up and down drafts. This occurs even though the layer is perfectly horizontal, contains no sources or sinks of fluids in a pattern that could cause the cellular structure shown, and has a constant temperature along the top and bottom surfaces. A necessary condition for self-organization to take place is that the system be maintained sufficiently far from equilibrium; this is attained in the Benard system via the applied temperature difference. The resulting Benard convection pattern has a velocity and cell size that is not imposed by nonuniformities in boundary temperatures or other variations along the top and bottom plates. Rather, pattern length and amplitude only depend on fluid-layer thickness, thermal expansivity, and viscosity, as well as the temperatures imposed at the top and bottom boundaries.

In addition to displacement from equilibrium, geochemical self-organization requires feedback in the reaction-transport network. Let us now discuss a number of examples of geochemical self-

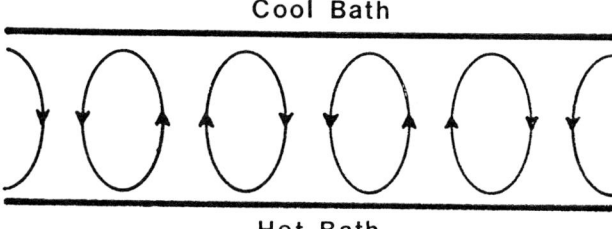

Figure 1. Benard flow cells in a horizontal fluid layer heated from below are formed when the temperature at the bottom exceeds that at the top by a critical value. The flow is driven by buoyancy caused by thermal expansion of the fluid.

organization, pointing out the feedback and special features of the reaction-transport dynamics that sustain them.

Diagenetic self-organization phenomena span the range of length scales from microns (in the case of oscillatory intracrystalline zoning) to kilometers (for basin compartments). A brief survey of observed or conjectured self-organizing diagenetic features is as follows (see also Ortoleva et al., 1987a; Dewers and Ortoleva, 1988; Ortoleva, 1990; Ortoleva et al., 1990).

Oscillatory Intracrystalline Zoning

The compositional zoning of solid solution crystals may be oscillatory, either in response to an oscillatory growth environment or to feedback effects involving the kinetics of crystal growth coupled to transport in the growth medium (Ortoleva et al., 1990). The latter "autonomous" oscillations can arise if the rate of construction of each end member depends on the fraction of each at the surface (Haase et al., 1980; Ortoleva et al., 1987a; Ortoleva, 1990a, b). For example, diagenetic oscillatory zoning has been observed in calcite (Reeder and Prosky, 1986; Reeder and Grams, 1987). Because zoning profiles can, in principle, yield detailed information about the fluids that flowed through a formation, it is important to be able to distinguish between autonomous and imposed zoning patterns.

Banded Authigenic Cements at Reaction Fronts

Banded patterns of mineral distributions of iron oxide cements, Mississippi Valley-type ores and skarns are common examples of alternating concentrations of minerals that appear to have been deposited at reaction fronts. These fronts arise when a fluid is imposed onto a rock that is out of equilibrium with at least one of the minerals in that rock. There are four known mechanisms wherein crossed gradients of coprecipitates can produce banded cements:

- the classic Ostwald (1925) mechanism of Liesegang banding (1913);
- unstable coarsening at reaction fronts, wherein a nucleation front of the cement-forming mineral is produced and becomes unstable to band genesis (Feinn et al., 1978; Lovett et al., 1978; Boudreau, 1987);
- autocatalytic nucleation of cement-forming minerals at reaction fronts (Flicker and Ross, 1974; Guy, 1990); and
- kinetic feedback not involving diffusion (Ortoleva, 1990b).

Stylolites and Compaction/Cementation Banding

Banded pressure solution features and stylolites appear in a variety of sedimentary rocks. Examples include:

- stylolites in limestones and sandstones (Tada and Siever, 1989);
- small-scale lamination in chalk (Ekdale and Bromley, 1988);
- diagenetic bedding (Ricken, 1986; Bathurst, 1986);
- mica laminations in very-fine-grained sandstones from the Pennsylvanian Coffeyville Formation (Z. Al-Shaieb, oral communication, 1988);
- evenly spaced, banded calcite and silica cements in sandstones and shales (Levandowski et al., 1973); and
- layers of enhanced intergranular pressure solution in sandstones (Heald and Anderegg, 1960).

In a companion chapter (Dewers and Ortoleva, 1990) we demonstrate how these phenomena may arise through the coupling of stress and dissolution, through the stress dependence of rock and mineral solubility, and the texture dependence of mechanical properties coupled to solute transport and grain growth/dissolution kinetics. As a result of this mechano-chemical coupling, the state of uniform compaction is unstable under some conditions, resulting in nonuniform distributions of cementation and dissolution in the form of stylolite arrays or bands. Rocks containing these diagenetically differentiated structures can serve as seals or reservoir rock (Powley, 1980, 1985, 1990; Bradley, 1975; Al-Shaieb et al., 1989; Tigert and Al-Shaieb, 1990).

Reaction-Front Fingering and Scalloping

The dependence of the Darcy permeability on porosity and other textural variables can lead to a flow self-focusing mechanism that can destabilize a planar flow-driven reaction front and lead to the development of fingered or scalloped reaction-fronts (Chadam et al., 1986, 1988; Ortoleva et al., 1987b; Chen and Ortoleva, 1990a, b). The flow-focusing instability is illustrated in Figure 2. A number of examples of reaction-front scalloping are cited in Ortoleva et al. (1987b). In the present paper we illustrate several reaction-front morphologies in carbonate-cemented sandstones that may occur because of this instability. As these phenomena can involve narrow channels of enhanced porosity and permeability, they can in principle be important in petroleum migration and trapping and may also be important during well stimulation procedures.

Reaction-Mediated, Buoyancy-Driven Convection

The dependence of fluid mass density on composition and temperature can lead to buoyancy-driven flows in a sedimentary basin. Classic studies on this effect were those of Bories and Combarnous (1973), Combarnous and Bories (1978), and Wood and Hewett (1982), wherein the effect of the geothermal gradient coupled to the thermal expansion of water was shown to lead to a density inversion that could drive kilometer-scale circulation currents. A number of mechanisms for creating analogous flows due to diagenetic reactions have been proposed (Dewers and Ortoleva, 1988). The decreasing mass density of a methane-water solution with increasing methane content can drive appreciable flows from methane generated by kerogen breakdown at depth (see Figure 3). Other proposed mechanisms include pressure solution, decrease of carbonate mineral solubilities with temperature increase, and overlying of domains containing more soluble minerals over domains with less soluble minerals. The most dramatic flows of the latter types are likely those associated with salt domes (Ranganathan and Hanor, 1988). In the absence of appreciable flows driven by "external forces" such as compaction or recharge flows, the buoyancy-driven flows can play a major role in determining the distribution of minerals and petroleum in a basin.

Basin Compartmentalization

Most sedimentary basins have apparently become compartmentalized hydraulically (Powley, 1980, 1985, 1990; Bradley, 1975). The bounding seals of these kilometer-scale compartments are typically 10 to 100 m in thickness and are characterized by low permeability. The low permeability commonly is caused by cementation and enables the seals to sustain high-pressure gradients. Compartment interiors typically have augmented permeability. Many compartments serve as petroleum reservoirs.

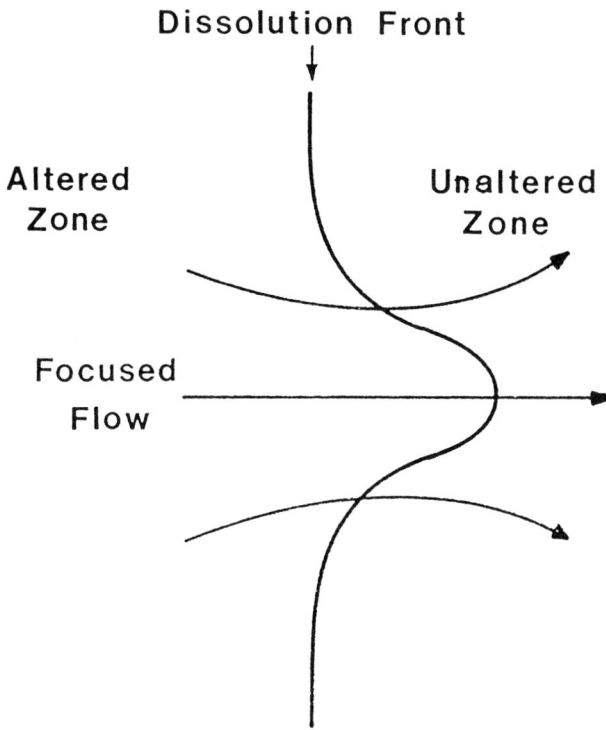

Figure 2. Flow self-focusing results when the altered zone of a reaction front has a higher porosity/permeability than that in the unaltered zone. This can lead to the instability of a planar reaction front to the formation of bumps or scallops.

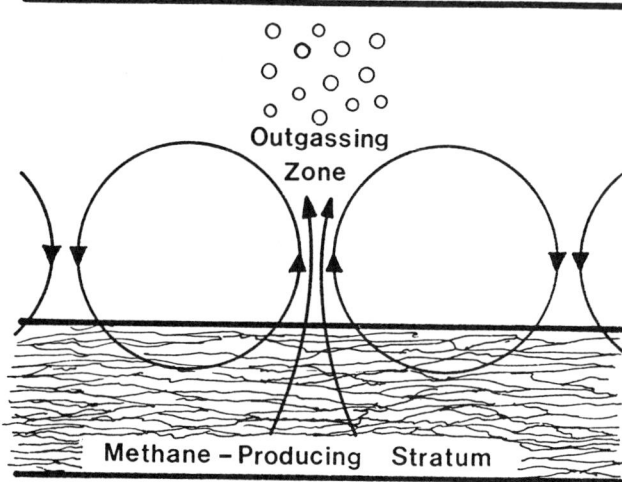

Figure 3. Production of methane can lead to buoyancy-driven flows.

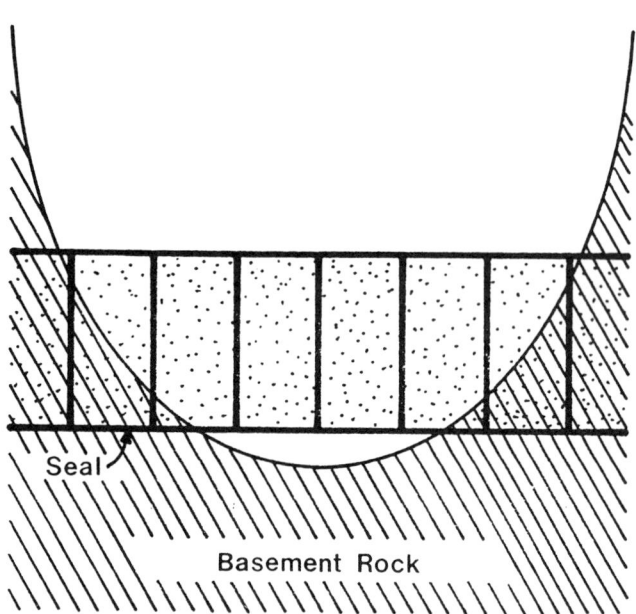

Figure 4. Schematic view of a sedimentary basin showing an array of sealed compartments. Evidence even exists for formation of compartments in the surrounding basement rock. Specific examples based on pressure-depth data have been given by Powley (1980).

The compartments commonly form arrays, dividing the basin into a boxwork of isolated kilometer-scale domains, as suggested in Figure 4; many examples have been cited in the work of Powley (1980, 1985, 1990; Bradley, 1975). It has been argued that in many cases compartment formation is caused by an internal diagenetic differentiation process and need not be the result of patterns of sedimentary origin or faulting (Dewers and Ortoleva, 1988).

Oscillatory Fluid Ejection from Source Rock or Overpressurized Compartments Through Fracturing/Healing Cycles

Compaction, thermal expansion, or release of components into the fluid phase can increase fluid pressure (Bradley, 1975; Palciauskas and Domenico, 1982; Domenico and Palciauskas, 1988). If pressure exceeds the fracture point, the fracture-induced flow pathways can release this pressure. When the pressure is reduced, the fractures may heal. If the pressurizing mechanisms continue to operate, the fracturing, release, and healing cycle can be repeated. In this way, for example, methane release from a low-permeability shale during kerogen breakdown or fluid release from a compartment through seal puncture can become oscillatory in time (Ghaith et al., 1990, in press).

The oscillatory release of fluids from sealed overpressurizing compartments was suggested by Bradley (1975). This mechanism involves coupling of the dynamics of fluid expulsion through the seal with seal fracturing and healing. As evidence of this concept, overpressures are commonly near the fracture pressure, and the top seals of overpressured compartments are commonly fractured (Powley, 1990).

Recent observations give circumstantial evidence for episodic fluid release from compartments. Baker et al. (1989) found evidence of two apparently independent fluid-ejection events in the Gulf of Mexico which, from their short duration and high implied flow rates, may have involved hydrofracturing to provide an efficient flow escape pathway. Also, estimated fluid temperature from the analysis of organic components (J. Whelan, oral communication, 1989) suggests that the temperatures within the seal from the Mill Creek graben, as identified by Tigert and Al-Shaieb (1990), was higher than that above or below this seal. This could be caused by the partial fracturing of a banded seal and consequent lateral spreading of fluids.

Formation of overpressure requires a seal (or low-permeability boundary) and a mechanism for increasing pore-fluid pressure. Seals may be related to sedimentary features or may be diagenetic in origin (Tigert and Al-Shaieb, 1990). A simple model of oscillatory fluid release is set forth and analyzed in the next to last section of this paper.

The above phenomena all apparently arise as repetitive variations of mineralization or petroleum content in space or time without the mediation of an externally imposed template. Here we attempt to make the point that such a self-organization dynamic may be operating on a range of length scales in a sedimentary basin and is responsible for a number of important diagenetic features.

As previously stated, a necessary condition for the self-organization of macroscopic patterns is that the system be maintained sufficiently far from equilibrium (Nicolis and Prigogine, 1977). It is also well understood that diagenesis of sediments occurs because of the existence of an overall state of fluid-rock disequilibrium. This state is sustained in a variety of ways, including the injection of fluids and heating and changes in stress with burial. In this way, a range of mechanisms exists for maintaining this equilibrium and therefore, on rather "general principles," we should expect a variety of self-organization phenomena in a basin.

Traditional experimental and theoretical geochemistry has focused on the identification and characterization of individual reaction, transport, or mechanical processes. However, the phenomena of interest here emerge from the strong coupling of processes. Understanding such phenomena in realistic systems then requires us to quantitatively integrate many effects. This is clearly not feasible without using quantitative models and numerical methods to simulate them. Thus, our group has emphasized the development of a new area—computational geochemistry—that will allow us to perform numerical "experiments" that would be unattainable with time and space scales available in the laboratory. Finally, the availability of efficient, robust numerical algorithms and high-speed, cost-effective computers are making this area of investigation feasible.

Let us now introduce the modeling approach we use and then discuss the properties and conditions favorable for some of the coupling and feedback they capture. Computer simulation techniques are then addressed. We emphasize our latest code, GENEX, which combines aspects of many of our earlier codes and introduces a number of new and powerful solution techniques. The remainder of the presentation focuses on three applications—natural convection driven by methanogenesis, fingering and scalloping of diagenetic reaction fronts, and oscillatory fluid release from overpressurizing compartments. In this way we hope to illustrate examples of coupling, feedback, and self-organization in systems of interest to the petroleum industry.

MODELING APPROACH

The great variety of processes operating in a sedimentary basin and the geometric and chemical complexity of the rocks make an attempt at realistic modeling quite formidable. Even if we have an accurate and complete model, we are faced with a very difficult problem in computer simulation. Thus it is essential to develop the minimal but complete model that captures the phenomena of interest, as well as codes that can efficiently simulate them on available hardware. In this section we review the progress we have made in this general area and then give actual examples in later sections.

Description

The rock is described in our approach via the variables fixing the state of the pore fluid and the solid matrix. If the pore fluid is a single phase, the composition, pressure, and temperature serve to characterize it. Because diagenetic changes are slow, gradients of these variables within a pore are usually negligible. The situation with multiple phases (water, oil, gas) is significantly more complex; each phase may constitute a spatially continuous medium or exist as micells or droplets. Clearly these cases require distinct treatments for reaction and transport processes. For the results presented here we restrict ourselves to the single pore filling phase. Thus the pores are described by

N = number of species in pore fluid;

$\underline{c} = \{c_1, c_2, ... c_N\}$ = set of N concentrations (moles/pore volume);

p = pore-fluid pressure; and

T = pore-fluid temperature.

The rock matrix is described by the distribution of grain sizes, orientations, mineral identity, and (for solid solutions) grain composition. Furthermore, a complete description must specify the geometry of grain contacts and (possible) coatings. Also, grain surfaces may have a degree of molecular adsorption coverage. Amorphous solid-phase (notably kerogen) and texturally complex biological material (shells, sponge spicules, coral, etc.) may be present. To simplify our modeling considerations, we have adopted the following description of the solid. For an assemblage of minerals we assume that grains are described by the following variables:

M = number of minerals;

i = mineral index ($i=1,2,...M$);

V_i = volume of a mineral i grain; and

n_i = number of grains of mineral i per rock volume.

Note that with these variables we can calculate the porosity ϕ as follows

$$\sum_{i=1}^{M} n_i V_i + \phi = 1 \tag{1}$$

Clearly this textural model is not sufficient for some applications. For mechano-chemical effects like pressure solution, one must have a more detailed model that allows for the distinction between grain-grain contact faces and "free faces" in contact with pore fluid. Results on such models are presented in a companion chapter (Dewers and Ortoleva, 1990a), where compaction and the genesis of stylolites and diagenetic bedding are modeled.

Other limitations arise when grain sizes or shapes occur throughout an appreciable range. Here, one needs to describe the probability of finding grains in each size or shape interval. Also, if solid solutions are present, we must allow for compositional zoning within a grain (Haase et al., 1980; Ortoleva, 1990a, b, and citations therein). For the present discussion we limit ourselves to the case of geometrically simple, compositionally homogeneous grains. Even within this simple model, we can describe significant redistribution of solids and porosity/permeability that takes place because of diagenesis.

These examples of mass redistribution raise the question of the "local" nature of the textural description of a rock. Of great interest is how the spatial distributions of the textural variables change with time. Our goal will be to determine $R_i(\vec{r},t)$, $i = 1, 2,...M$ where

$\vec{r}$ = vector specifying position in the system; and

t = time.

Processes, Coupling, and Feedback

The operating processes advance the state of the rock in time and also lead to coupling between the descriptive variables that allow for interesting and important phenomena. Let us now list some processes that we have considered and demonstrate qualitatively how they can lead to coupling and feedback.

Processes may be "local" or "nonlocal." Let the system be divided conceptually into an array of "macrovolume elements." Each such element is assumed to contain a statistically significant number of grains but is smaller than the scale of the phenomena of interest. For example, the latter could be the length scale over which the porosity varies appreciably. That is, whereas local processes operate within a volume element, nonlocal processes allow for the interaction between volume elements by transport of mass or energy. The separation of local and nonlocal processes is illustrated in the following list:

Local Processes
- fluid phase reactions
- grain growth/dissolution/nucleation
- kerogen breakdown
- pressure solution
- fracturing and healing

Nonlocal Processes
- diffusion and dispersion
- Darcy flow thermal conduction and convection force balance
- compaction

This list is clearly limited, but processes such as adsorption/desorption of molecules at grain surfaces, kinetics of solid-solution grain growth leading to intracrystalline zoning, and gouge or multiphase flow, can also be included.

Evolution Equations and Rate Laws

The starting point for most geochemical modeling studies attempting to predict the temporal course of diagenesis is a set of evolution equations (usually arising from conservation of mass, momentum, and energy) and a set of phenomenological laws expressing the rates of all processes in terms of the variables used to describe the system. The usual form of these equations is

$$\left\{\begin{array}{c}\text{Net rate of change} \\ \text{of a given} \\ \text{descriptive variable}\end{array}\right\} = \left\{\begin{array}{c}\text{Sum of rates of all} \\ \text{processes affecting} \\ \text{that variable}\end{array}\right\} \tag{2}$$

Perhaps the most central of such laws in geochemical modeling are the equations for mass conservation of pore-fluid species and for grain growth/dissolution/nucleation. Let us examine these equations in more detail to further illustrate aspects of the modeling approach and its limitations.

Conservation of mass for the pore-fluid species is taken in the form

$$\frac{\partial \phi c_\alpha}{\partial t} = \vec{\nabla} \cdot \vec{J}_\alpha + \sum_{\kappa=1}^{N_a} \upsilon_{\alpha\kappa} W_\kappa + \sum_{j=1}^{N_m} \mu_{\alpha j} U_j \quad (3)$$

where

- α = index labeling species in the pore fluid;
- $\vec{J}_\alpha$ = vector flux (moles/area-time) for pore-fluid species α;
- $-\vec{\nabla} \cdot \vec{J}$ = net rate of change of moles of α per rock volume due to all transport processes;
- k = index labeling each of the N_a pore-fluid reactions;
- $\upsilon_{\alpha k}$ = stoichiometric coefficient of α in the reaction k;
- W_k = rate of reaction k (mole equivalents/volume-time);
- j = index labeling one of the N_m reactions affecting at least one mineral (j = 1, 2,...N_m);
- $\mu_{\alpha j}$ = stoichiometric coefficient of α in the reaction j; and
- U_j = rate of production of mole equivalents due to mineral reaction j.

This set of N equations ($\alpha = 1, 2,...N$) must be supplemented by a set of equations specifying the functional form of the rates W_k and U_j and the fluxes $\vec{J}_\alpha$ for the processes one believes to operate.

Even if all the rate laws are known, equation 3 presents itself as a very difficult equation to solve on the computer for a number of distinct reasons:

- some reactions proceed on a very short time scale (say 10^{-6} sec) although we need to make simulations on the 10^{+6} sec time scale;
- the number of species N can be very large and hence equation 3 constitutes a very large set of coupled equations that must be solved simultaneously;
- furthermore, equation 3 must be solved simultaneously and self-consistently with energy balance and stress mechanical equations along with equation 4 (below) to determine the evolution of the solid matrix, creating an even larger coupled set of differential equations of various types;
- the density of solids is much greater than that of components in solution, so that the time scale for significant change in mineralization is typically much longer than that for the establishment of equilibrium with respect to the pore-fluid reactions;
- sharp (cm-scale or finer) reaction fronts may develop, although we may be interested in km-scale problems, and furthermore these fronts are constantly moving and being created or annihilated; and
- for practical research we typically need to perform many simulations to test system sensitivity to variations of parameters.

These technical issues are dealt with in Ortoleva et al. (1987a) for our one-dimensional program REACTRAN and in Chen et al. (1990) for a new, two-dimensional code GENEX.

Another family of equations arises from the consideration of grain growth/dissolution/nucleation and possible plastic or other deformation processes. Clearly the form of these equations depends strongly on the specific model of grain geometry and packing used. At the crudest level, mass balance for a given mineral yields one condition as in the following:

$$\frac{\partial}{\partial t}(n_i \rho_i V_i) = -\vec{\nabla} \cdot (n_i \rho_i V_i \vec{u}) + \sum_{j=1}^{N_m} \omega_{ij} U_j \quad (4)$$

where we define

- ω_{ij} = stoichiometric coefficient of i in the reaction j;
- $\vec{u}$ = velocity of rock deformation (due to compaction, shearing, or other deformation) and due to tectonic movement; and
- ρ_i = molar density of solid i.

The first term on the RHS (right-hand side) accounts for the possible motion of the rock past a point of observation due to rock deformation or subsidence and other tectonic motion. A shortcoming of the above model is that it does not provide enough information to calculate all the variables needed to describe the (changing) grain geometry and packing, and that the rates U_j will typically depend strongly on these descriptive variables.

The effective sphere model is the simplest textural description. One assumes that all grains are approximately spherical. Hence

$$V_i = \frac{4}{3} \pi R_i^3 \quad (5)$$

for the effective grain radius R_i. If the rock has high porosity, one might assume that most of the grain surface area ($\approx 4\pi R_i^2$) is in contact with the pore fluid. In this case we may take

$$U^j = 4\pi R_{i(j)}^2 n_{i(j)} \rho_{i(j)} F_j \quad (6)$$

where

- i(j) = the mineral label for the j-th mineral reaction (assumed here to affect mineral i(j) only);

F_j = radial growth rate measured for an isolated grain surrounded by fluid.

If we ignore nucleation, then

$$\frac{\partial n_i}{\partial t} = -\vec{\nabla} \cdot (n_i \vec{u}). \tag{7}$$

Combining equations 4,7 we then obtain, assuming ρ_i is constant,

$$\frac{\partial R_i}{\partial t} = -\vec{u} \cdot \vec{\nabla} R_i + \sum_{j=1}^{N_m} \omega_{ij} F_j \tag{8}$$

If we neglect stress corrections, then F_j depends only on pore-fluid composition ($\underline{c}$), temperature, and pressure. Studies accounting for pressure solution and other mechano-chemical effects, as well as nonspherical and changing grain geometry, are discussed further in Dewers and Ortoleva (1990a).

That there is a sum of terms in equation 8 can be quite essential, because more than one dissolution/precipitation reaction can take place for each mineral. For example, the mechanism of calcite dissolution changes dramatically with pH (Sjoberg and Rickard, 1984; Busenberg and Plummer, 1986). Therefore, by specifying a number of reactions with different proton stoichiometry one may capture such pH-dependent reaction rates and other analogous multimechanism effects.

The spherical approximation cannot describe a number of effects such as the following:

- pressure solution at contacts;
- grain coatings (as clay on quartz);
- clays more commonly occur as platelets, needles, or other high-surface-area morphologies; and
- preferred dissolution of, say, quartz at quartz-mica contacts.

All these effects require more complex grain geometry models with an attendant increase in the number of textural variables needed to characterize the state of each mineral grain.

To complete the description of the state of the pore fluid, we must calculate the velocity of the various fluid phases that fill the pores. For single-phase systems the most commonly used model is Darcy's law:

$$\vec{v} = -\kappa \{\vec{\nabla} p + \Delta g \vec{z}\} \tag{9}$$

where

κ = (permeability/fluid viscosity);

$\vec{\nabla} p$ = pressure gradient;

Δ = fluid mass per volume;

g = gravitational acceleration;

$\vec{z}$ = upward-pointing unit vector. If the fluid has one dominant component (dominantly H_2O), the major mode of transport for that component will usually be by flow. In that case we have

$$\vec{J}_s \approx \phi c_s \vec{v} \tag{10}$$

where c_s is the concentration of the solvent. Then equation 3 for the solvent becomes

$$\frac{\partial \phi c_s}{\partial t} = -\vec{\nabla} \cdot (\phi c_s \vec{v}) + \phi \sum_{\kappa=1}^{N_a} \upsilon_{s\kappa} W_\kappa + \sum_{j=1}^{N_m} \mu_{sj} U_j. \tag{11}$$

This still leaves open the question of the pressure, p. The additional condition needed to determine p is usually obtained from the thermodynamic equation of state (p as a function of $\underline{c}$ and T); one may rewrite the equation of state in the form, letting $\underline{c}$ represent the set of solute concentrations only,

$$c_s = c_s(\underline{c}, p, T). \tag{12}$$

This is then inserted into equation 11 to obtain an equation for p.

The resulting pressure equation is quite instructive. Assume that $c_s(p,T,\underline{c})$ can be linearized about a reference pressure p, temperature $\overline{T}$, and composition $\overline{\underline{c}}$. Then we have

$$c_s \approx c_s(\overline{p},\overline{T},\overline{\underline{c}})$$

$$\left[1 + \beta(p-\overline{p}) - \beta\gamma(T-\overline{T}) + \beta \sum_{\alpha=1}^{N} \Gamma_\alpha (c_\alpha - \overline{c}_\alpha) \right] \tag{13}$$

for compressibility β and other constants γ and Γ_α that can be determined from partial molar volume and compressibility properties. With this, equation 11 becomes

$$\frac{\partial p}{\partial p} = \Gamma + \frac{1}{\phi \beta \overline{c}_s} \left[-\vec{\nabla} \cdot (c_s \vec{v}) + \phi \sum_{\kappa=1}^{N_a} \upsilon_{s\kappa} W_\kappa + \sum_{j=1}^{N_m} \mu_{sj} U_j \right], \tag{14}$$

$$\Gamma = -\frac{c_s}{\overline{c}_s \beta} \frac{\partial \ln \phi}{\partial t} + \gamma \frac{\partial T}{\partial t} - \sum_{\alpha=1}^{N} \Gamma_\alpha \frac{\partial c_\alpha}{\partial t} \tag{15}$$

From this we can identify the following factors contributing to fluid pressures:

- compaction or other porosity-altering process (the first contribution to Γ);
- thermal expansion of the fluid (the γ term in Γ);

- changes in solute concentrations due to reaction or transport processes affecting these solutes (the ∂c/∂t terms in Γ);
- solvent transport (the first term on the RHS of equation 11); and
- reactions affecting the solvent (the last two terms on the RHS of equation 11).

All these factors can affect the temporal evolution of the spatial distribution of basin pressures. This has been discussed in the context of abnormal pressures (Palciauskas and Domenico, 1982; Domenico and Palciauskas, 1988).

The above fluid dynamics can couple strongly to the chemical dynamics through the dependence of k on p,T,c (through the viscosity) and on mineral growth/dissolution (through the dependence of permeability on the state of the rock matrix; that is, grain size, orientation, fractures, etc.).

The temperature can couple strongly to other variables in a number of ways, and hence a full integration of the evolution of its spatial distribution into a geochemical model is often crucial. This coupling can lead to feedbacks as follows:

- fluid viscosity can depend on temperature, whereas the rate and distribution of flow can change temperature;
- the equilibrium and rate constants for mineral reactions can depend on temperature which, in turn, can affect temperature by changing fluid flow through permeability variation; and
- changes in temperature can affect the rates of reactions, changing fluid density and therefore fluid flow through buoyancy and thereby the distribution of temperature.

To evaluate mechano-chemical effects, we must solve force-balance equations on both the macroscopic (lithologic, stratigraphic, and basin) scale to account for tectonic forces and overburden, and the microscopic (grain-size) scale to describe pressure solution effects. Dewers and Ortoleva (1990a) demonstrate the development of such integrated "two-level" mechanical models and the implications of these effects. Some discussion of the effects of natural fracturing is given in the next to last section below.

Computer Simulations

To extract an understanding of the genesis, migration, and trapping of petroleum and diagenetic phenomena that determine reservoir quality arising from the reaction-transport mechanisms outlined above, we must solve the equations numerically. Computationally, this process involves solving a large set of equations recursively using numerical techniques, such as finite elements or finite-difference methods. This is now quite feasible for even very complex models because of:

- the rapid improvement in the computational capacity and affordability of computer hardware and
- improvements in techniques for solving the equations. Both points bear more detailed examination.

Computer Hardware

The solution techniques used to solve the above models involve numerical methods, through which the partial differential equations are reduced to a large set of algebraic equations. In turn, these equations are solved by iteration schemes. This is a standard technique used in other fields such as climatic, aerodynamic, and fluid-dynamic modeling.

New types of computers are not only capable of faster processing but also can perform vector and parallel processing. By taking advantage of these features, very complex modeling can be carried out using cost-effective mini-super computers such as STARDENT Titan or the IBM3090.

Parallel processing allows one to simultaneously compute independent blocks of a program, whereas vector processing allows one to simultaneously carry out many similar calculations. Through the use of these techniques, we are now able to model very complex chemical systems, where previously this was not feasible.

To interpret the results of these simulations, we must organize the data by using graphical techniques. Most ideally we require not only that the results be displayed after the simulation is completed, but also that the graphics be integrated into the program in order to monitor the progress of the simulations. Integrated graphics provide for early termination of uninteresting simulations, and also can increase the efficiency of studies on the sensitivity of the system to variations in parameters.

The graphical analysis of geochemical problems in one and two dimensions is fairly standard. In one spatial dimension, one simply requires cartesian plots of various descriptive variables as functions of space and time. In two dimensions, we may use contour or surface maps and, for fluid flows, vector or stream line diagrams. The problem is more challenging in three dimensions. There we may display serial sections, stream lines, or contour surfaces. To capture a sense of the temporal evolution of the system, movies can be made during the execution of a code and later viewed. Because a complex system will require hours and days of processing, the videotaping of results can be a very useful option. Our geochemical modeling group develops simulators in which these computational and graphical techniques are integrated.

Available Reaction-Transport Codes

A number of reaction-transport simulators have been developed by our group to analyze reaction, transport, and mechanical aspects of diagenesis or other fluid rock interaction systems. A code GENEX (Geological Engineering and Exploration) has been developed and is available through Geo-Chemical Research Associates, Inc. (1990). This code can model two-dimensional flow/reaction systems using most of the modeling features described above. Intended for general reaction/transport modeling, the program is a physico-chemical exploration tool, wherein the specific chemical and geometric model may be specified by the user. To enhance its flexibility, we incorporated user-friendly approaches to data entry and graphic display. The program is highly modular, enabling easy upgrading or customization.

GENEX uses a self-adapting spatial grid, allows for multiple reactions for each mineral, accounts for activity coefficients, and calculates the effects of changing temperature on equilibrium and rate constants. We are currently installing modules that calculate the effects of pressure solution, fracturing (Dewers and Ortoleva, 1990a), and multiple pore-fluid phases and organic matter decomposition.

METHANOGENIC BUOYANCY-DRIVEN FLOWS

We now consider a simple model of methanogenic convection, as suggested in Figure 3. Such buoyancy-driven flows illustrate a number of important features of basin self-organization. In particular, we demonstrate the following:

- symmetry-breaking instability;
- development of coherent patterns from noisy nonuniformity;
- oscillatory convection;
- differentiating of diagenetic patterns in initially nonuniform media; and
- pattern nucleation and propagation.

We make these points by using a simple reaction-transport model.

We limit our treatment to systems for which methane only exists in the aqueous phase. The factor allowing for flow is the dependence of the mass density of the aqueous fluid on methane concentration. Below we give some details on the construction of the mathematical model used (Park et al., 1990).

Reaction-Transport Model

Our treatment is based on conservation of mass equations for methane and water and an equation for the decomposition of kerogen. Kerogen particles of radius R are allowed to decompose into methane and inert residue by the reaction

$$\text{kerogen} \rightarrow \text{methane} + \text{inert residue}. \qquad (16)$$

The decomposition of kerogen is assumed to proceed at rate G (cm/sec):

$$\frac{\partial R}{\partial t} = -G \qquad (17)$$

Conservation of mass (as expressed in equation 3) yields for the concentration m of methane

$$\frac{\partial \phi m}{\partial t} = -\vec{\nabla} \cdot \{\phi D \vec{\nabla} m - \phi \vec{v} m\} + 4\pi \rho n R^2 G \qquad (18)$$

where D is the diffusion coefficient, n is the number of kerogen particles per rock volume, and ρ is the molar density of kerogen. The m dependence of the fluid mass density a(m) provides the essential coupling needed to generate patterns of methanogenesis-driven convection. It is also assumed that methane exists in aqueous phase only.

Conservation of mass for water yields an equation for p. To simplify the model, we neglect changes in porosity and permeability with R and furthermore neglect the dependence of the molar density c_s of water on m and p (through the equation of state $c_s = c_s(m,p)$). With this we obtain

$$\vec{\nabla} \cdot (\phi \vec{v}) = 0. \qquad (19)$$

$$\vec{v} = -\kappa \left[\vec{\nabla} p + \Delta g \vec{z} \right]$$

Where κ is the ratio of permeability to fluid viscosity, g is the gravitational constant, and $\vec{z}$ is an upward-pointing unit vector; the mass density $\Delta(m)$ was taken to be in the form $\Delta = \Delta_0(1 + \alpha m)$ for constants Δ_0 and α.

Thus the model consists of solving equations 17, 18, and 19 subject to boundary and initial data. The burial history was reflected in the rate of reaction (equation 16) through an Arrhenius law

$$G = \kappa_0 e^{-T^*/T} \qquad (20)$$

where the distribution of the absolute temperature T in time and space was determined by the geothermal gradient and subsidence rate of the sediment (Park et al., 1990). In equation 20, κ_0 is the pre-exponential factor and T^* is the activation energy divided by the gas constant.

Boundary and Initial Data

To study methanogenic convection, we consider a rectangular domain taken through a specified burial history. The system was assumed to be closed to mass transport. The system was initialized with a distribution of kerogen particle radius ($R(\vec{r},0)$) that differed for each simulation. The initial methane was assumed to be $m(\vec{r},0) = 0$; that is, no methane was present before reaction (equation 16) was initiated. The values of the transport, chemical, and other parameters used are summarized in Table 1.

Table 1. Parameters used to generate simulations of methane buoyancy driven flows

Figures 5, 7, 8, 9

Porosity	25%
Permeability	0.1 Darcy
Fluid viscosity	0.001 poise
Diffusion coefficient of methane	$1.0 \times (10^{-6})$ cm²/sec
Geothermal gradient	30°C/km
Rate of subsidence	30 m/my
Surface temperature	26.85°C
Fluid mass density	1.0 g/cm³
Pre-exponential factor, $\kappa_0{}^1$	$1.0 \times (10^{16})$ cm/sec
Activation energy	70.0 kcal/mol
Kerogen particle number density, n	150 grains/cm³
Organic matter molar density, ρ	0.01 moles/cm³
ω	0.24 g/mol

Figures 9, 10

Sandstone porosity	20%
permeability	0.1 Darcy
Shale porosity	30%
permeability	0.0001 Darcy
Pre-exponential factor, $\kappa_0{}^1$	$1.0 \times (10^{17})$ cm/sec
Activation energy of diffusion[2]	15.0 kcal/mol
Pre-exponential factor of diffusion[3]	
sandstone	100.0 cm²/sec
shale	10.0 cm²/sec

[1] After Ungerer et al. (1987)
[2] After Krooss and Leythaeuser (1988).
[3] After Park et al. (1990).

Simulating Cellular and Oscillatory Convection

The above reaction-transport model was simulated using a finite-difference method as implemented in the ELLPACK software package (Rice and Boisvert, 1984). Simulations were carried out to demonstrate various effects as follows.

The Basic Convection Cell

A flow pattern evolving from a uniform initial kerogen distribution with a small excess of kerogen in the lower left corner illustrates half of a convection cell (the full cell is a mirror image of the pattern shown in Figure 5 with respect to the left vertical boundary). During burial the kerogen decomposes progressively bottom-upward as T increases with time because of subsidence. The low values for permeability adopted in our study (0.1 Darcy) represent typical deep sandstone permeabilities, and hence velocities are small (less than 1 cm/yr).

Oscillatory Convection

The methanogenic convection cells are subject to an oscillatory instability. In Figure 6 we see the temporal course of the fluid velocity in the updraft at point A of Figure 5. A tendency for oscillation can be argued as follows. At first there is no methane and hence no flow. With subsidence and methane

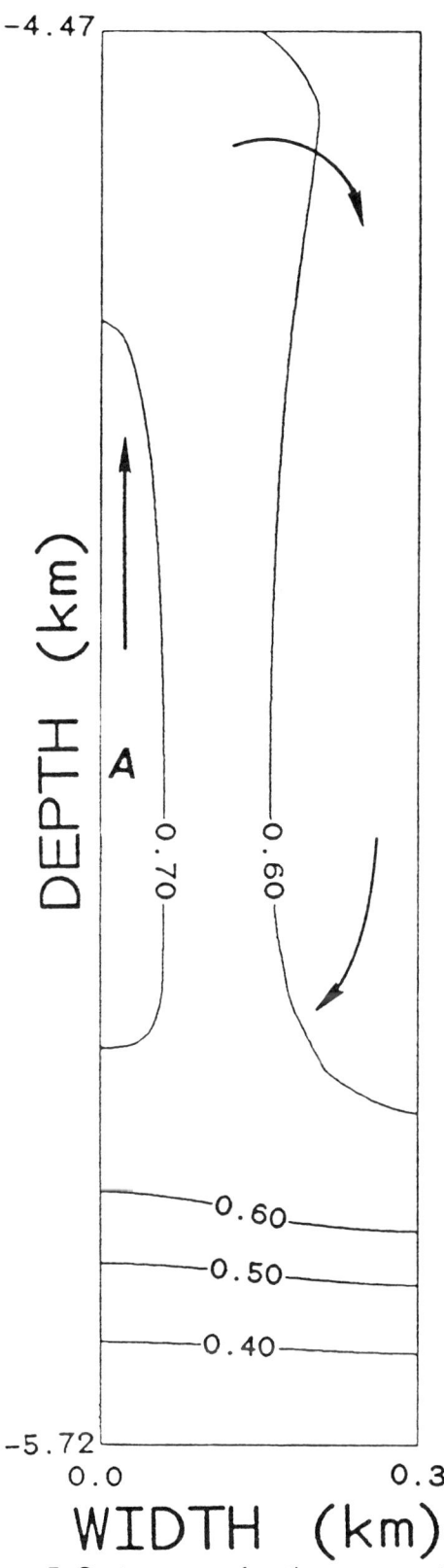

Figure 5. Contour map of methane concentration and flow directions in a convection half-cell. Methane is being generated in the lower regions and carried up by the updraft flow. Arrows indicating flow (here and in Figures 7, 8, 9, 10) are hand-drawn from vector maps of flow velocities.

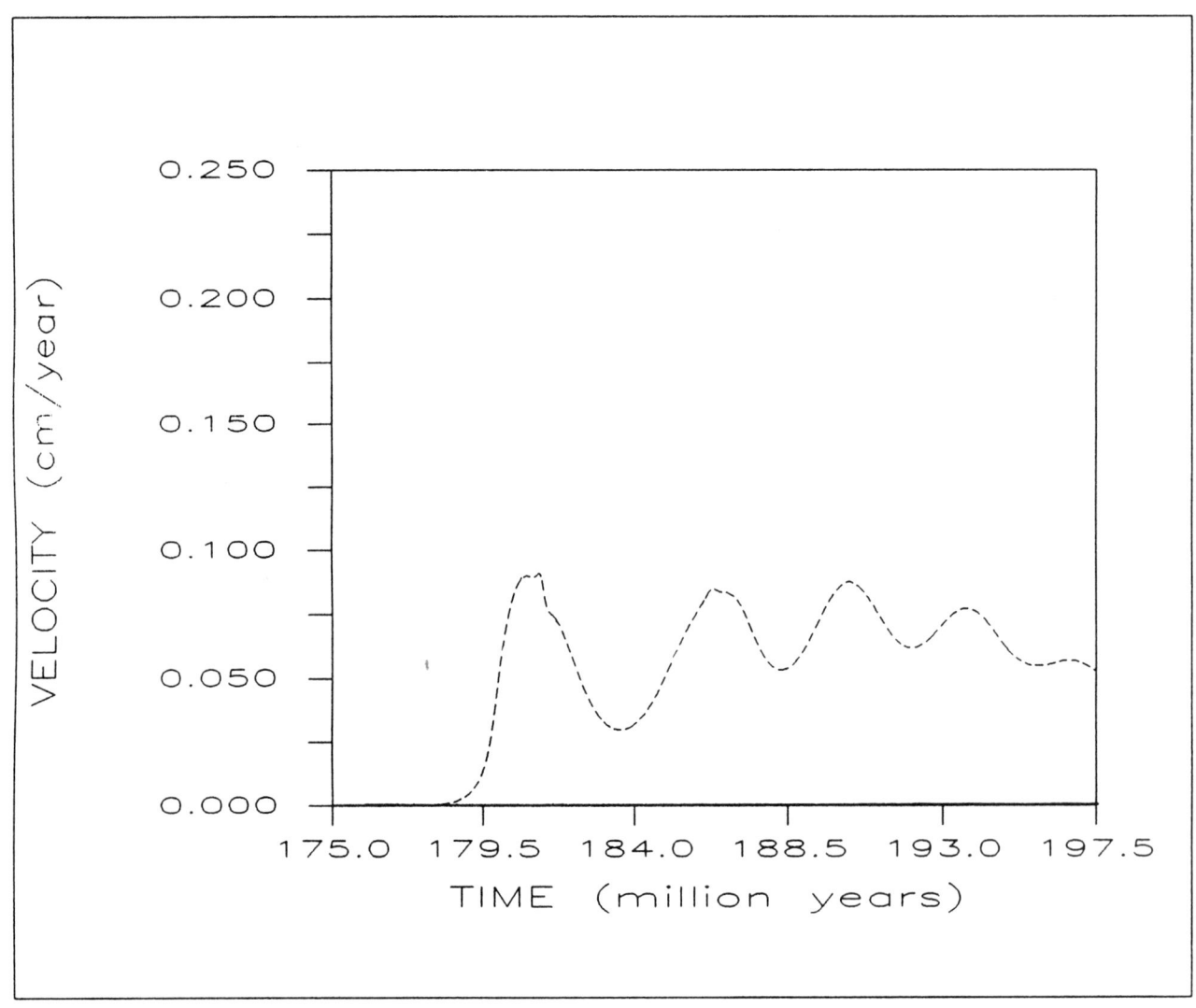

Figure 6. Updraft velocity of flow at point "A" on Figure 5. Oscillation is due to periodic pulsing updrafts of methane-rich fluid. Decreasing overall velocity with time is due to depletion of kerogen.

enrichment of the aqueous phase near the bottom of the domain, a critical point of methane saturation is achieved, thus initiating an upward flow. After one overturning, the methane-rich fluid is swept from the methane-producing region. Some time must then lapse for the decomposing kerogen to recharge the fluid at depth with methane and initiate the next overturning. This cycling is illustrated in Figure 6, wherein the updraft velocity is shown as a function of time. Oscillation continues until all kerogen is decomposed.

From Random Kerogen Distributions to Orderly Cellular Convection

A well-ordered array of convection cells can emerge from an initially inhomogeneous distribution of kerogen. This demonstrates the robustness of the self-organizing tendency of a convective system even in the presence of imposed nonuniformity. We tested this by assigning a random initial kerogen particle radius distribution in space, whereby we obtained spatially varying rates of methane production. Figure 7a-c shows a typical simulation result. Figure 7a shows small, disordered arrays of convection cells, which develop early and which reflect the local variations of G imposed by the nonuniform initial kerogen distribution. In Figure 7b and c we see the development of larger scale coherent flows and the repression of early-stage, smaller scale patterns. Once this ordered pattern is achieved, it persists until the source of methane becomes depleted. Depending on the initial random kerogen distribution pattern, as well as the geothermal gradient, subsidence rate, activation energy of kerogen decomposition, and other physico-chemical parameters, different final convection patterns can be achieved. However, the system tends to select only a few of the infinite variety of flow patterns the initial kerogen distribution can impose on the system. Given the parameters used

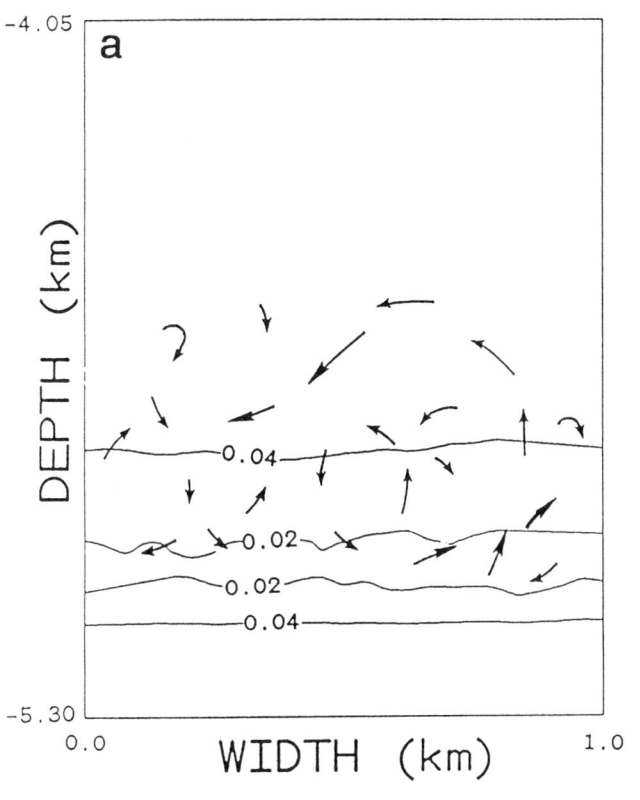
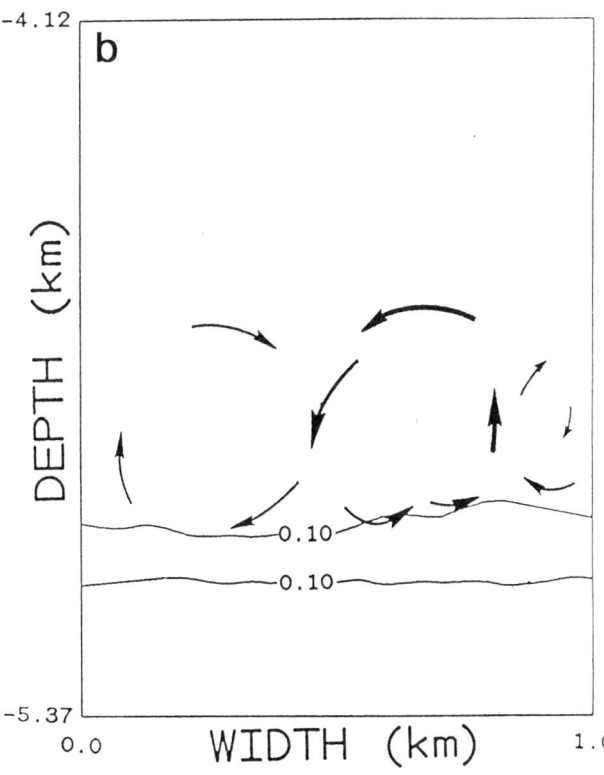
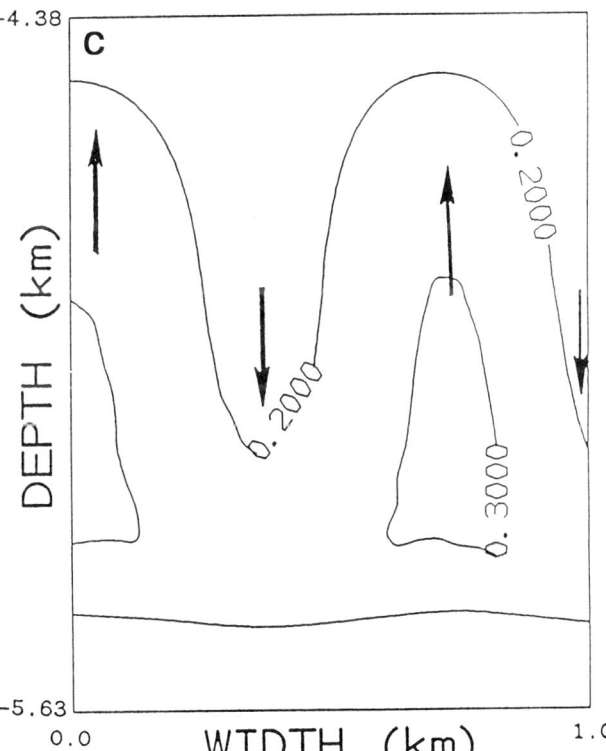

Figure 7. Evolution of patterned array of convection cells from a noisy initial kerogen distribution. (a) Early phase, where small convection cells arise from local inhomogeneities in methane production rate; (b) coalescence of small cells to form a large-scale coherent flow pattern; and (c) late-stage convection cell pattern.

here (Table 1), we usually found three or four half-cells eventually forming in the simulation domains.

Induction and Propagation of Cellular Convection

A front of convection cell induction can propagate in space. In the case shown in Figure 8, a small excess of kerogen is placed at the lower left of an otherwise uniform, elongate system. The initial kerogen nonuniformity starts an associated convection cell at the left. Rather than causing an overall elongate convection pattern, the system chooses to evolve into an array of shorter scale convection cells. This array is generated sequentially with the induction of an increasing number of up and down drafts from left to right. The induction front will not advance indefinitely in a real system, because omnipresent kerogen distribution nonuniformity can start local cells ahead of the propagating disturbance. Given sufficient time, the state of the system evolves into an ordered cellular convective regime.

Convection in the Presence of Hydraulic Heterogeneity

Bedding, lenses, and other depositional features lead to nonuniformity in porosity and permeability. As a result, the idealized convection cell structure as in Figure 5 can be strongly distorted. This effect has been investigated through a simulation of kerogen-rich, low-permeability shale beds and lenses placed within a more permeable sandstone. To account for the increase in diffusion coefficient with temperature, we used an Arrhenius form for the

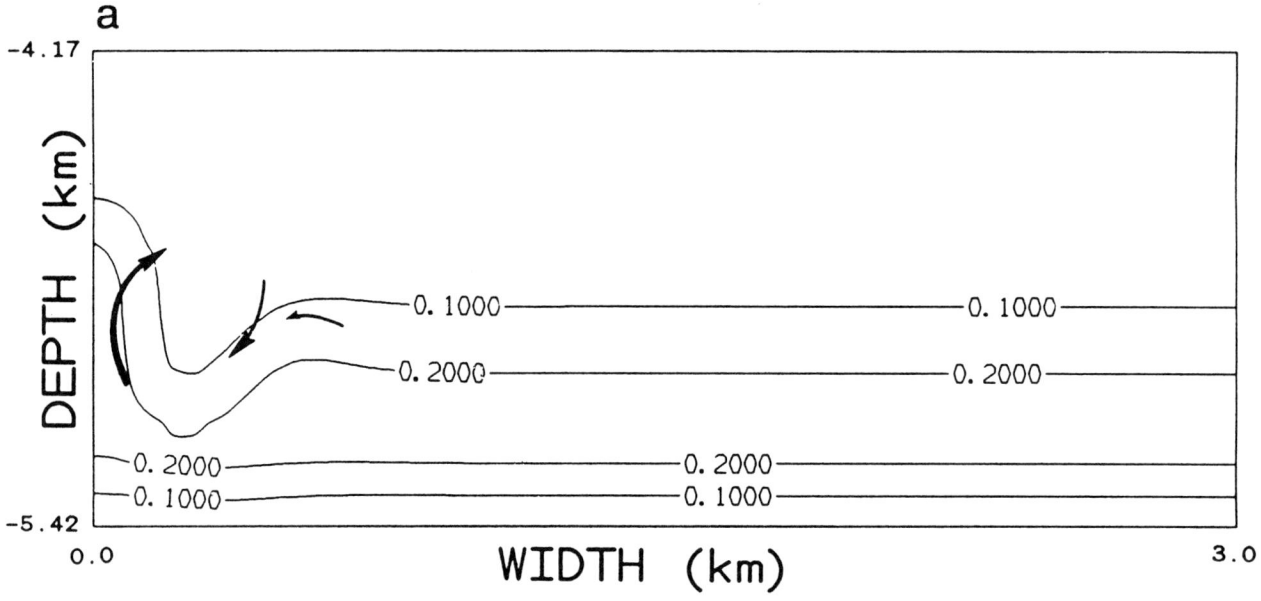

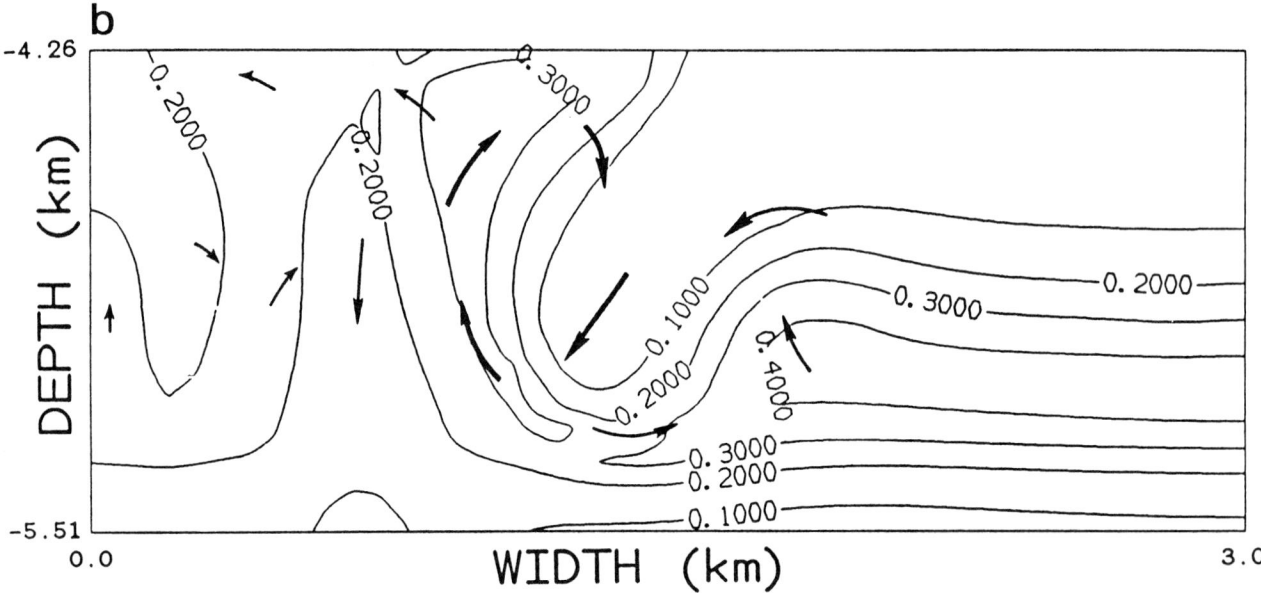

Figure 8. (a-c) Propagating wave of convection cell induction emanating from a local small initial excess kerogen content in the lower left corner at three consecutive times.

diffusion coefficient similar to that of equation 20. As the system subsides, kerogen decomposition is successively activated in the kerogen-rich strata at increasingly higher elevations. The resulting flow is highly irregular, involving a number of reversals as methanogenesis in higher lying beds is initiated while kerogen in lower beds becomes depleted. As methane winds its way upward by following more permeable pathways, new circulations are initiated, old ones can be reversed, and overall a very complex fluid migration pattern evolves (Figure 9).

A result of a simulation in another initially inhomogeneous system is illustrated in Figure 10. Here, two layers of kerogen-rich shaly strata were separated by more permeable sandstone, and a repetitious antiform structure was added. Kerogen decomposition causes a series of plumes to rise. However, the locations of rising plumes do not necessarily coincide with the antiform structure. That is, not all stratigraphic traps associated with the antiform become filled with methane-rich fluid. An implication of this result is that not necessarily

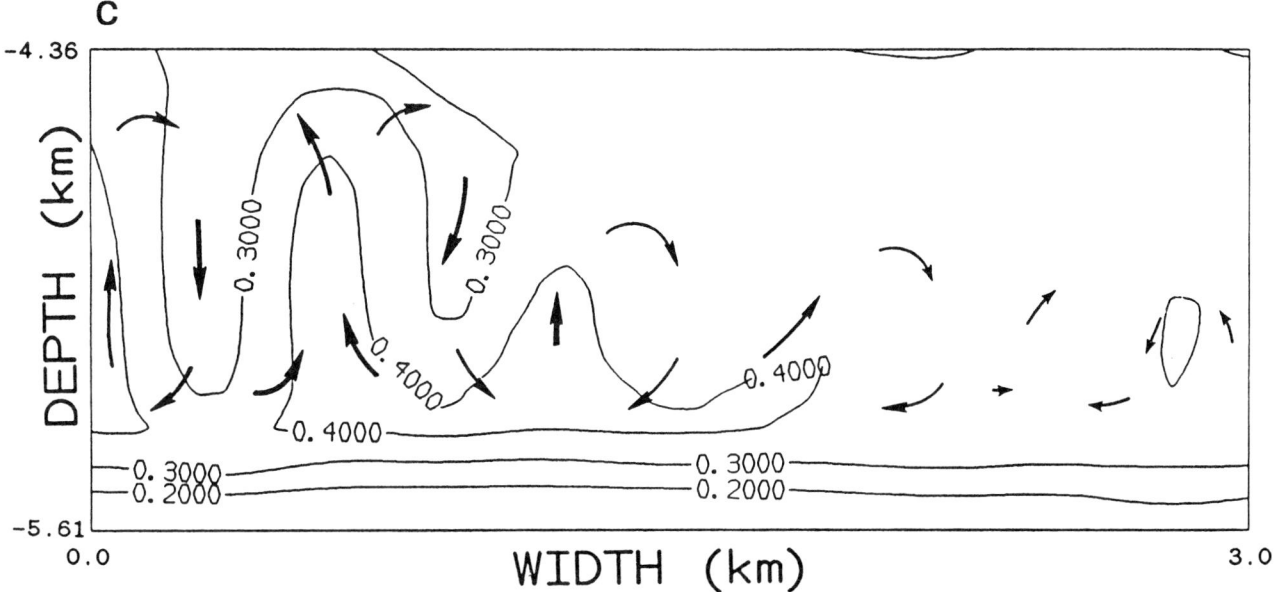

Figure 8. (Continued)

all stratigraphic traps may produce petroleum, and that the spatial distribution of trapped petroleum may strongly depend on both the lithologic and fluid flow patterns of the basin.

REACTION-FRONT MORPHOLOGY: FINGERED CALCITE DISSOLUTION AND DOLOMITIZATION FRONTS

Model

We now consider alteration fronts in a calcite-cemented sandstone with resulting dolomitization and/or calcite dissolution. We assume that the carbonates are present as cements in an inert quartz matrix. The aqueous reactions considered are

$$H^+ + CO_3^{2-} \rightleftarrows HCO_3^- \quad (21)$$

$$H^+ + HCO_3^- \rightleftarrows H_2CO_3 \quad (22)$$

$$H_2O \rightleftarrows H^+ + OH^- \quad (23)$$

$$CaCO_3 \rightleftarrows Ca^{2+} + CO_3^{2-} \quad (24)$$

and the mineral dissolution reactions are

$$\text{calcite} \rightleftarrows Ca^{2+} + CO_3^{2-} \quad (25)$$

$$\text{dolomite} \rightleftarrows Ca^{2+} + Mg^{2+} + 2CO_3^{2-} \quad (26)$$

We have omitted complexes such as CaCl (for saline fluids), Ca(OH)$_2$ (for high pH), or organic complexes with Ca^{2+} and Mg^{2+}.

The central feature of reactive porous media that allows for the destabilization of a planar reaction front (suggested in Figure 2) through flow self-focusing is the dependence of the Darcy permeability on porosity and other textural variables. For example, the Fair-Hatch or Kozeny-Carman relation for the ratio of the Darcy permeability to fluid viscosity, κ, may be written

$$\kappa = \kappa_0 \phi^3 / F^2 \quad (27)$$

where $\kappa_0 = (J\mu_w\Theta^2)^{-1}$, J is a packing factor ($\simeq 5$), Θ is a geometric factor (7.7 for angular grains and 6 for spherical grains), μ_w ($\sim .001 - .01$ Poise) is the fluid viscosity when water is the solvent, and ϕ is porosity. The factor F is defined by:

$$F = \sum_{i=1}^{3} \phi_i / L_i. \quad (28)$$

Here ϕ_i ($=4/3\pi R_i^3 n_i$) is the volume fraction of the rock occupied by mineral i; and L_i^3 ($\equiv 4/3\pi R_i^3$) is the grain volume. Because ϕ and F depend on grain size, the permeability is affected by mineral growth/dissolution reactions.

The reaction-transport equations for the above problem were solved in a rectangular domain with lossless top and bottom boundaries, "inlet" end at the left and "outlet" at the right. Solute concentrations and pressure were fixed at the inlet end. The system was assumed to be isothermal. Results obtained were as follows (Chen and Ortoleva, 1990a, b).

A small local region of calcite cement depletion at the left (inlet) end in Figure 11a evolves into a fingered calcite dissolution front that advances steadily and stably into an initially uniformly cemented bed. If

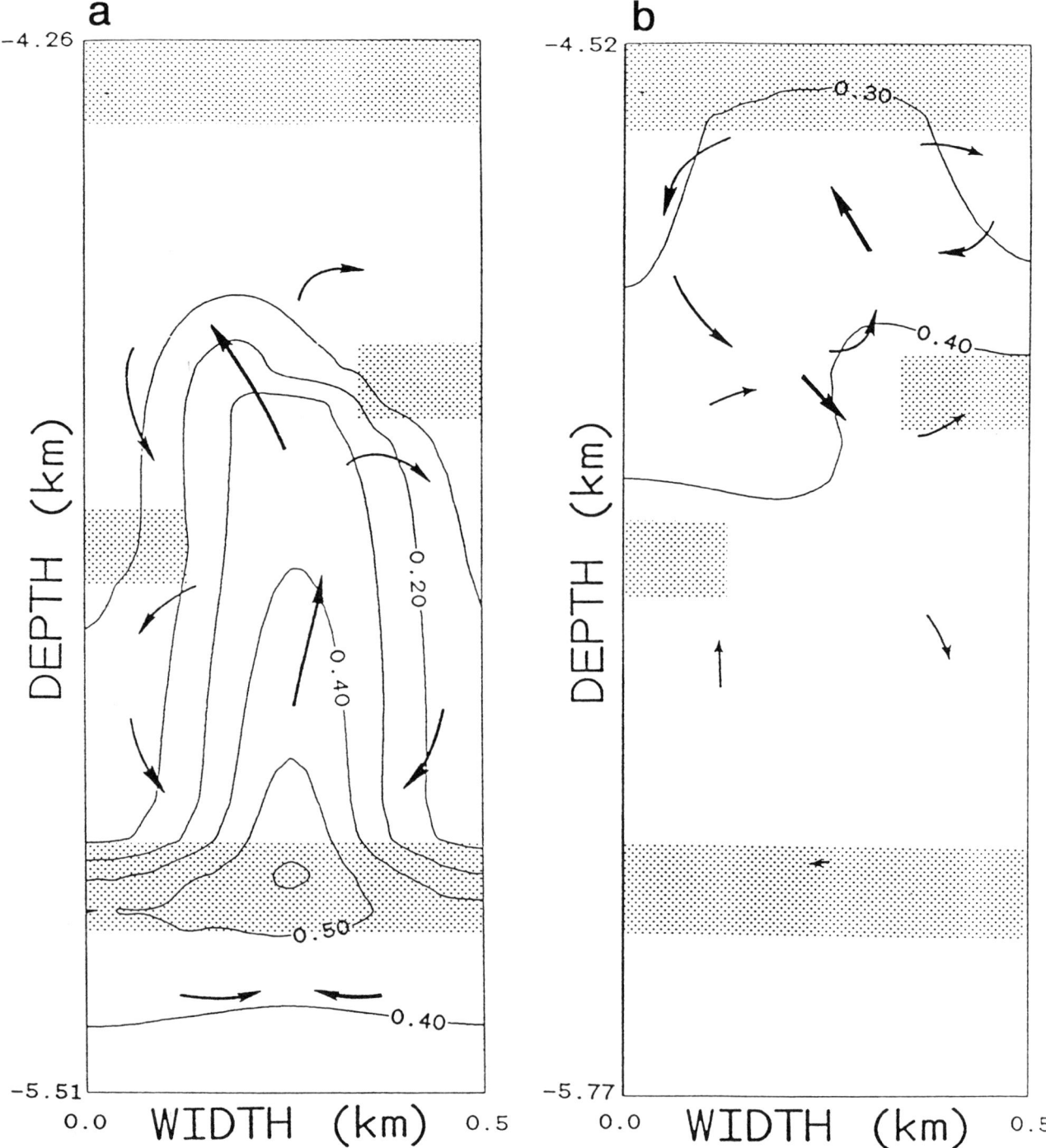

Figure 9. In a lithologically complex domain the temporal and spatial pattern of methane buoyancy-driven flows can be highly complex. In this simulation, the initial kerogen was concentrated in the low-permeability shales (shaded regions). Note distortion of the cell flow pattern, as well as changes in direction that take place as kerogen increasingly higher up in the domain decomposes during subsidence.

the imposed fluid velocity does not exceed a critical value, the disturbance is removed and the front becomes planar, as in Figure 11b. As the inlet flow velocity, or domain width, or inlet fluid undersaturation are increased the morphology established by the Darcy flow-reaction coupling can take on exotic geometries (Figures 12, 13). These patterns are inherent in the system dynamics and do not reflect initial nonuniformity.

The inherent nonplanar reaction front morphologies that are allowed in the initially uniform system can be highly distorted in the presence of pre-existing nonuniformity. In Figure 14 we see that the initial nonuniformity (frame a) causes the dissolution finger

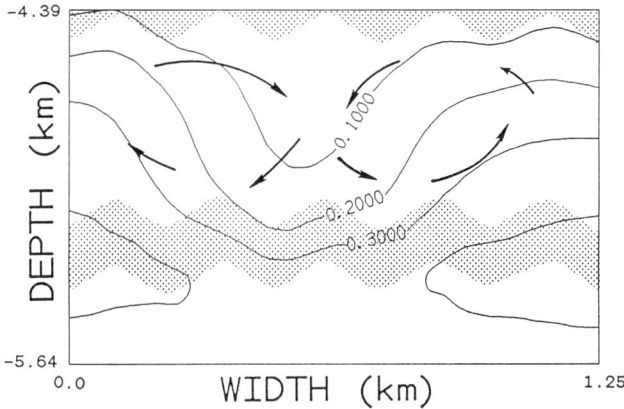

Figure 10. Typically, the pattern of structural trap distribution arising, for example, from a periodic array of antiforms will not match the natural pattern of buoyancy-driven flows. As a result, not all stratigraphic traps become charged with petroleum. This is illustrated here in terms of a periodic array of antiforms that do not conform to the methanogenic buoyancy-driven flow pattern.

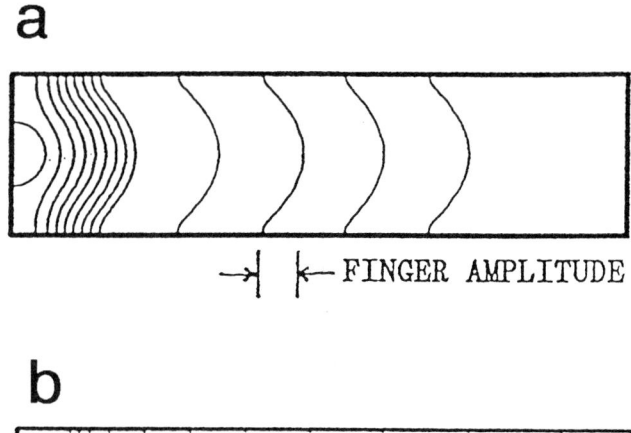
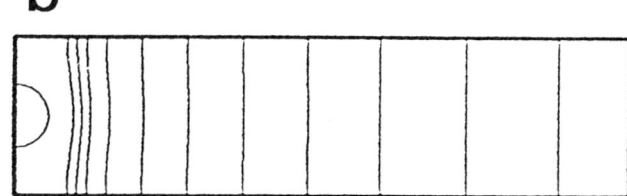

Figure 11. (a) Temporal development of a steady finger-shaped dissolution front for a system of 7.5 cm × 2 cm. The time interval between contours is 10^8 sec for the first 9 contours from the left and 8×10^8 sec for the last 5 contours. The imposed flow along the horizontal direction is 1.4×10^{-3} cm/sec. (b) Similar to (a), except that the imposed flow velocity is 0.5×10^{-3} cm/sec, in which case no dissolution finger develops. The time interval is 10^9 sec.

to steer around regions of low porosity and to be attracted by those of initially higher porosity. The higher inlet velocity case of frame c differs from the lower velocity of frame b in that the finger tends to split around obstacles in the former case. This splitting is a direct consequence of the tendency for a single finger to become unstable to the formation of branches as the imposed velocity increases.

Reaction Front Fingering in a Calcite/Dolomite System

A key element of the flow self-focusing leading to the above phenomena is that permeability in the altered zone exceeds that ahead of the reaction zone. A classic case of such a permeability and porosity increase is the dolomitization of a limestone by Mg^{2+}-rich, Ca^{2+}-poor fluids. Such appears to be the case in the Arabian Shelf (Broomhall and Allan, 1987). Figure 15 shows a simulation of a fingering phenomenon in a calcite/dolomite-cemented sandstone.

Each frame in Figure 15 is a set of three contour maps, showing the spatial distribution at a given time of (from top to bottom) the volume percent of the rock occupied by dolomite, calcite, and pore space, respectively. The calcite dissolution is observed to be uneven in space because of the initial high-permeability bump which focuses the inlet fluid through it, leading to a fingered calcite dissolution front that advances downstream in a way similar to that shown in Figures 2 and 11. At time 4.5×10^8 sec, only about 1% of calcite remains at the outlet end of the system (frame d). Dolomitization is also uneven, although the initial dolomite volume fraction was uniform. Frame b shows a noticeable dolomite precipitation at 5×10^7 sec. The faster dissolution of calcite at the central region of the domain causes a faster dolomite precipitation there. This is evidenced by the crescent-shaped dolomite deposition cap at the tip of the calcite dissolution finger. In the region next to the dolomite cap, there was less dolomite precipitation because less flow passed through that region as a result of the focusing effect of the calcite finger. Farther from the cap, dolomite precipitation forms two side islands at the top and bottom boundaries of the domain. These islands represent the residue of what would be the planar front if the initial high-permeability bump were not present in the system. The precipitation cap builds up with time (see frame c) and eventually terminates the flow-focusing of the initial permeability bump. The flow is then diverted to the sides of the cap. The regions between the precipitation cap and the two side islands are favorable for flowthrough. As a result, flow is funnelled there and produces two new precipitation "islands" (frame d).

The simulation of the calcite/dolomite system demonstrates that the planar multimineral reaction front can be unstable to a local permeability perturbation. In general, the morphological stability of planar reaction fronts can lead to complex patterns in a multiple mineral system, involving rather

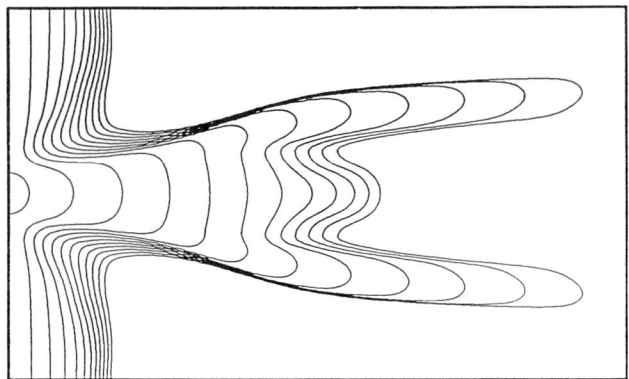

Figure 12. Triply fingered reaction front ("branching tree structure") in a domain of 20 × 12 cm. The imposed horizontal flow is 5.3 × 10^{-3} cm/sec. The time interval between adjacent contours is 5 × 10^7 sec.

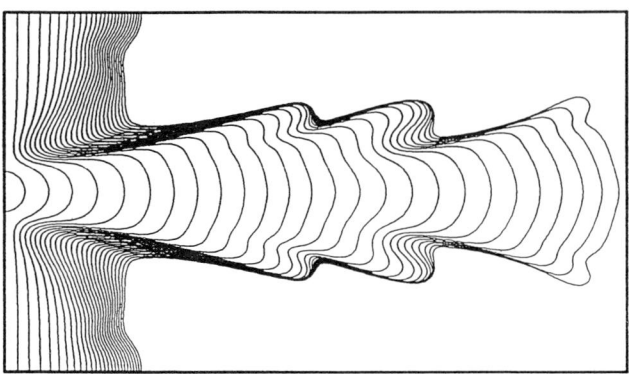

Figure 13. Triply fingered reaction front ("branching tree structure") in a domain of 20 × 12 cm. The imposed flow is 7.2 × 10^{-3} cm/sec. The time interval between adjacent contours is 2 × 10^7 sec. Buds are seen to emerge periodically in time in a frame moving with the advancing main shoot but are ultimately drawn back to the main flow channel.

a

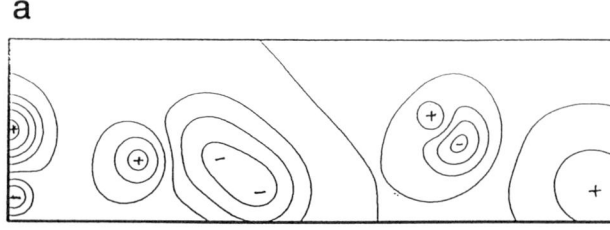

b

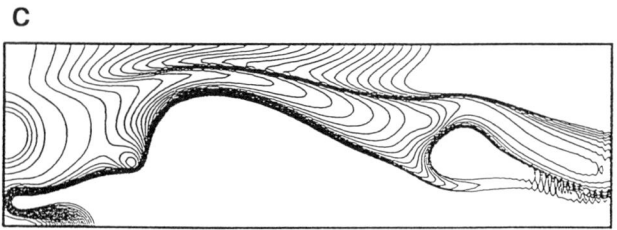

c

Figure 14. Temporal development of a calcite dissolution front in an initially nonuniformly cemented sandstone. The size of the system is 10 × 3 cm. (a) Porosity contour map at the initial time. The contour interval is 2%, and the plus (minus) sign means that the porosity there is higher (lower) than the average porosity of the system. (b) Temporal development of a front under an imposed flow of 1 × 10^{-3} cm/sec. The time interval is 5 × 10^7 sec. (c) Same as in frame (b), except the imposed flow is 1 × 10^{-2} cm/sec. The time interval is 2 × 10^6 sec.

unpredictable events wherein fingers develop in the dissolution front of one mineral and are capped by the precipitation of another mineral, with subsequent development of alternative pathways.

OSCILLATORY FLUID RELEASE FROM OVERPRESSURIZED COMPARTMENTS

From the discussion in the introduction, it is clear that fluid release from overpressurized compartments can in principle be oscillatory. However, qualitative arguments do not capture the properties and range of existence of the oscillations. Key properties are the timing of fluid release pulses and the amount of fluid release per pulse. Parameters that affect the oscillation include burial rate and fluid chemistry, geothermal gradient, mineralization within the overpressured internal volume (compartment) and its surrounding seal, and compaction dynamics, as well as the thickness, tensile strength, and permeability of the seal in healed and fractured states. Here we attempt to place the discussion of these factors within a simple quantitative framework. A successful quantification could yield valuable information on the relationship of the timing of petroleum genesis and migration relative to that of fluid release from sealed compartments for a given geologic context.

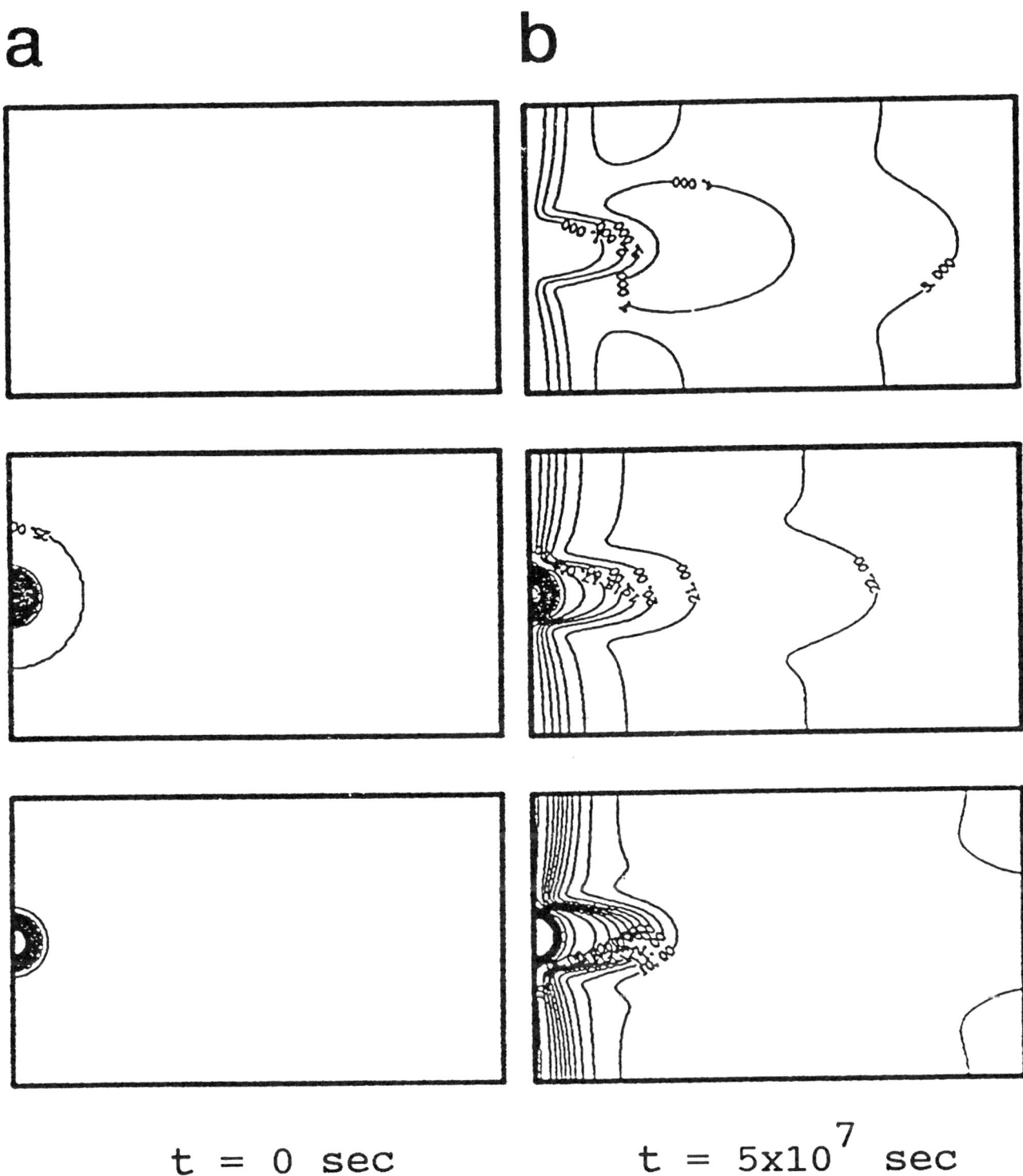

Figure 15. Temporal development of a reaction front and resulting patchwork structure for a system initially cemented uniformly with calcite and dolomite in an insoluble quartz matrix. The initial mineral modes are: 65% quartz, 25% calcite, and 0.5% dolomite. The mineral texture is initially uniform except for semicircular perturbation at the inlet wall, which makes the calcite volume fraction lower and porosity higher in that region than elsewhere in the system. Each of the frames is a set of three contour maps, showing the space distribution at given time of (from top to bottom) dolomite, calcite, and porosity. The size of the domain is 20 × 12 cm; the imposed flow is 3×10^{-2} cm/sec. The times are (a) 0 sec, (b) 5×10^7 sec, (c) 2×10^8 sec, and (d) 4.5×10^8 sec.

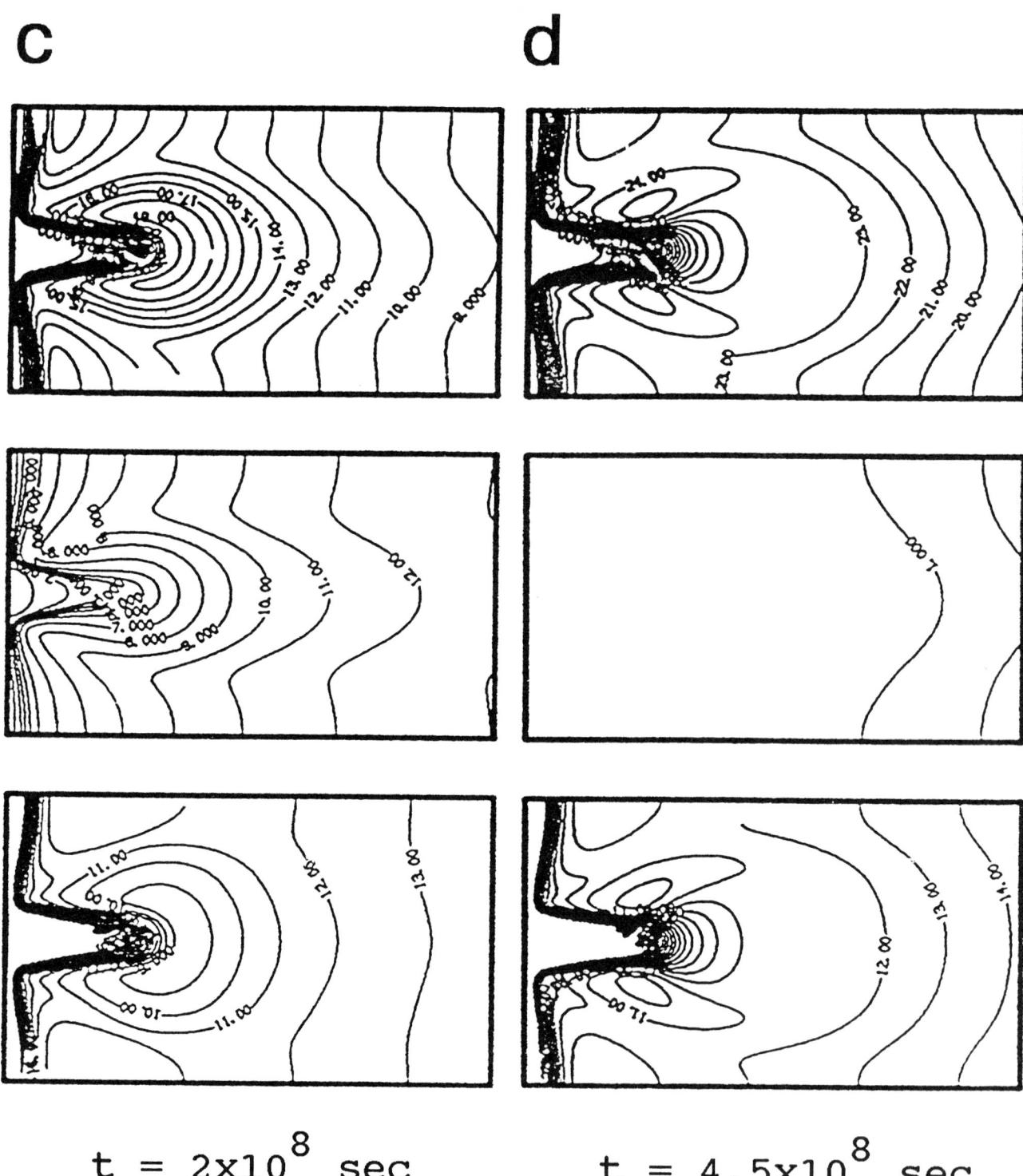

$t = 2 \times 10^8$ sec $t = 4.5 \times 10^8$ sec

Figure 15. (Continued)

First we present the formulation of the model. Results on simple estimates and numerical experiments based on the model are then discussed. We conclude with implications for petroleum genesis and containment.

Simple Mathematical Model

Consider the model suggested in Figure 16. An overpressurized compartment is seen bounded by a top seal (and bottom and side seals, not shown)

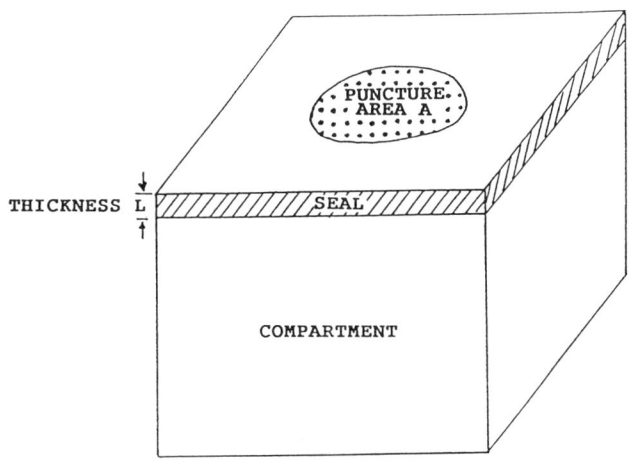

Figure 16. Schematic diagram of a portion of an overpressured, sealed compartment. (Side and bottom seals not shown.)

(Ghaith et al., in press). We consider the possibility that the top seal can be "punctured" by local fracturing, as in Figure 16. The compartment is pressurized because of subsidence and other factors. The key to the model is in the kinetics of seal puncturing based on phenomenological considerations of the behavior of seal permeability during fracturing and healing.

Before developing the quantitative model, the results may be summarized as follows. A typical simulation is seen in Figure 17. In frame a the pressure in the compartment oscillates as the compartment descends at a depth of about 3 km. During oscillation the fluid pressure within the compartment fluctuates between values at which the seal can become fractured and become healed. Frame b shows the accompanying oscillation in seal permeability in terms of the ratio (κ) of the permeability to fluid viscosity.

A simple quantitative model of the dynamics of seal puncture during overpressuring may be set forth as follows. The fluid velocity within the seal is taken to be given by Darcy's law, equation 9. Note that κ is not a fundamental variable but rather reflects the local state of texture (grain shape, packing, and fracturing) within the seal. However, for simplicity we assume that we can obtain a dynamical equation for it in the form

$$\frac{\partial \kappa}{\partial t} = f \qquad (29)$$

The rate function f depends on κ and three pressure parameters:

p_c = pressure in the overpressured compartment just below the seal;

p_f = value of p_c at which the seal will fracture; and

p_h = value of p_c below which the seal will heal in time.

Both p_f and p_h depend on lithostatic stress, tensile strength, and p_0, the fluid pressure at the top boundary of the seal.

The essence of the fracture-healing dynamics is captured by the form of the null curve $f = 0$. Figure 18 illustrates such a case, which embodies the fracture-healing dynamics of interest. In our approach an explicit form for f is set forth and the parameters are chosen to fit the observed data. The form used in this study is

$$f = q \left[p - p^* + a(\kappa - \kappa) + b \frac{(\kappa - \kappa^*)^3}{[(\kappa - \kappa_a)(\kappa - \kappa_b)]^2} \right] \qquad (30)$$

where K^* and p^*, κ_a, and κ_b are defined in Figure 18. The parameters a and b can be obtained in terms of p_h and p_f. These parameters along with K^*, p^*, and q can be used to fit the fracturing-healing dynamics to agree with field or experimental data. The parameter q may be varied to change the rate of healing and fracturing; if the rate of fracturing is much greater than that of healing, q can be taken to increase with p_c.

Fluid mass conservation within the overpressured zone and the equation of state of the fluid yield an equation (equation 14) for the fluid pressure p_c within the compartment. Here we take a simple approach whereby the pressure within the compartment is assumed to be uniform but, as a result of overpressuring, can differ from the pressure above the seal. Let V_c and ϕ_c be the compartment volume and porosity. Because the thermal expansion of the aqueous fluid volume for the temperature range 10 to 80°C is 2 orders of magnitude higher than that for a typical rock matrix, the latter can be neglected (Palciauskas and Domenico, 1982); V_c and ϕ_c are assumed to only vary because of compaction. Assuming that the aqueous fluid has a composition independent of temperature, the molar density of water ρ (c_s in equation 11) is taken to be a function of fluid pressure p and absolute temperature T only. We take the simple linear equation of state

$$\rho(p,T) = \rho_0 [1 + \beta(p - \bar{p}) - \alpha(T - \bar{T})] \qquad (31)$$

for fluid compressibility β, and thermal expansivity α, and reference pressure and temperature $\bar{p}$ and $\bar{T}$. The total amount of water (moles) in the compartment is $p\phi_c V_c$. Its rate of change is then

$$\frac{d}{dt}\left[\rho\phi_c V_c\right] = \rho_0 \beta \phi_c V_c \left[\frac{dp_c}{dt} - \Gamma\right]. \qquad (32)$$

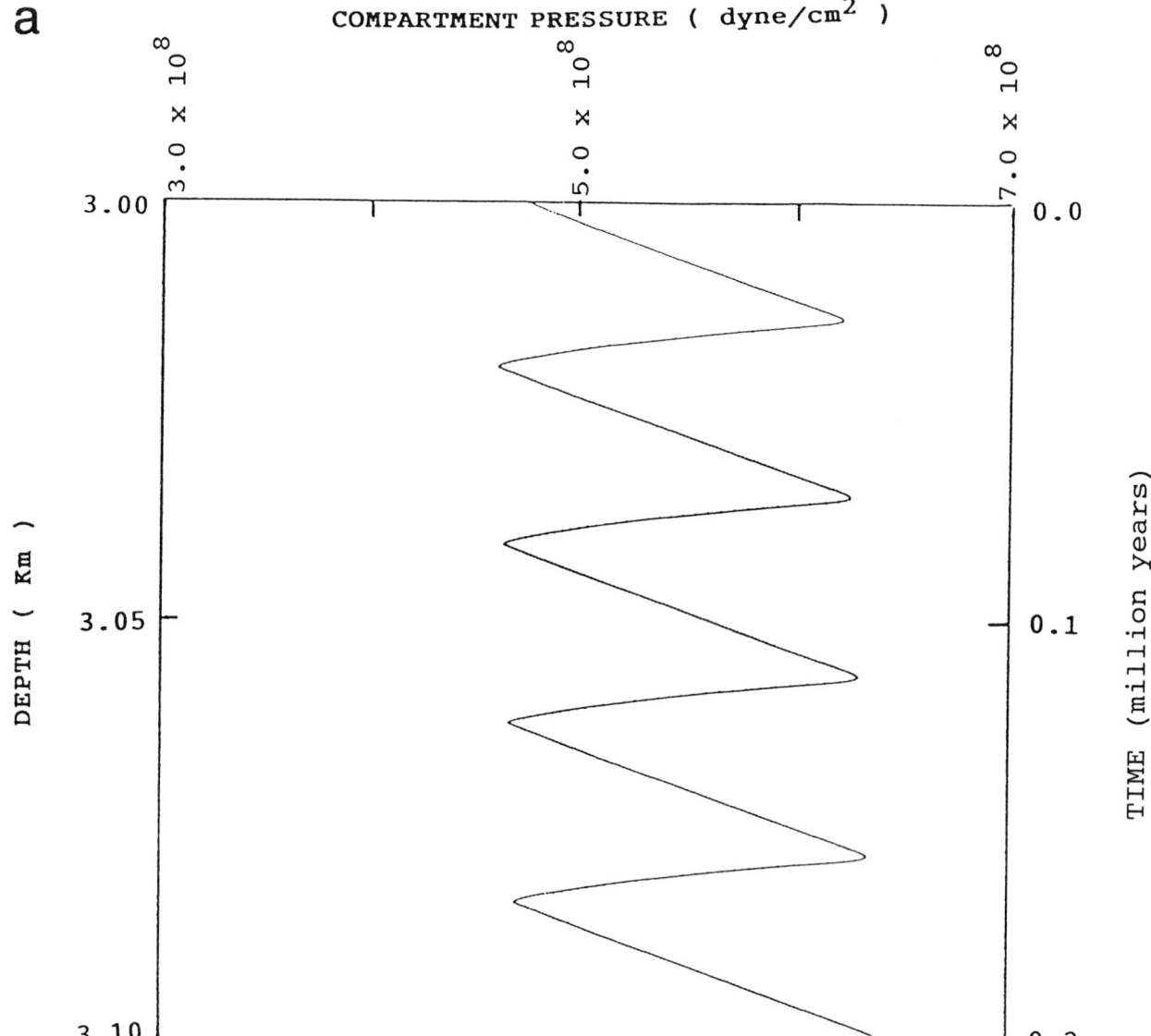

Figure 17a. Time-dependence of the fluid pressure within the compartment just below the seal during subsidence between 3.0 and 3.1 km.

In equation 32 the "pressurizing rate" Γ is given by

$$\Gamma = \frac{\alpha}{\beta}\frac{dT}{dt} - \frac{1}{\beta\phi_c V_c}\frac{d(\phi_c V_c)}{dt}. \quad (33)$$

We assume the temperature within the compartment is constant in space but increases in time due to burial. We assume Γ to be a known function of time that depends on specified compaction rate, burial rate, and geothermal gradient.

Conservation of mass within the compartment accounting for loss through the seal yields

$$\beta\phi_c V_c \left[\frac{dp_c}{dt} - \Gamma\right] = \int dxdy \left[\phi\kappa\frac{\partial p}{\partial z}\right]_b \quad (34)$$

The subscript b indicates evaluation just inside the seal at the bottom of the seal; here z is the vertical coordinate and x and y fix position along the top seal in the horizontal plane. The x,y integration is over the area A of the putative puncture suggested in Figure 16.

Neglecting all reactions, conservation of water within the seal (equation 11) yields

$$\frac{\partial\phi\rho}{\partial t} = -\vec{\nabla}\cdot(\phi\rho\vec{v}) \quad (35)$$

We expect that most of the water balance within the putative puncture area of the seal will be accounted for by flowthrough and not by change in seal porosity ϕ or fluid molar density ρ. Thus the left-hand side of equation 35 may be neglected. If

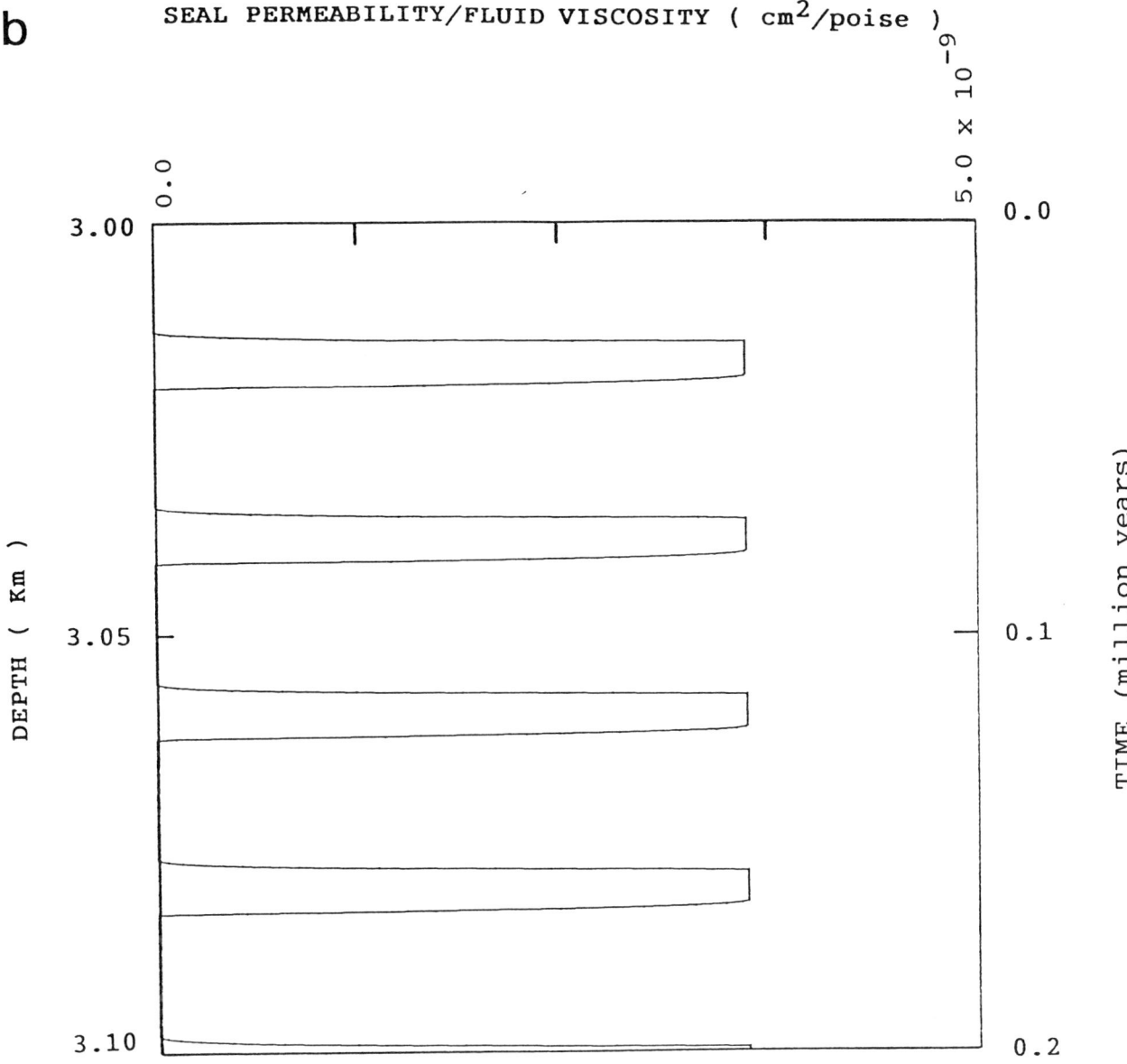

Figure 17b. Variations in the Darcy permeability (for the seal) divided by the fluid viscosity coordinated with pressure variations as in frame (a).

we consider that ϕ within the seal does not vary appreciably during any fluid ejection event (that is, fracturing adds connectivity between pores but not much porosity), then equation 35 takes the form

$$\vec{\nabla} \cdot \vec{v} = 0. \tag{36}$$

The boundary conditions applied to the seal are that at its bottom, $z = 0$, and p is p_c, its value within the compartment; at its top, $z = L$, and p is some specified value denoted p_0 that may be hydrostatic or some abnormal pressure outside the compartment. Because seals are usually relatively thin (about 100 m), we can neglect the gravitational contributions to the Darcy flow (see equation 9), and thus we have

$$p = p, z = 0 \tag{37}$$
$$p = p_0, z = L.$$

The above approximations imply

$$p = p_c + (p_0 - p_c)z/L. \tag{38}$$

With this we obtain

$$\frac{dp_c}{dt} = \Gamma + \Lambda\kappa(p_0 - p_c) \tag{39}$$

where

Prediction of Reservoir Quality Through Chemical Modeling

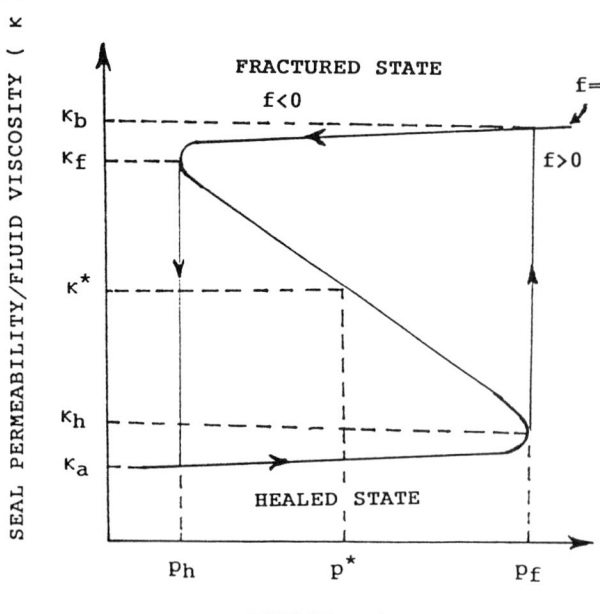

Figure 18. S-shaped null curve (f = O) for permeability dynamics capable of supporting the cycle of alternating fracturing, depressurizing, healing, and pressurizing phases of compartment dynamics.

$$\Lambda \equiv \frac{A\phi}{V_c\phi_c\beta L} \quad (40)$$

is a loss rate/compartment capacity parameter that depends on the puncture and compartment size. With this we obtain equation 39, an ordinary differential equation coupling p_c to the seal permeability-fluid viscosity ratio κ.

Range of Existence of Oscillation

The range of existence of the oscillation in terms of system parameters can be obtained analytically for our simple model. The parameter Λ (equation 40) serves as a measure of the efficiency for the puncture to relieve pressure relative to compartment capacity.

The range of existence of oscillation in the Λ-Γ plane can be delineated as follows (Ghaith et al., in press). If Γ is sufficiently small, the healed seal can relieve pressure and a steady (nonoscillatory) state can be set up whereby p_c is kept below p_f. The marginal state is attained when p_c just equals p_f. In such a state the rate of pressurization Γ is just balanced by the rate of pressure loss $\Lambda\kappa_h(p_f-p_0)$, κ_h being κ in the healed state. Thus we obtain the condition

$$\Lambda > \Lambda_h \equiv \Gamma/\kappa_h(p_f-p_0), \text{ no oscillation.} \quad (41)$$

Similarly, if Γ is large, the seal will remain open in a fractured state. This yields

$$\Lambda < \Lambda_f = \Gamma/\kappa_f(p_h-p_0), \text{ no oscillation,} \quad (42)$$

κ_f being κ in the fractured state. The marginal case ($\Lambda = \Lambda_f$) is when p_c can just be sustained at p_h when the seal is fractured; thus, p_h appears on the right-hand side of equation 42.

Any state of oscillation must occur in what remains of the Γ-Λ plane when equations 41 and 42 are eliminated (Figure 19).

Numerical Experiments on Subsiding Compartments

The above results were obtained under the artificial assumption that p_h and p_f were constants. Actually, they change with depth. To study this effect we simulated the model numerically. The simulations were performed by assuming a subsidence and thermal history. This was implemented in the model by making a number of specific parametric and phenomenological assumptions as follows. The present model reduces to the simultaneous solution of equations 29 and 39.

Data for Numerical Simulations

First we assumed a phenomenology for p_f and p_h by taking an empirical approach. Of the more than 400 abnormally pressured gas reservoirs in Louisiana reported by the petroleum industry in 1972, compartment pressures just under the top seals varied with depth and, with few exceptions, were in the interval from 1.6×10^3 to 2.0×10^3 (dyne/cm²/cm). We assumed that if the compartments in Louisiana were undergoing oscillatory fluid release, the oscillations would take p_c between p_f and p_h, as in Figure 17. We interpreted this to imply that

$$\frac{dp_f}{dz} = -2.0 \times 10^3 \text{ dyne/cm}^3 \text{ (-0.9 psi/ft)} \quad (43)$$

$$\frac{dp_h}{dz} = -1.6 \times 10^3 \text{ dyne/cm}^3 \text{ (-0.7 psi/ft)}$$

The former is consistent with estimates of p_f based on the formula $p_f = p_0 + \sigma_h + n$, where p_0 is the pore fluid pressure above the seal, σ_h is the horizontal principal lithostatic stress, and n is tensile strength.

Simulations were then performed by using specified geothermal gradient and subsidence and compaction rates; α and β values used were from Straus and Schubert (1977). The compaction rate, given by

$$\frac{-1}{\beta(1-\phi_c)} \frac{d(\ln\phi_c)}{dt} \quad (44)$$

was calculated by taking ϕ_c to change exponentially with depth, and therefore $d(\ln\phi/dt)$ is constant. As seen on Figure 17, the period of oscillation is much longer than times for seal fracturing and even likely for seal healing (fractures may be cemented on the time scale of decades; Boles and Ramseyer, 1987).

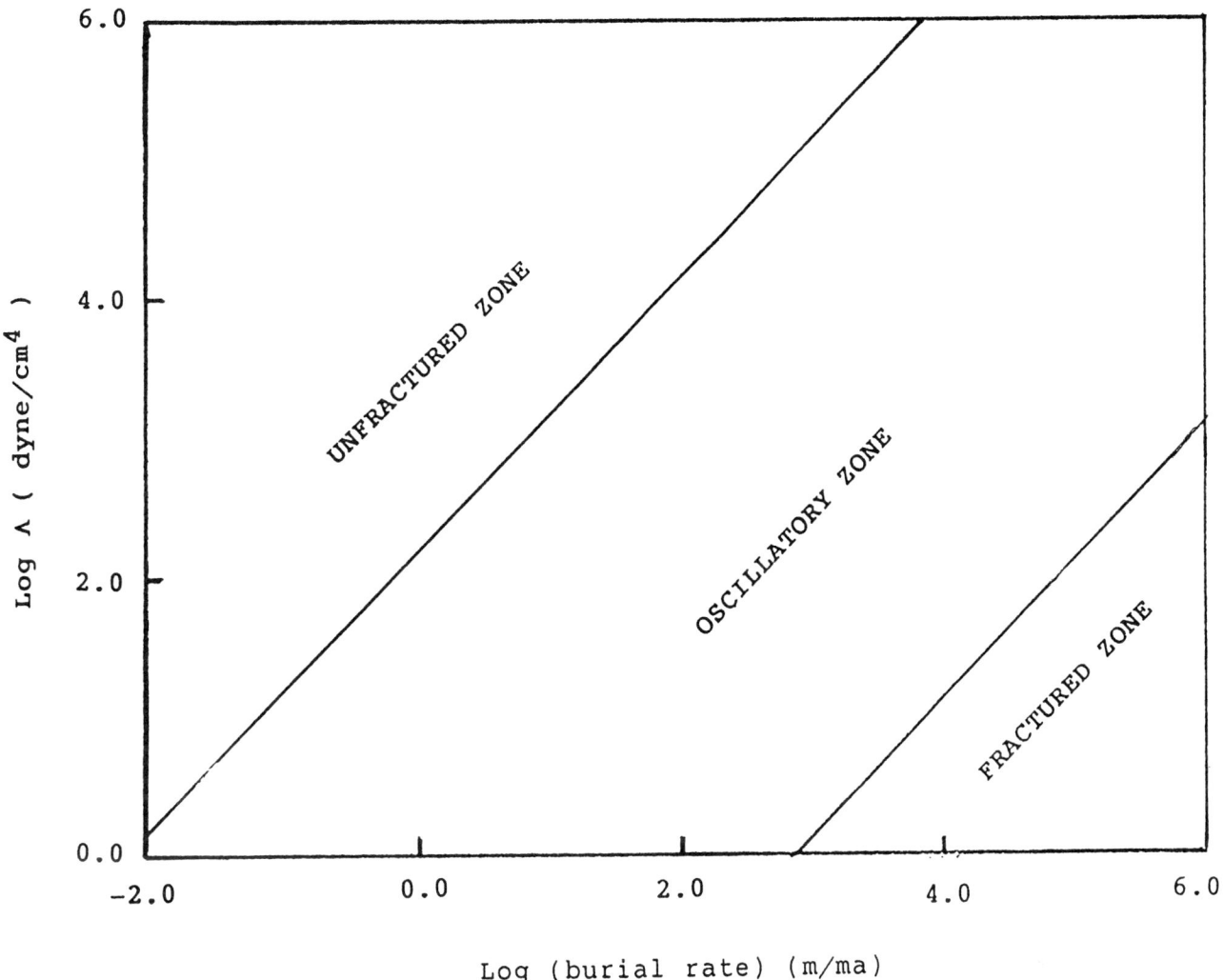

Figure 19. Plane of Λ and subsidence rate showing domain of oscillatory fluid release.

Sustained and Transient Oscillatory Release

A number of simulations illustrating oscillatory fluid release were shown in Ghaith et al. (in press). The range of existence of oscillation may change with burial because of changing Γ or the rate of increase of p_f and p_h (see equations 41, 42). Therefore, oscillation may commence or terminate as the compartment descends. In Figure 20 we illustrate a case with two switching events. Oscillation gives way to a steady healed state, because the rate of pressurization cannot keep up with the increase of p_f with depth. The compartment starts off in the steady fractured state, but as compaction slows down, it enters the oscillatory regime.

Oscillation in seal permeability and compartment pressure are also accompanied by fluctuations in fluid release. Analytical estimates for fluid release per cycle as well as the period of a cycle may be obtained for the present model (Ghaith et al., in press). A typical number, as for the oscillations shown on Figure 17, is 0.12 km³ for the pulse occurring near 3 km in the case of a 10-km³ compartment. Thus, considerable fluid can be released in each cycle.

CONCLUSIONS AND DISCUSSION

Reaction-transport modeling has great potential in the petroleum and related industry. This type of modeling can provide guidelines for exploration by predicting conditions favorable for the creation of reservoir rock and the timing and geometry of petroleum migration. For petroleum engineering, it can help in the design of well completion, acid stimulation, and EOR (enhanced oil recovery) strategies. Related questions of hazardous chemical waste disposal in geologic repositories and diagenetic

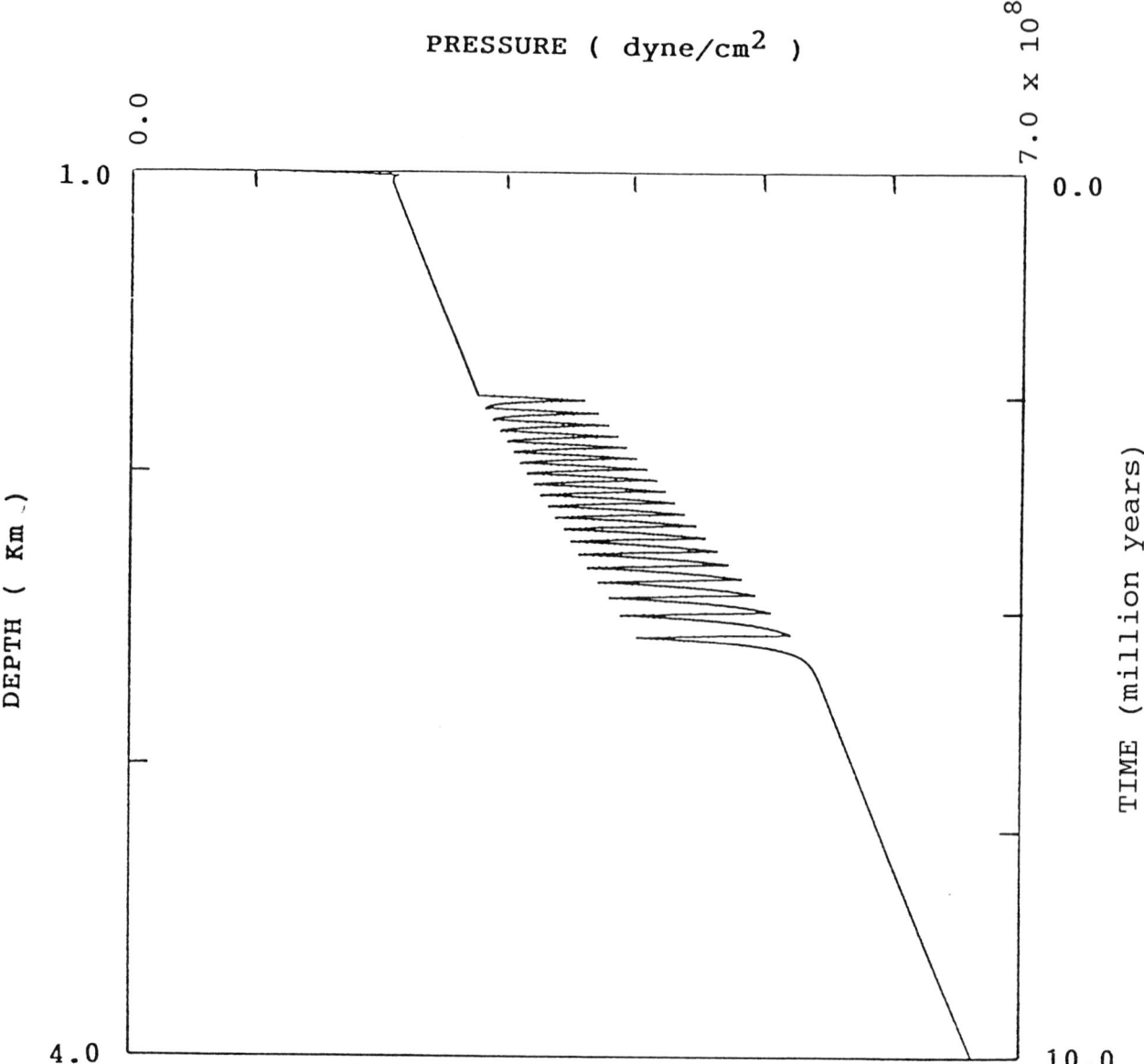

Figure 20. Simulation of sinking compartment showing switching from steady fractured state to oscillatory state and then to steady healed state.

orebodies can also be modeled for engineering or exploration studies.

For models to be realistic, they must be fully coupled. This allows for the description of phenomena such as reaction-front fingering, pulsatile fluid motion, or other geochemical self-organizing effects. These self-organization phenomena are of particular interest because they force us to have a new approach to the study of the distributions of porosity, permeability, mineralization, and petroleum not in the realm of more classical types of geological analysis.

In order to make this type of modeling practical, rather general reaction-transport codes and asso- ciated graphical analysis packages must be developed. We have developed a number of such codes, the most notable of which is GENEX. It incorporates a large range of phenomena, works in one and two spatial dimensions, and is easily adapted to include a range of factors such as pressure solution, fracturing, burial, subsidence, and thermal effects.

To illustrate the power of our approach, we have focused on three phenomena arising from the strong coupling of processes—reaction-front fingering, buoyancy-driven cellular methane migration, and oscillatory fluid release from overpressurized compartments. In reality, these phenomena may act individually or may operate constructively or destructively with each other. For example, a positive chemical feedback may arise that promotes local seal puncturing. If fluids deep within a compartment tend to dissolve minerals comprising its top seal, a small region of elevated permeability within the seal can become increased in size and permeability as fluid flows through it. This increased permeability attracts even more flow, thus completing a constructive feedback loop. This dynamic may also couple to seal fracturing because of the enhancement of the fracture size through dissolution and because dissolution tends to decrease tensile strength. Such a process may underlie the puncturing of an anhydrite/calcite seal in the Arabian Shelf, as discussed by Broomhall and Allan (1987).

A significant feature of quantitative modeling is its potential for predicting the timing of key diagenetic events relative to that of petroleum genesis and migration. For example, under some conditions a compartment can pass through the regime of oscillatory seal fracturing before petroleum genesis and migration into the top of the compartment. Thus, further sinking of such a compartment will lead to deep compartment-sequestered petroleum. The model also suggests that multiple distinct phases of petroleum injection into a reservoir rock should be commonplace. We believe that future work in refining the model through improved descriptions of texture and transport and grain growth rate laws will bring these considerations into the realm of practical exploration strategies.

ACKNOWLEDGMENTS

The authors are grateful to Grant Garven and John Bradley for their thorough and insightful reviews of an earlier draft. The studies presented also benefited greatly from frequent discussions with Tom Dewers. The technical support by Ken DeHart on computer hardware and Alicia Wilbur's organizational abilities are also gratefully acknowledged.

This research was supported by a contract from the Gas Research Institute and a grant from the Basic Energy Sciences Program of the U.S. Department of Energy.

REFERENCES

Al-Shaieb, Z., V. Tigert, J. Puckette, and R. Ely, 1989, Genesis of pressure compartments and seals: OKINTEX Group, Progress Report IA for the Gas Research Institute, 15 November 1989.

Baker, E. T., J. W. Lavelle, R. A. Feely, G. J. Massoth, and S. L. Walker, 1989, Episodic venting of hydrothermal fluids from the Juan de Fuca Ridge: Journal of Geophysical Research, v. 904, p. 9237-9250.

Bathurst, R. G. C., 1986, Diagenetically enhanced bedding in argillaceous platform limestones: stratified cementation and selective compaction: Sedimentology, v. 34, p. 740-778.

Boles, J. R., and K. Ramseyer, 1987, Diagenetic carbonate in Miocene sandstone reservoir, San Joaquin basin, California: AAPG Bulletin, v. 59, p. 1475-1487.

Bories, S. A., and M. A. Combarnous, 1973, Natural convection in a sloping porous layer: Journal of Fluid Mechanics, v. 3, p. 63-79.

Boudreau, A. E., 1987, Pattern formation during crystallization and the formation of fine-scale layering, in J. Parsons, ed., Origins of igneous layering, I.: Dordrecht, Reidel, NATO ASI Series, Ceries C, p. 453-471.

Bradley, J. S., 1975, Abnormal formation pressure: AAPG Bulletin, v. 59, p. 957-973.

Broomhall, R. W., and J. R. Allan, 1987, Regional caprock destroying dolomite on the Middle Jurassic to Early Cretaceous Arabian Shelf: Society of Petroleum Engineers Formation Evaluation 13697.

Busenberg, E., and L. N. Plummer, 1986, A comparative study of the dissolution and crystal growth kinetics of calcite and aragonite, in F. A. Mumpton, ed., Studies in diagenesis: USGS Bulletin 1578, p. 139-168.

Chadam, J., A. Sen, and P. Ortoleva, 1986, Reactive infiltration instabilities: Institute of Mathematics and Its Applications, Journal of Applied Mathematics, v. 36, p. 207-221.

Chadam, J., A. Sen, and P. Ortoleva, 1988, A weakly nonlinear stability analysis of the reactive infiltration interface: Society for Industrial and Applied Mathematics, Journal of Applied Mathematics, v. 48.

Chen, W., and P. Ortoleva, 1990a, Self-organization in far-from-equilibrium reactive porous media subject to reaction front fingering, in D. Walgraefand and N. M. Ghoniem, eds., NATO ASI Series, Patterns, defects and materials instabilities: v. 183, p. 203-220.

Chen, W., and P. Ortoleva, 1990b, Reaction front fingering in carbonate cemented sandstones, in P. Ortoleva, B. Hallet, A. McBirney, I. Meshri, R. Reeder, and P. Williams, eds., Proceedings of the Workshop on Self-Organization in Geological Systems: Earth Science Reviews (Amsterdam, Elsevier).

Chen, W., A. Park, and P. Ortoleva, in press, GENEX: a general two-dimensional reaction/transport simulator for geochemical engineering and exploration analysis: AICGE November 1990 meeting in Chicago.

Combarnous, M. H., and S. A. Bories, 1978, Hydrothermal convection in saturated porous media: Advanced Hydroscience, v. 10, p. 231-307.

Dewers, T., and P. Ortoleva, 1988, The role of geochemical self-organization in the migration and trapping of hydrocarbons: Applied Geochemistry, v. 3, p. 287-316.

Dewers, T., and P. Ortoleva, 1990a, Diagenesis through coupled processes: modeling approach, self-organization, and implications for exploration, this volume.

Dewers, T., and P. Ortoleva, 1990b, Nonlinear hydrologic phenomena in deep sedimentary basins: a coupled stress, fracturing, reaction and transport model: OKINTEX Group, Second Quarterly Report for the Gas Research Institute.

Domenico, P. A., and V. V. Palciauskas, 1988, The generation and dissipation of abnormal fluid pressures in active depositional environments, in W. Back, J. S. Rosenshein, and P. R. Seaber, eds., The Geology of North America: Hydrogeology: GSA, v. 2.

Ekdale, A. A., and R. G. Bromley, 1988, Diagenetic microlamination in chalk: Journal of Sedimentary Petrology, v. 58, p. 857-861.

Feinn, D., P. Ortoleva, W. Scalf, S. Schmidt, and M. Wolff, 1978, Spontaneous pattern formation in precipitating systems: Journal of Chemical Physics, v. 67, p. 3771.

Flicker, M. R., and J. Ross, 1974, Mechanism of chemical instability for periodic precipitation phenomena: Journal of Chemical Physics, v. 60, p. 3458.

Geo-Chemical Research Associates, Inc., 1990, GENEX: A fluid-rock interaction simulator and analysis system for geochemical engineering and exploration: Concepts and User's Manual, 78 p.

Ghaith, A., J. Bradley, D. Powley, and P. Ortoleva, in press, Oscillatory fluid release from overpressurized compartments: AAPG Bulletin.

Ghaith, A., W. Chen, and P. Ortoleva, 1990, Oscillatory methane release from shale source rock, in P. Ortoleva, B. Hallet, A. McBirney, I. Meshri, R. Reeder, and P. Williams, eds., Proceedings of the Workshop on Self-Organization in

Geological Systems: Earth Science Reviews (Amsterdam, Elsevier).
Guy, B., 1990, Banded skarns: an introduction, *in* P. Ortoleva, B. Hallet, A. McBirney, I. Meshri, R. Reeder, and P. Williams, eds., Proceedings of the Workshop on Self-Organization in Geological Systems: Earth Science Reviews (Amsterdam, Elsevier).
Haase, S., D. Feinn, J. Chadam, and P. Ortoleva, 1980, Oscillatory zoning in plagioclase feldspar: Journal of Sedimentary Petrology, v. 30, p. 568-577.
Heald, M. T., and R. C. Anderegg, 1960, Differential cementation in the Tuscarora Sandstone: Journal of Sedimentary Petrology, v. 30, p. 568-577.
Krooss, B. M., and D. Leythaeuser, 1988, Experimental measurements of the diffusion parameters of light hydrocarbons in water-saturated sedimentary rocks II. Results and geochemical significance: Organic Geochemistry, v. 12, p. 91-108.
Levandowski, D. W., M. E. Kaley, S. R. Silverman, and R. G. Smalley, 1973, Cementation in Lyons sandstone and its role in oil accumulation, Denver basin, Colorado: AAPG Bulletin, v. 57, p. 2217-2244.
Liesegang, R. E., 1913, Geologische diffusionen: Steinkopff, Dresden.
Lovett, R., J. Ross, and P. Ortoleva, 1978, Kinetic instabilities in first order phase transitions: Journal of Chemical Physics, v. 69, p. 947.
Nicolis, C., and G. Nicolis (eds.), 1987, Nonlinear phenomena and dynamical systems analysis in geosciences: Proceedings of the NATO Conference in Crete, July 1985, Reidel.
Nicolis, G., and I. Prigogine, 1977, Self-organization in non-equilibrium systems: New York, Wiley.
Ortoleva, P., 1990a, Role of attachment kinetic feedback in the oscillatory zoning of crystals grown from melts, *in* P. Ortoleva, B. Hallet, A. McBirney, I. Meshri, R. Reeder, and P. Williams, eds., Proceedings of the Workshop on Self-Organization in Geological Systems: Earth Science Reviews (Amsterdam, Elsevier).
Ortoleva, P., 1990b, Geochemical self-organization: Princeton University Press.
Ortoleva, P., J. Chadam, E. Merino, and C. H. Moore, 1987a, Geochemical self-organization I: Feedback mechanisms and modeling approach: American Journal of Science, v. 287, p. 979-1007.
Ortoleva, P., E. Merino, J. Chadam, and A. Sen, 1987b, Geochemical self-organization II: the reactive infiltration instability: American Journal of Science, v. 287, p. 1008-1040.
Ortoleva, P., B. Hallet, A. McBirney, I. Meshri, R. Reeder, and P. Williams (eds.), 1990, Proceedings of the Workshop on Self-Organization in Geological Systems: Earth Science Reviews (Amsterdam, Elsevier).
Ostwald, W., 1925, Kolloid Z., v. 36, p. 330.
Palciauskas, V. V., and P. A. Domenico, 1982, Characterization of drained and undrained response of thermally loaded repository rocks: Water Resources Research, v. 18, p. 281-290.
Park, A., T. Dewers, and P. Ortoleva, 1990, Cellular and oscillatory self-induced methane migration, *in* P. Ortoleva, B. Hallet, A. McBirney, I. Meshri, R. Reeder, and P. Williams, eds., Proceedings of the Workshop on Self-Organization in Geological Systems: Earth Science Reviews (Amsterdam, Elsevier).
Powley, D. E., 1980, Normal and abnormal pressures: AAPG Petroleum Exploration School Lectures.
Powley, D. E., 1985, Pressures, normal and abnormal: Lecture Notes, "Techniques of Petroleum Exploration II": AAPG School, South Padre Island (Texas), Sept. 16-19, 1985.
Powley, D. E., 1990, Pressures, hydrogeology and large scale seals in petroleum basins, *in* P. Ortoleva, B. Hallet, A. McBirney, I. Meshri, R. Reeder, and P. Williams, eds., Proceedings of the Workshop on Self-Organization in Geological Systems: Earth Science Reviews (Amsterdam, Elsevier).
Ranganathan, V., and J. S. Hanor, 1988, Salinity driven groundwater flow near salt domes: Chemical Geology, v. 74, p. 173-188.
Reeder, R. J., and J. L. Prosky, 1986, Compositional sector zoning in dolomite: Journal of Sedimentary Petrology, v. 56, p. 237-247.
Reeder, R. J., and J. C. Grams, 1987, Sector zoning in calcite cement crystals: implications for trace element distributions in carbonates: Geochimica et Cosmochimica Acta, v. 51, p. 187-194.
Rice, J. R., and R. E. Boisvert, 1984, Solving elliptic problems using ELLPACK: New York, Springer-Verlag, 497 p.
Ricken, W., 1986, Diagenetic bedding: Berlin, Springer-Verlag, 210 p.
Sjoberg, E. L., and D. T. Rickard, 1984, Temperature dependence of calcite dissolution kinetics between 1 and $62°$ C at pH 2.7 to 8.4 in aqueous solutions: Geochimica et Cosmochimica Acta, v. 48, p. 485-493.
Straus, J. M., and G. Schubert, 1977, Thermal convection of water in a porous medium: effects of temperature- and pressure-dependent thermodynamic and transport properties: Journal of Geophysical Research, v. 82, p. 325-333.
Tada, R., and R. Siever, 1989, Pressure solution during diagenesis: a review: Annual Review of Earth and Planetary Sciences, v. 17, p. 89-118.
Tigert, V., and Z. Al-Shaieb, 1990, Pressure seals and their diagenetic zebra structure patterns, *in* P. Ortoleva, B. Hallet, A. McBirney, I. Meshri, R. Reeder, and P. Williams, eds., Proceedings of the Workshop on Self-Organization in Geological Systems: Earth Science Reviews (Amsterdam, Elsevier).
Ungerer, P., F. Behar, M. Villalba, O. R. Heum, and A. Annie, 1987, Kinetic modelling of oil cracking: 13th International Meeting on Organic Geochemistry, 22 p.
Wood, J. R., and T. A. Hewett, 1982, Fluid convection and mass transfer in porous sandstones—a theoretical model: Geochimica et Cosmochimica Acta, v. 46, p. 1707-1713.

Effects of Fluid and Rock Compositions on Diagenesis: A Modeling Investigation

Craig H. Moore
Peter J. Ortoleva
Geo-Chemical Research Associates
Bloomington, Indiana, U.S.A.

In order to demonstrate the potential use of geochemical modeling using nonequilibrium, coupled reaction-transport codes in predicting reservoir quality, we have carried out simulations of four generalized systems using the REACTRAN computer code. The systems chosen contained "clean" and "dirty" arkoses and were subjected to alteration by fresh and saline waters at a constant temperature of 110°C. We show the results of these simulations at various times during the first 10 million years of diagenesis and describe the salient geochemical features leading to the spatial and temporal distributions of authigenic minerals.

The timing and extent of diagenetic events simulated here is a function of initial rock composition, the rate of fluid flow, and the rates of mineral dissolution and precipitation assumed here. When appropriate rate data is included as a function of temperature and pH, one anticipates a good comparison of simulated and observed diagenetic events. The prediction of the timing and the extent of diagenetic events is the key to the prediction of reservoir quality.

INTRODUCTION

Diagenesis in most sedimentary basins is commonly accompanied by nonequilibrium conditions and fluid flow, as suggested by reaction fronts and zones in existing sedimentary rocks. Consequently, realistic simulation of diagenesis requires models that consider at least these two factors. In addition, the models should allow for the complete disappearance of minerals as well as nucleation of new minerals in the system, because both situations are observed or inferred from petrographic examinations of many reservoir rocks.

REACTRAN, a computer code for coupled/flow-reaction simulations, has these features and was used to perform a preliminary investigation of the controls that rock and fluid composition impose on diagenetic evolution of rocks under various conditions. In general, the program takes a stated geochemical system and evolves it in both space and time. The program uses a solution technique that follows events on the time scale which is associated with mineral reactions (slow), rather than the time scales which are associated with fluid flow or reactions among aqueous species (fast). Nucleation and the finite rate of growth of minerals, as well as finite rate and fast (equilibrated) reactions among pore-fluid species, are allowed. Porosity and permeability changes, attendant to grain growth and dissolution, are accounted for; permeability is calculated by using the Fair-Hatch formula (Bear, 1972). Therefore, by including Darcian flow equations, mineral growth and dissolution are coupled to fluid flow.

The mathematics of the physical processes pertinent to the REACTRAN code are discussed in Ortoleva et al. (1987). REACTRAN as used here as well as a new more general and two-dimensional reaction-transport code GENEX (which includes adaptive gridding and other accelerating and stabilizing numerical techniques) are available commercially (Ortoleva et al., 1990).

Simulations of this type can be used in two basic ways—as "diagnostic" tools or as "predictive" tools. In the diagnostic mode, they can provide data to answer questions about which pre-existing conditions could be responsible for the present state of the rocks in which we are interested. In the predictive mode, these simulations can be used to speculate on the final condition of rocks subjected to a variety of environments. In this paper, we present applications of the modeling through several examples of predictive simulations. In order to limit the number of parameters that need be considered in this introduction, we have chosen a simplified scenario for the temporal and spatial evolution of the system. We will consider a system that is isothermal (in both time and space), thus eliminating the complexities introduced by the time-temperature history of the basin. Including the time-temperature history, however, can have a significant influence on diagenesis. The results obtained are not meant to apply to any specific site; the goal of the present study is to gain an understanding of the general processes that lead to the formation of various authigenic minerals. Figure 1 is a schematic view of the system being considered.

Initial Data

A variety of data is needed to perform a simulation. The mineralogy of the initial rock must be provided, including chemical composition of the minerals, their abundance, spatial distribution, shape, and size. The chemical composition of the introduced fluid and its velocity must be specified. The mechanisms for each reaction must be given, as well as the equilibrium constants and rate constants (for slow reactions) for these mechanisms, over the temperature range to which the simulation domain will be subjected. For systems whose temperature will vary with time, the time-temperature history for the basin must also be stated. In addition to these "geological" data, parameters for controlling program execution (such as time step and space grid size) must be provided.

Output Data

REACTRAN simulations provide data on the spatial distribution of the amounts of aqueous and solid species in the system at different times in its evolution. In this paper, such distributions are presented as graphs of mineral or porosity percentage on the abscissa and space on the ordinate at specified times. Fluid flow is from bottom to top. A simple example of this type of plot is shown in Figure 2. Such graphs represent a time "snapshot" of the system. The data are quantitative but are subject to error resulting from the numerical approximations used. This numerical error can be reduced by using smaller time steps and more grid points in the simulation. The data presented here are from relatively "coarse" simulations, at least for the longer-time results, and should be considered "semiquantitative" for those times. The raw (unsmoothed) data are presented, including likely numerical errors, some of which are mentioned in the text. In general, "single point" errors seen as very sharp spikes probably were generated by enhancement of numerical errors from earlier times, when rates were high and "transient" minerals were present. For all data presented here for times up to 1 m.y. (million years), there are 200 space points/

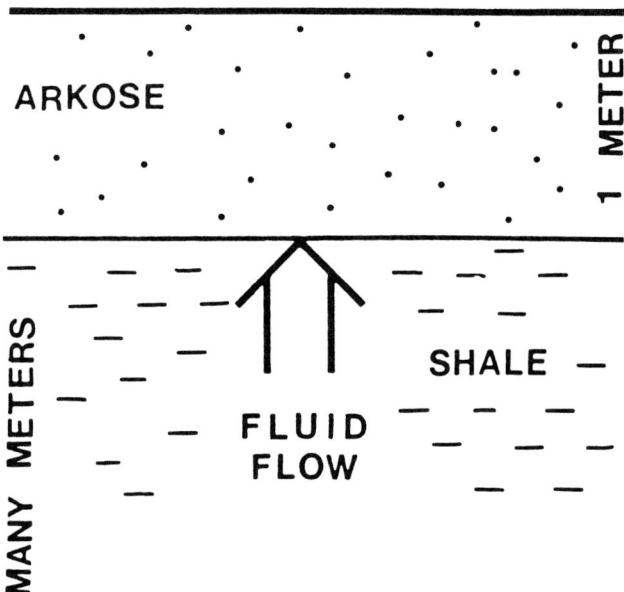

Figure 1. Schematic of the system being simulated. Fluid equilibrated with shale (see Table 2) flows upward through a 1-meter-thick arkose.

meter and the maximum time step was 6E10 seconds, or approximately 2 k.y. (thousand years). The actual time steps taken are variable and are chosen by the program to minimize the errors from extrapolation in time. Problems such as low resolution and buildup of numerical noise have been overcome in GENEX (which was not available at the time of the present study).

The results of simulations can be compared with observable field data in at least two ways—the mineral distribution in the rock and the paragenetic sequences observed. The comparison of mineral distribution is relatively straightforward. Comparisons of paragenetic sequences are more difficult, in part because ambiguous textures are commonly seen in thin sections and in part because one must decide what amount of mineral predicted by the simulations is likely to be noticed and recorded in routine petrographic work.

An additional problem in comparing predictions of simulations with natural systems is the relatively few thermodynamic and kinetic data available for use in the simulations, thus restricting the mineral phases that can be reliably accounted for. For example, very little data are available for illite and its solid solutions. Because of the wide range of illites found in nature, limiting all simulations to considering only one illite composition is bound to result in some discrepancies between simulations and observations. Fortunately, considerable work is being done by experimentalists to determine many of these data as well as by theorists to develop ways to estimate them, and more and more data are becoming

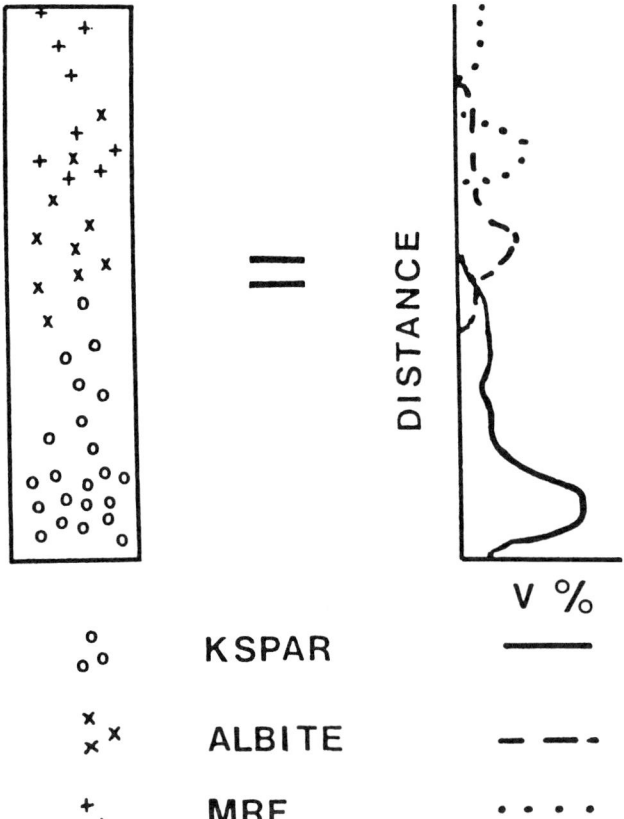

Figure 2. Data presented as plots of volume % of mineral vs. position in space. The left-hand diagram represents actual distributions of minerals in the column 1 m high and with the cross-sectional area of 100 cm². On the right is the graphical depiction. MRF = meteoric rock fragments.

Table 1. Mineral composition/porosity of initial "dirty" and "clean" arkoses and authigenic minerals considered

	Volume %	
Mineral/Porosity	Dirty Arkose	Clean Arkose
Quartz	45	60
K-feldspar	10	10
Albite	5	—
Muscovite	6	—
Annite	2	—
Chlorite	2	—
Porosity	30	30
Authigenic Minerals Considered		
Hematite	—	—
Calcite	—	—
Dolomite	—	—
Siderite	—	—
Illite	—	—
Kaolinite	—	—
Na-nontronite (smectite)	—	—
Analcime	—	—
Pyrite	—	—

available. Even with these limitations, many of the salient features of diagenesis can be captured and predicted using coupled flow/reaction simulations.

EXAMPLES OF SIMULATIONS OF ARKOSE DIAGENESIS

We have simulated the diagenesis of two different arkoses, one "clean" and one "dirty." The arkose was initially assumed to be uniform in composition, as shown in Table 1. The mineral grains were assumed to be spherical, with initial radius of 0.25 mm. One-meter-thick units of these arkoses were allowed to react with fluids flowing through them. Two fluids were considered—the first was average seawater (saline water) equilibrated with an "average" shale + dolomite at the temperature of interest, and the second was "fresh" groundwater from the Ohio Shale equilibrated with the same assemblage + siderite. The compositions of these two fluids at 110° C are summarized in Table 2, along with the saturation index of each mineral considered in the simulations. In the actual simulations, speciation among 55 aqueous species was calculated and these values were used to obtain the reaction rates of 15 minerals. In all simulations, the fluid velocity was assumed to be 0.5 m/yr. The inlet fluid composition was held constant and was at equilibrium with or undersaturated with respect to all initial minerals in the arkose (Table 2). The nucleation threshold for newly nucleated species was chosen to be 10% above saturation. Thermodynamic data used were those in the EQ3/6 data base (Wolery, 1984). Rate constants were chosen so as to produce rates similar to those reported by Lasaga (1984).

Because the composition of the inlet water is held constant through time, the ultimate (equilibrium) mineral assemblage one would find would be quartz, chlorite, calcite, and dolomite in the case of saline water at the inlet; and quartz, chlorite, calcite, dolomite, and siderite in the case of fresh water at the inlet. This assumes that the nucleation thresholds for the carbonates (and chlorite in the case of clean arkose) would be reached, and this is quite likely because of the small nucleation threshold used in the simulations. We will see, however, that even after 50 m.y. at 110°C this equilibrium assemblage has not been achieved in some systems.

SIMULATION RESULTS

General Remarks

The simulation results are presented in graphic form. Only data for minerals present in significant

Table 2. Inlet fluid compositions (totals of each element, moles/liter fluid) and saturation indices with respect to the minerals in Table 1

FLUID I

Saline Water: pH = 5.87

K 7.1E-4	Na 6.4E-1	Fe 2.1E-6	Mg 2.4E-3	Ca 4.0E-2
Al 4.3E-6	SiO_2 1.0E-4	S 1.0E-3	CO_3 1.0E-3	Cl 5.3E-1

Saturation Indices: 1.00 = equilibrium

Quartz 1.0	K-feldspar 8.1E-2	Albite 5.6E-1	Muscovite 9.6E-1
Annite 1.9E-6	Chlorite 1.0	Hematite 1.9E-1	Calcite 1.0
Dolomite 1.0	Siderite 3.0E-10	Illite 3.2E-2	Kaolinite 9.7E-1
Na-nontronite 9.5E-1		Analcime 3.6E-1	Pyrite 9.5E-1

FLUID II

Fresh Water: pH = 6.4

K 1.9E-4	Na 7.0E-3	Fe 1.3E-7	Mg 2.1E-4	Ca 2.4E-3
Al 4.5E-6	SiO_2 9.9E-4	S 3.6E-3	CO_3 1.9E-3	Cl 5.9E-4

Saturation Indices: 1.00 = equilibrium

Quartz 1.0	K-feldspar 8.0E-2	Albite 3.3E-2	Muscovite 9.6E-1
Annite 1.2E-5	Chlorite 1.0	Hematite 5.0E-1	Calcite 1.0
Dolomite 1.0	Siderite 1.0	Illite 3.2E-2	Kaolinite 9.7E-1
Na-nontronite 9.5E-1		Analcime 2.1E-2	Pyrite 9.4E-1

amounts (greater than approximately 0.3 volume % of the rock) are shown. The distribution of the species is given at various times during the evolution of the system, with time zero being the beginning of diagenesis.

A common problem in many simulations using REACTRAN and other codes that use thermodynamic data in the literature is the stability of muscovite with respect to illite in many situations where observations show that illite is the stable phase. This may be a result of poor thermodynamic data for either illite or muscovite; alternatively, it may be due to the fact that the illite forming in nature has a composition and thermodynamic properties different from those for which data are available. This problem may be approached in one of two ways—disallow muscovite in the system in favor of illite, or perform the simulation with both muscovite and illite as possible minerals, but keep in mind the problem. We have taken the second approach. Consequently, we see muscovite as a common diagenetic mineral in the simulations. We suspect that this really represents illite (and/or sericite) growth, but cannot state so with certainty; we will refer to this as muscovite* in the remainder of the text and MUSC in the figures.

Dirty Arkose Altered by Saline Water

Figures 3 through 7 show the evolution of the distribution of mineral species in space and time when the dirty arkose of Table 1 is subjected to a flow of saline fluid (Fluid I, Table 2) at 110°C. These processes, as well as phenomena that occur but are not shown because of space limitations, are discussed below.

The first significant process that occurs in this simulation is the "replacement" of annite by chlorite (although REACTRAN cannot show individual mineral textures, some processes are clearly coupled spatially and temporally, and we sometimes refer to this as replacement). This process forms approximately one-quarter of the authigenic chlorite precipitated in the entire system over 50 m.y. Also, at this time, smectite (represented in the simulations by sodium nontronite*, abbreviated as NANONT by the computer) forms, once again in association with annite dissolution. Locally, as much as 0.4% (of the total of 2%) of the annite has dissolved after 200 years of diagenesis. All annite has dissolved from the system by approximately 25 k.y.; as much as 0.5% quartz, precipitated at some space points, is derived from the dissolution of K-feldspar and plagioclase; silica released by annite dissolution was incorporated in chlorite and smectite growth and was not a contributor to quartz growth. Although carbonates have formed (dolomite near the inlet and calcite toward the center of the system), their connection to dissolving silicates is not absolutely clear. However, it seems that calcite growth is most closely connected with K-feldspar dissolution, whereas albite dissolution is related to dolomite growth (Figures 3,

---

*Sodium nontronite is used instead of a smectite because of availability of thermodynamic data on clay minerals of specific composition. Similarly, muscovite abbreviated on the figures as "musc" is used instead of an illite, because of the availability of the thermodynamic data.

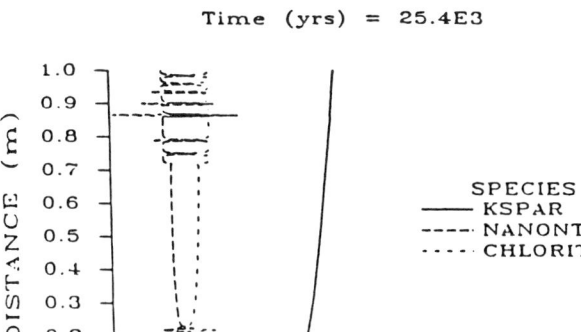
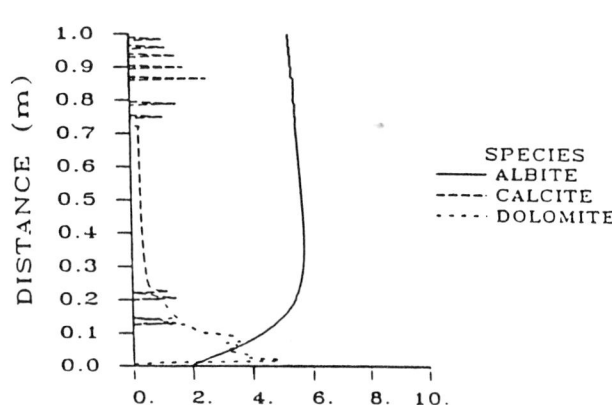

Figure 3. Distribution of indicated minerals in dirty arkose altered by saline water after 25.4 k.y. (See text for discussion of this and subsequent figures.)

4). In both cases, the likely cause of carbonate precipitation is a rise in pH resulting from silicate dissolution. Kaolinite occurs in small amounts near the inlet, apparently as a product of the dissolution of the feldspars, and where it is growing, dolomite is the stable carbonate rather than calcite. Muscovite* growth (not shown) is also associated with feldspar dissolution. Albitization is occurring near the center of the system, and results from the dissolution of albite near the inlet and its reprecipitation in the different chemical environment present downflow. Smectite and chlorite amounts show an inverse relationship. This distribution is shown in Figure 3. Figure 4 shows the system configuration at about 127 k.y. Features to note are (1) the nearly complete dissolution of K-feldspar, (2) the growth of calcite associated with this dissolution (especially downflow where no albite is dissolving), (3) the downflow movement of the albite dissolution front and reprecipitation of the albite downflow, and (4) the association of calcite + chlorite that is exclusive to the association dolomite + smectite.

K-feldspar has dissolved out of the system by about 225 k.y., and albite is gone by about 775 k.y. Figure 5 shows the distribution of minerals in the system at about 952 k.y. By this time, as much as 4% chlorite and as much as 8% muscovite* have been added at some points; some of this growth is directly attributed to elements gained from feldspar dissolution, but much comes from replacement of smectite, which is no longer stable at about 0.5 m into the system and is replaced by the chlorite + muscovite* assemblage. While albite was still in the system, sodium derived from its dissolution also locally enhanced the albitization process. Dolomite persists near the inlet, and calcite occurs in various amounts throughout the rest of the system; kaolinite (not shown) occurs near the inlet and two-thirds of the way through the system but is slowly being dissolved, and muscovite* and chlorite are growing. Smectite is also being superseded by this assemblage. A net increase in silica from these silicate reactions results in quartz growth. Trace amounts of pyrite are present. Porosity has decreased to as low as about 24% everywhere except the first space point. Features to note in Figure 5 are (1) the continued inverse relationship in the amounts of smectite and chlorite, (2) the positive correlation between muscovite* and calcite, and (3) the absence of large amounts of calcite and dolomite in association with each other.

Further evolution of the system primarily enhances these earlier diagenetic features. Figure 6 shows how smectite has been completely replaced in the second two-thirds of the system by chlorite (and by kaolinite; see Figure 7) by 31.5 m.y. It also shows the distribution of quartz (originally 45%) and porosity (originally 30%) at this time. For the most part, overall porosity has remained nearly the same, although it may have locally decreased and quartz growth has occurred nearly everywhere (the first space points in Figures 6 and 7 are likely to have accumulated a significant amount of numerical error by this time).

Overall equilibrium (that is, all minerals at a point in the system in equilibrium with the fluid there) has not been achieved by 50 m.y., and slow dissolution of muscovite* and smectite is still occurring. Figure 7 shows the distribution of all the significant mineral species and the porosity at 9.52 m.y. Because the minerals remaining are very near equilibrium with the incoming fluid, their dissolution is quite slow and this distribution is essentially the final assem-

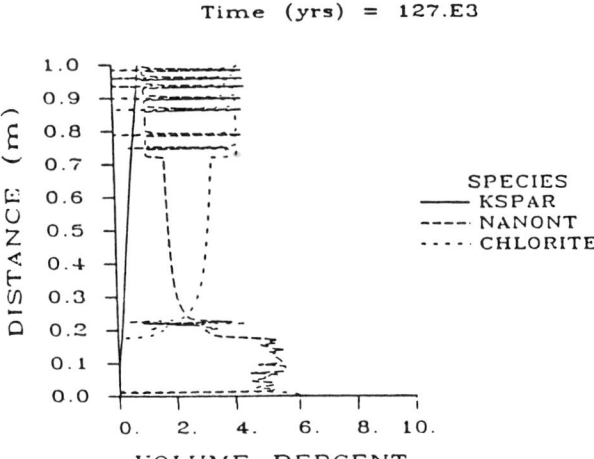
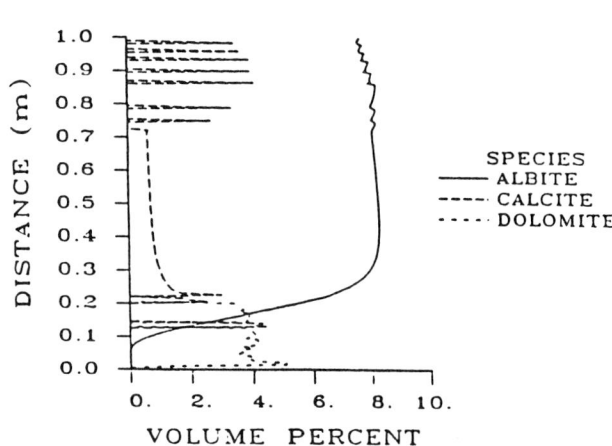

Figure 4. Distribution of indicated minerals in dirty arkose altered by saline water after 127 k.y.

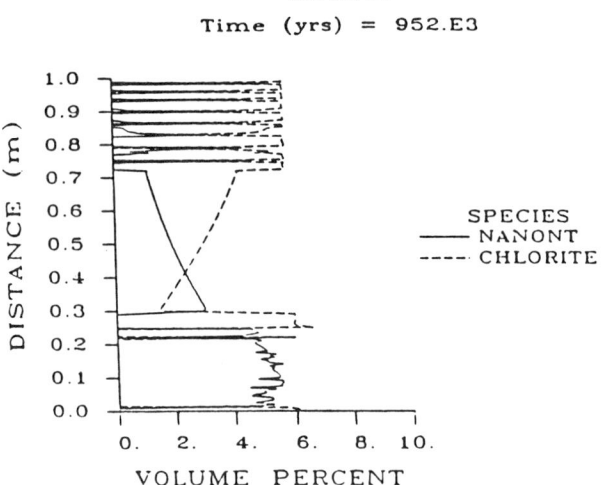
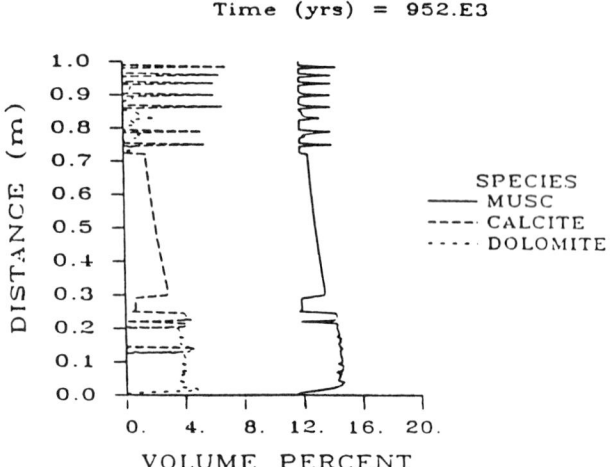

Figure 5. Distribution of indicated minerals in dirty arkose altered by saline water after 952 k.y.

blage (although the smectite near the center of the system will have disappeared by 50 m.y.). The overall effect of diagenesis on porosity has been to produce a general, small decrease in porosity. The factors likely to have the most significant effect on reservoir quality are the form of occurrence and the location in which the diagenetic muscovite* and chlorite are found (pore filling, bridging, etc.) and whether the quartz has formed as cement or overgrowths.

Clean Arkose Altered by Saline Water

The early stages of the diagenesis of clean arkose by saline water are marked by the formation of muscovite*, albite, and calcite associated with the dissolution of K-feldspar as shown in Figure 8. Muscovite* forms throughout the system, in greater abundance near the inlet, whereas albite forms toward the end of the system, and calcite occurs from about 0.2 m to the end of the system. Dolomite occurs nearer the inlet, where it is associated with the muscovite*, and may also be spatially a result of the dissolution of previous authigenic albite. At this point in time, only minor quartz growth has occurred, and the porosity is essentially unchanged.

Our next view of the system is at 219 k.y. By this time, K-feldspar has dissolved from the system and the albite is nearly gone (Figure 9). A zone of kaolinite appears to be associated with the albite dissolution,

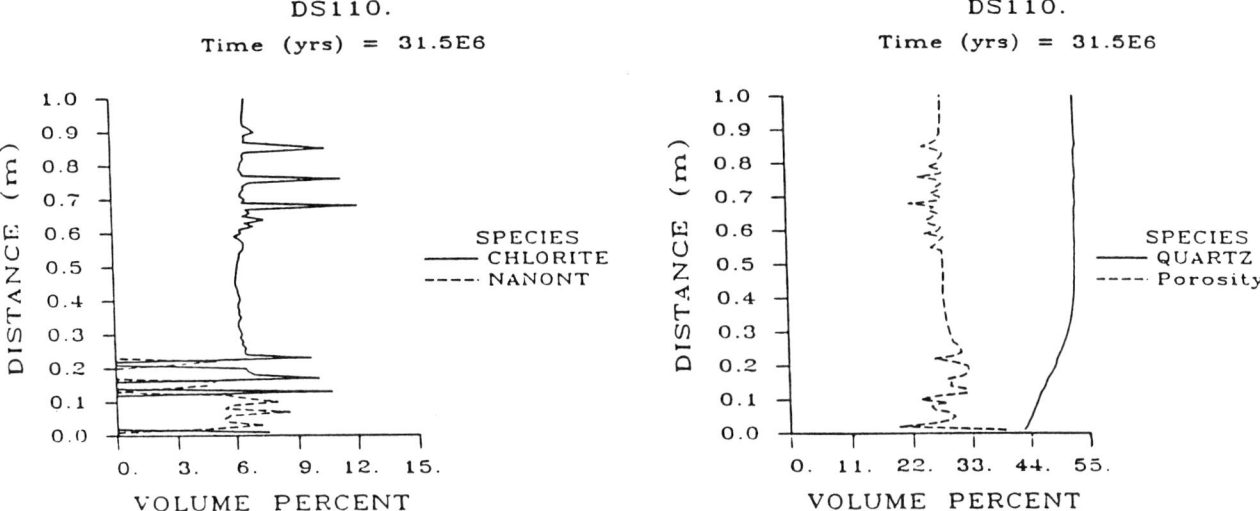

Figure 6. Distribution of indicated minerals in dirty arkose altered by saline water after 31.5 m.y.

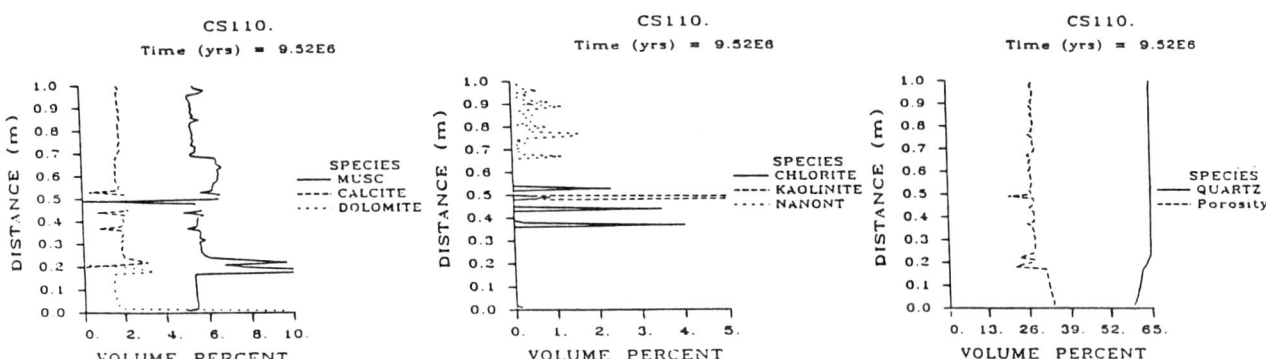

Figure 7. Distribution of the significant minerals remaining in dirty arkose altered by saline water after 9.52 m.y.

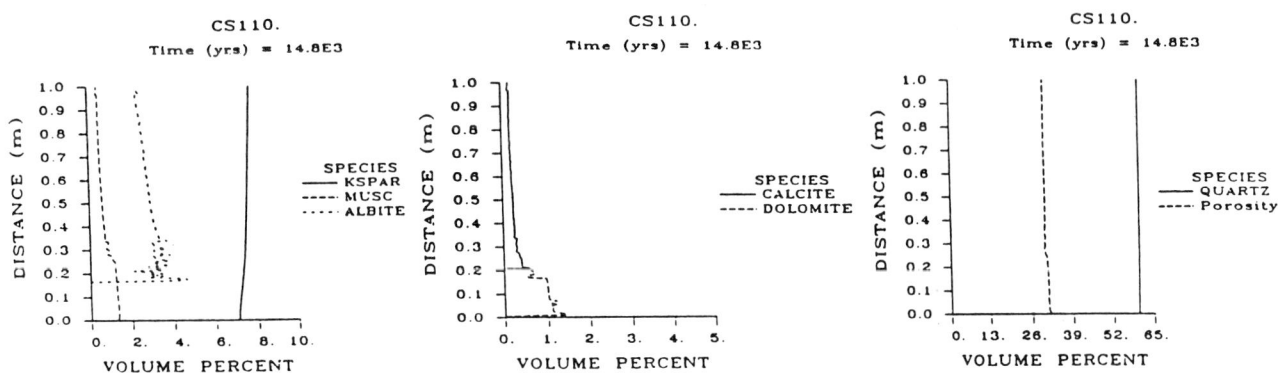

Figure 8. Distribution of indicated minerals in clean arkose altered by saline water after 14.8 k.y.

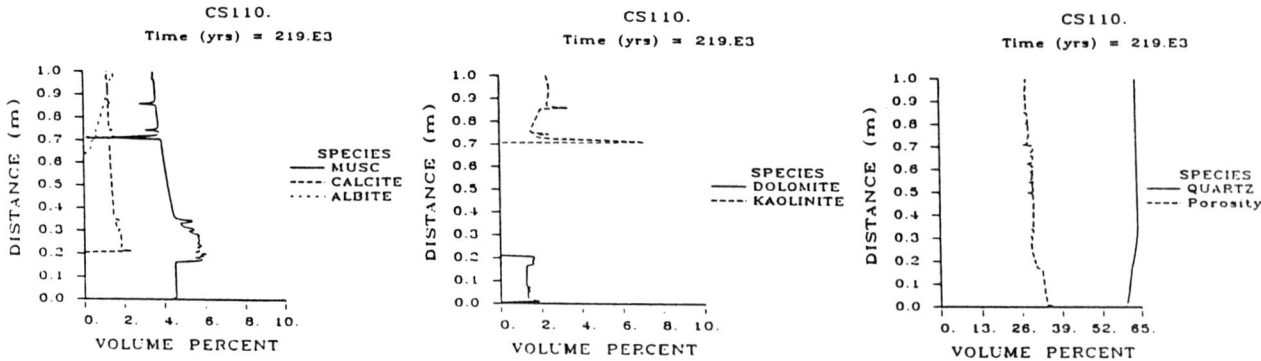

Figure 9. Distribution of indicated minerals in clean arkose altered by saline water after 219.3 k.y.

probably representing kaolinization of the albite. Quartz has increased by as much as 3% as a result of feldspar dissolution, and porosity has decreased by as much as 3.5% near the 0.2-m mark.

Figure 10, for the time of 9.52 m.y., shows that the overall mineral distribution has changed little from the previously shown time. Albite has disappeared by dissolution, possibly leading to an increase in kaolinite, which was then "replaced" by muscovite* (illite/smectite) and calcite (examine the area from 0.6 m to the end of the system and compare with Figure 9). The chlorite and kaolinite contents plotted in the center of the graph probably represent a numerical error, although small amounts of these minerals may be present. Quartz abundance has increased to about 65% from the initial 60% in most of the system. Overall porosity has decreased by about 3.5% everywhere except near the inlet, where it has remained essentially the same since the beginning of the simulation. Once again, the primary control on reservoir quality is probably the form of occurrence and position of the clay minerals.

Dirty Arkose Altered by Fresh Water

Figure 11 shows mineral distributions in this system at 63.5 k.y. Early diagenesis has resulted in the dissolution of the feldspars, annite, and chlorite to produce muscovite*, smectite, and carbonates. Chlorite has been removed from most of the first half of the system (the chlorite at the inlet may be a result of numerical error). Although kaolinite appears near the inlet, its possible relationship to the dissolution of a particular silicate is unclear. Smectite and dolomite growth show strong positive coupling. At this time, there has been little change in the amount of quartz or the porosity.

By 254 k.y., albite has been removed from the system (Figure 12). Chlorite is nearly gone, but K-feldspar and annite persist. Muscovite*, smectite, and dolomite are the main authigenic minerals in the system. Kaolinite has been removed except at the inlet. The strong positive coupling between the carbonates and smectite is still seen; dolomite is the dominant carbonate, except perhaps near the center of the system. The volume of quartz has increased by about 3%, and the overall porosity has decreased by about the same amount.

By 952 k.y. (Figure 13), smectite and muscovite* volumes have increased as a result of the dissolution of annite, and K-feldspar is now gone from the system. The distribution of the carbonates has changed very little from that at 254 k.y. The amounts of quartz and porosity also have undergone negligible change, suggesting that the replacement of annite and K-feldspar by muscovite* and smectite was nearly volume-for-volume. Except for the removal of the small amount of annite that remains in the system, this is essentially the same configuration as the system at 9.52 m.y. (Figure 14). An interesting feature to note is the "banded" nature of the authigenic calcite. This banding apparently is a result of variations in the upstream mineralogy, which affect the downstream pH and, consequently, control calcite precipitation. Banded carbonate, silica, and other cements in arkoses have been reported by Tigert and Al-Shaieb (1990) and have been given an alternative interpretation based on a mechano-chemical model (Dewers and Ortoleva, 1990). The bandings observed in the present study differ from those cited above in that each carbonate band seen in Figures 12 and 13 is associated with a change in other mineralogy, each change being distinct for each band.

Clean Arkose Altered by Fresh Water

The most simple of the diagenetic histories is that of clean arkose altered by fresh water. The earliest minerals formed are muscovite*, dolomite, and quartz, all associated with the dissolution of K-feldspar. As a result of these reactions, quartz is also formed. The distribution of minerals at approximately 190 k.y. is shown in Figure 15. By 100 k.y., some calcite has also formed, from about 0.82 m to the end of the system. Evolution of the system continues along this path until the K-feldspar has dissolved (at about 575 k.y.). This authigenic

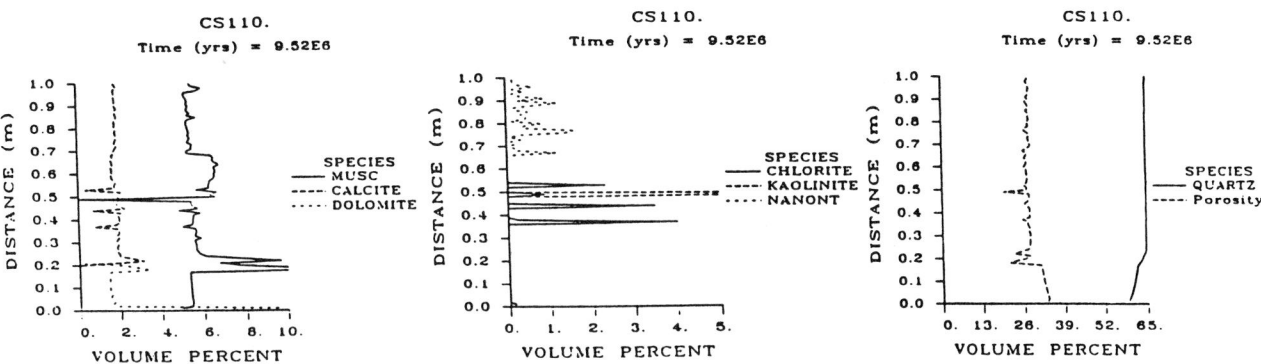

Figure 10. Distribution of the significant minerals remaining in clean arkose altered by saline water after 9.52 m.y.

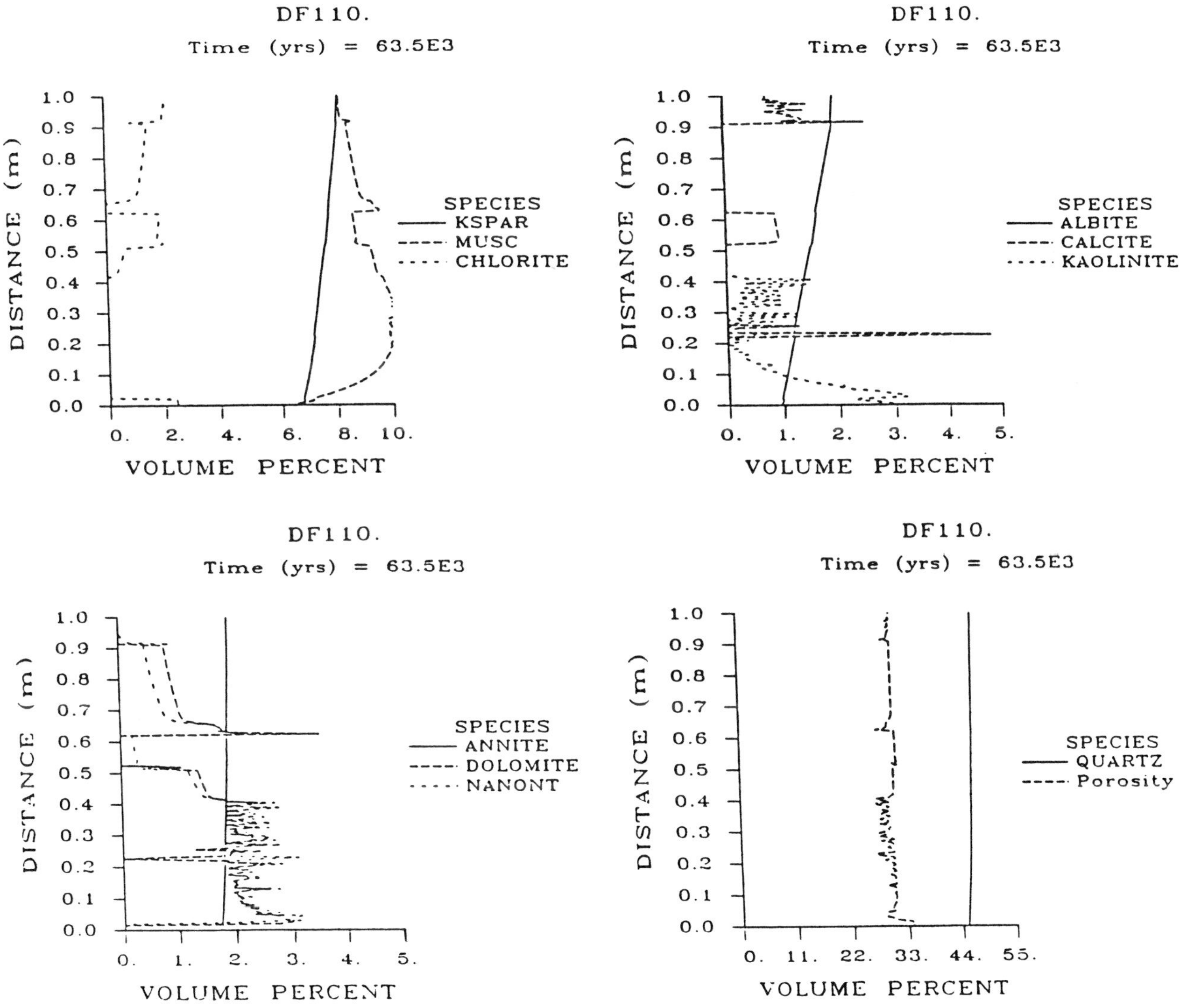

Figure 11. Distribution of indicated minerals in dirty arkose altered by fresh water after 63.5 k.y.

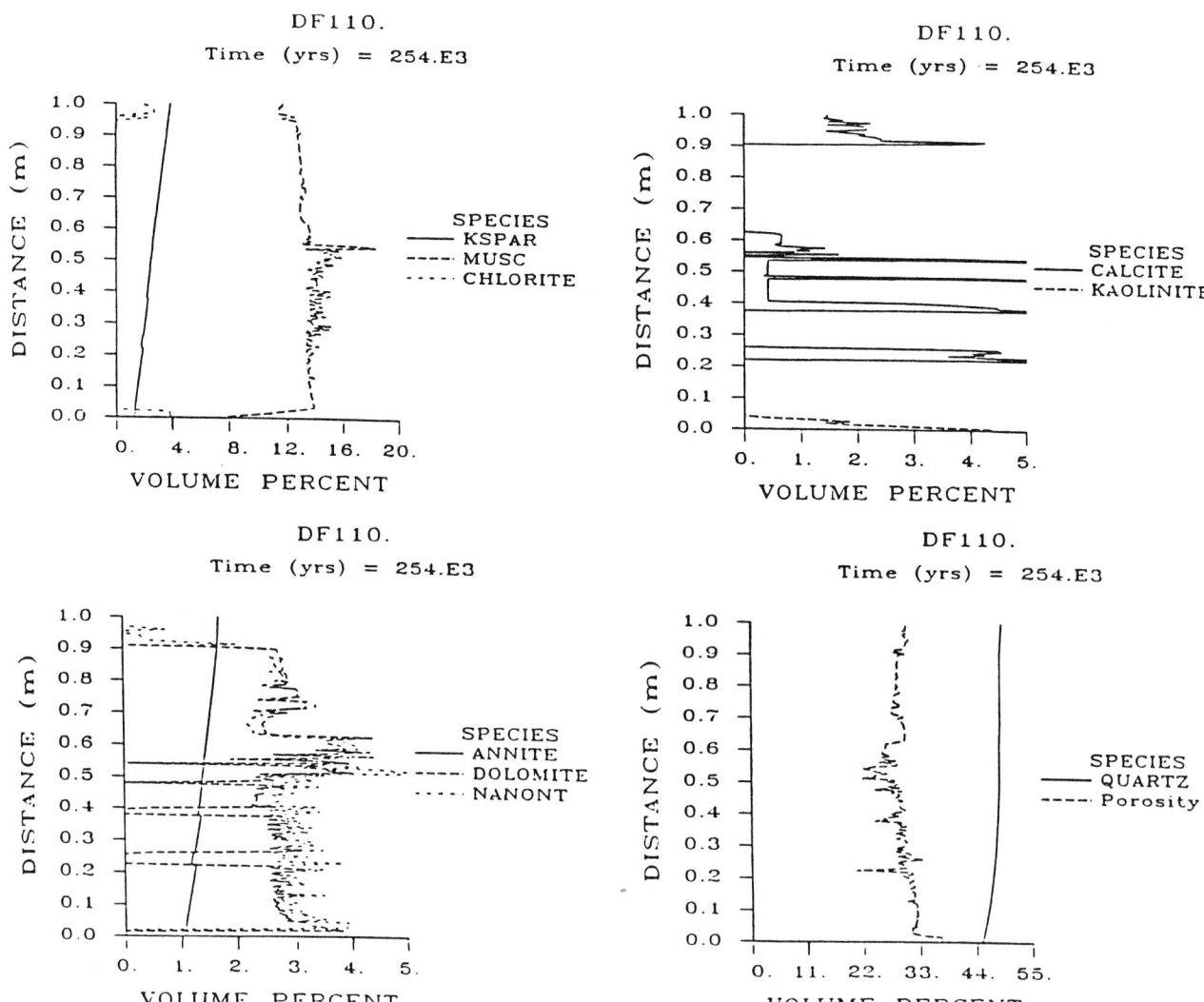

Figure 12. Distribution of indicated minerals in dirty arkose altered by fresh water after 254 k.y.

assemblage then persists, and these minerals are in equilibrium with the inlet fluid. The distribution of minerals at 9.52 m.y. is shown in Figure 16. This diagenetic evolution results in very little modification of the original porosity.

Comparison of Simulation Results

In this section we give a brief comparison of the results of the four simulations. We begin with a comparison of the four systems as they appear at 9.52 m.y. In the dirty arkoses, alteration by saline waters results in a slightly higher percentage of authigenic quartz than alteration by fresh water. The same relationship is true for the clean arkose (Figure 17). The porosities of the dirty arkoses are quite similar for alteration by both fresh and saline waters. The porosities of clean arkoses, however, differ by a few percentage points, the least porosity being in the arkose altered by saline water (Figure 18). Comparison of the two arkoses altered by the same type of water shows what one would expect— the dirty arkose shows a slightly smaller porosity change, and the effect is most pronounced in the clean arkoses.

In all four simulations, calcite and dolomite were both formed as authigenic minerals. Siderite was not formed except in trace amounts early in the system evolution. For the clean arkose, calcite and dolomite occupy spatially well-defined zones; dolomite is present in the part of the system nearer the inlet, and calcite in the part away from the inlet. The width of the dolomite zone is much smaller in the arkose altered by saline water. In the dirty arkoses, calcite and dolomite are also spatially separated, but are separated into many zones throughout the system, not just two. As with clean arkoses, the first zone is dolomite, and it is wider in the case of freshwater

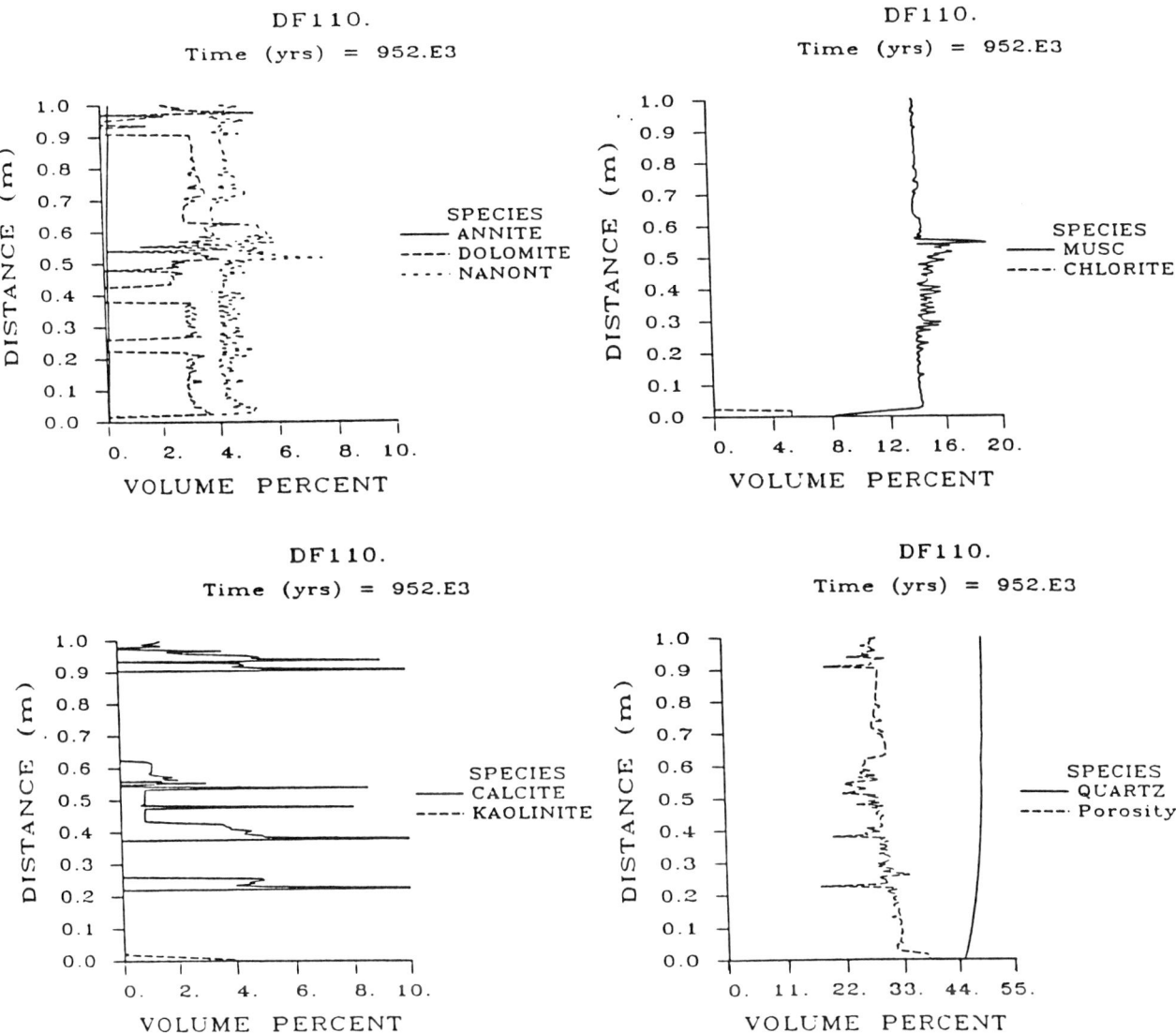

Figure 13. Distribution of indicated minerals in dirty arkose altered by fresh water after 952 k.y.

alteration. Both arkoses contain more authigenic carbonate in the system when they are altered by fresh water. These relationships are shown in Figure 19.

The greatest contrast in authigenic mineralogy is the difference in phyllosilicate content and distribution among the four simulations. Where clean arkose is altered by saline water, muscovite* is the only authigenic phyllosilicate present. For clean arkose altered by saline water, small amounts of chlorite and kaolinite may remain, but this could be caused by numerical error. Some smectite may persist but will slowly disappear, leaving only muscovite*. Dirty arkose altered by fresh water contains predominantly muscovite* and smectite, with only a small amount of chlorite near the outlet end. Dirty arkose altered by saline water also contains muscovite* and smectite, with only a small amount of chlorite near the outlet end. Dirty arkose altered by saline water also shows an interesting zoning of smectite-chlorite-kaolinite (with some overlap) from the inlet toward the outlet. This zoning does not persist, however, and kaolinite is dissolved by about 32 m.y.; smectite dissolves very slowly. Distributions of the nonmuscovite* phyllosilicates are shown in Figure 20. For the dirty arkoses, muscovite* is slightly more abundant when the arkose is altered by fresh water; in the clean arkose, however, the opposite is found (Figure 21).

Discussion

The simplifications made to allow easier understanding of geochemical relationships may make specific comparison with observed examples of arkose diagenesis seem untenable. However, several general

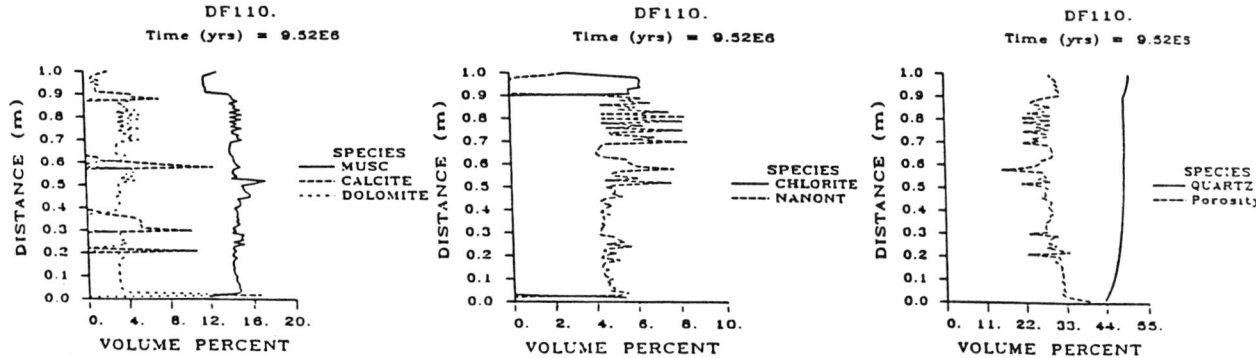

Figure 14. Distribution of the significant minerals remaining in dirty arkose altered by fresh water after 9.52 m.y.

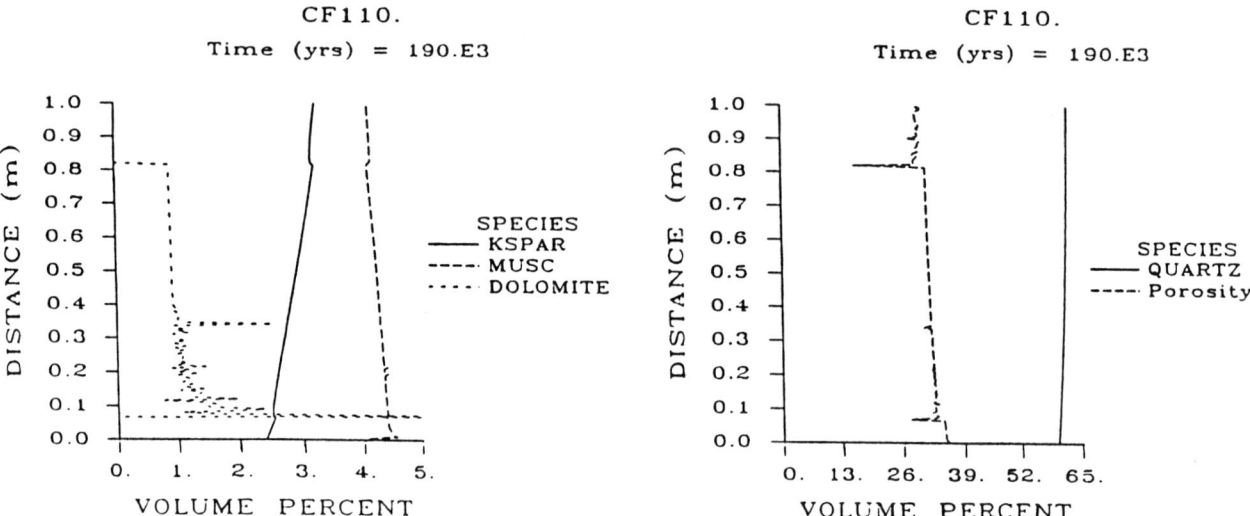

Figure 15. Distribution of indicated minerals in clean arkose altered by fresh water after 190 k.y.

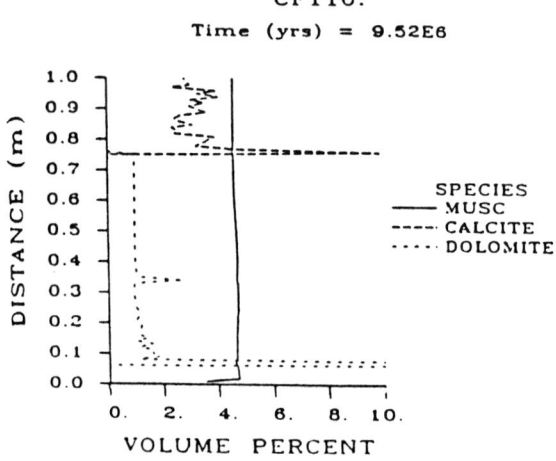
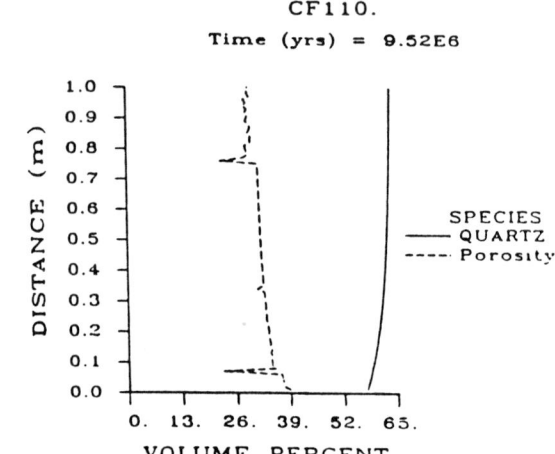

Figure 16. Distribution of the significant minerals remaining in clean arkose altered by fresh water after 9.52 m.y.

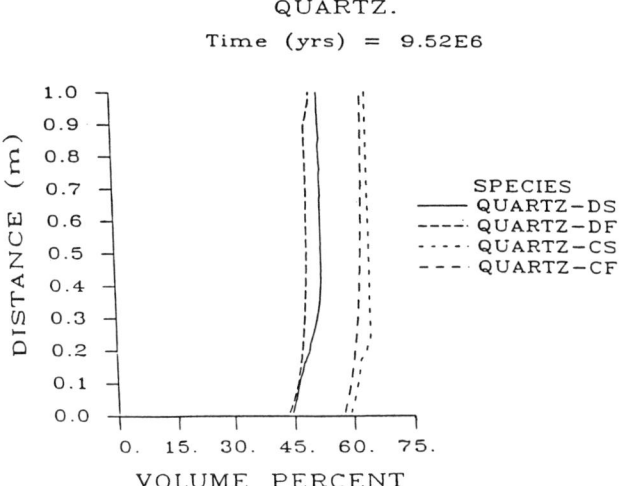

Figure 17. Quartz distribution after 9.52 m.y. in the four arkose-fluid systems simulated. DS = dirty arkose, saline water; DF = dirty arkose, fresh water; CS = clean arkose, saline water; CF = clean arkose, fresh water.

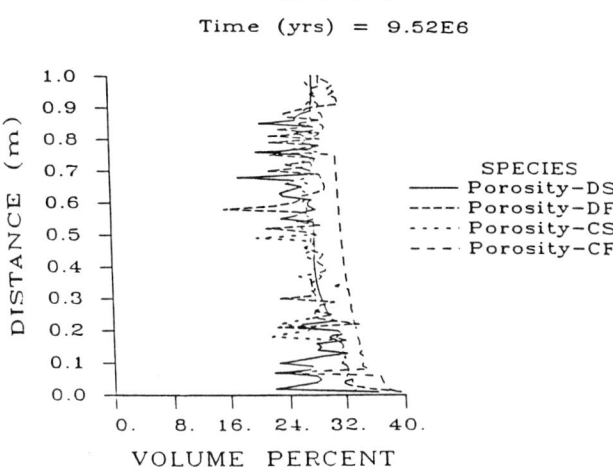

Figure 18. Porosity distribution after 9.52 m.y. in the four arkose-fluid systems simulated. Labels are as in Figure 17.

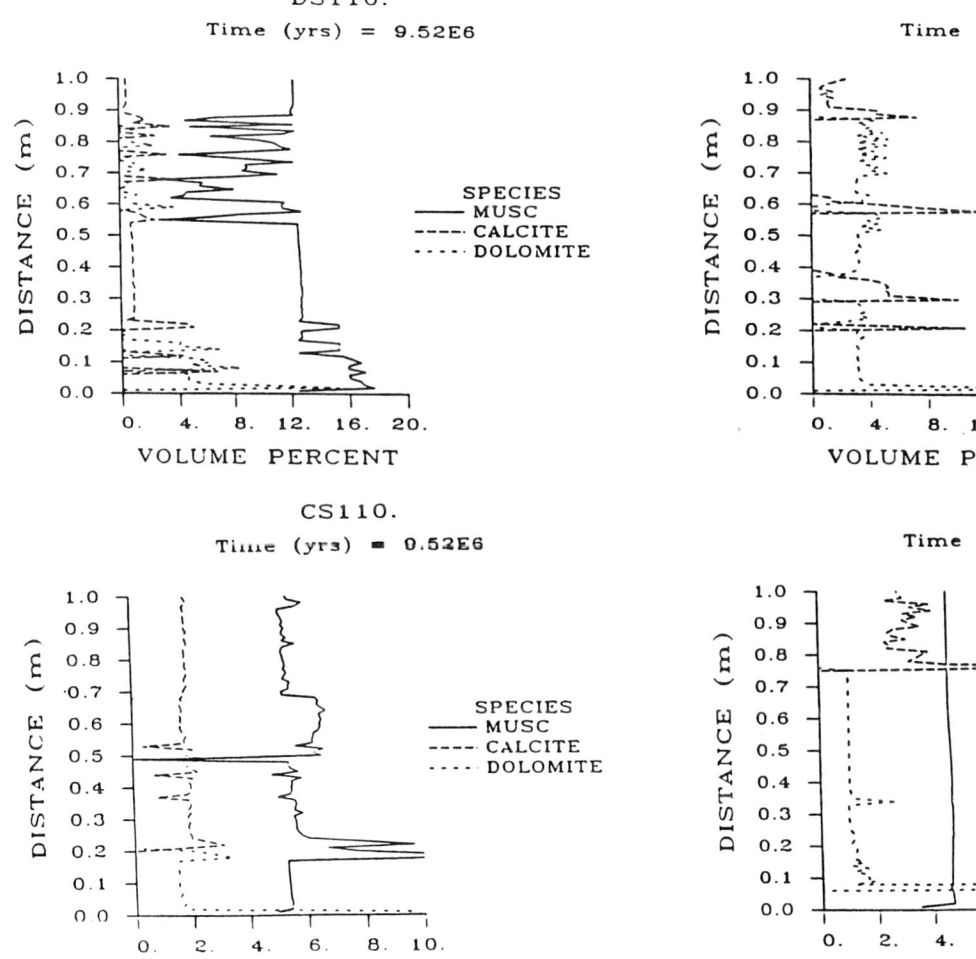

Figure 19. Distributions of muscovite*, calcite, and dolomite after 9.52 m.y. in the four arkose-fluid systems simulated. Labels are as in Figure 17.

Prediction of Reservoir Quality Through Chemical Modeling

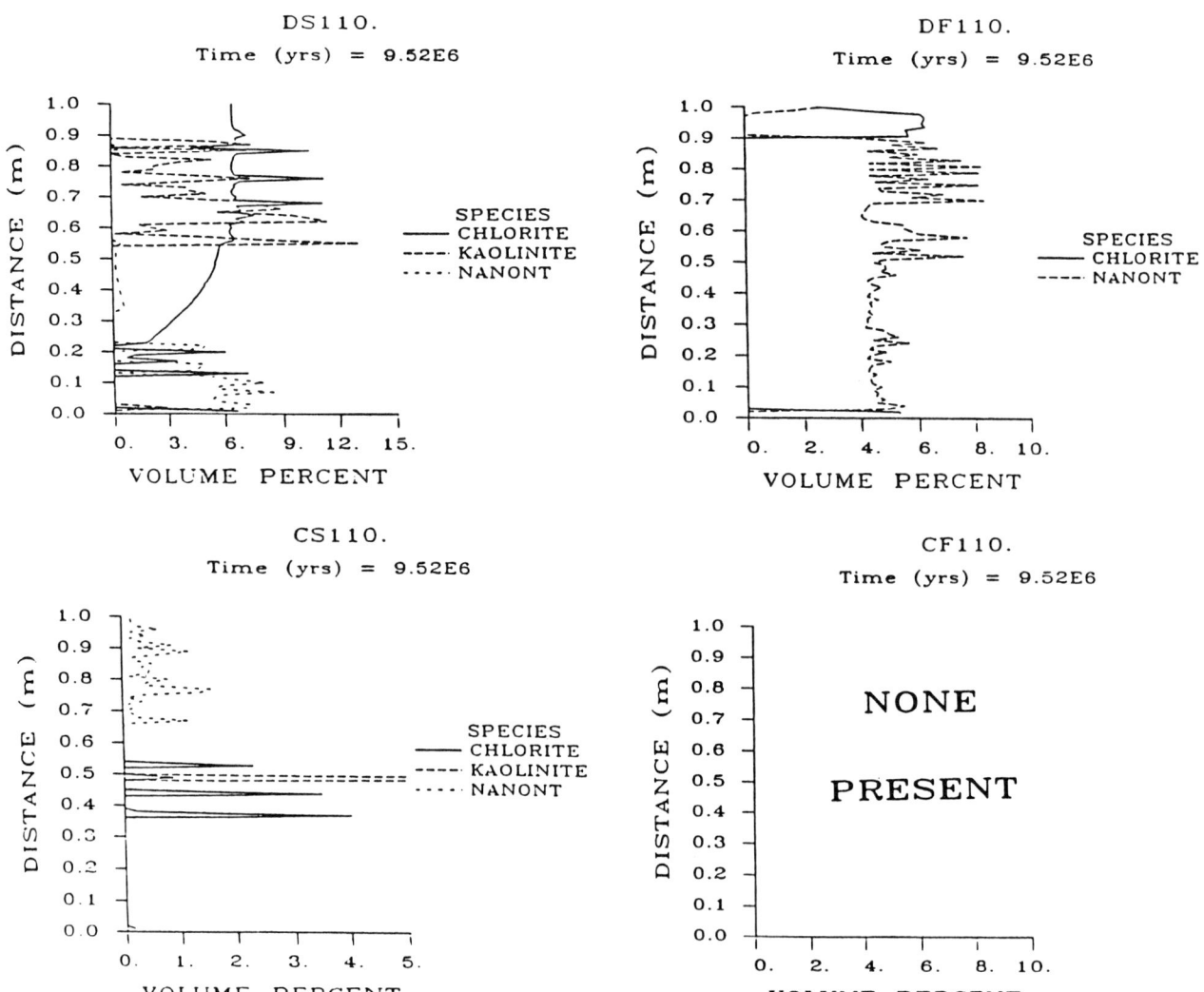

Figure 20. Distribution of chlorite, kaolinite, and smectite (NANONT) after 9.52 m.y. in the four arkose-fluid systems simulated. Labels are as in Figure 17. Minerals in the legend but not plotted occurred in amounts too small to plot on the graph.

observations can be compared with currently accepted concepts of diagenesis. The most clearcut is that albitization occurs only in systems with saline water. Chlorite does not form (or forms in very small amounts) in clean arkoses, regardless of the type of fluid imposed. Similarly, smectite is rare or absent in clean arkose but common in dirty arkose. However, smectite eventually disappears in deference to muscovite*; this takes place more quickly in saline waters.

Although relatively small porosity decreases were predicted in these simulations, other simulations show much larger changes. Moreover, reservoir quality also strongly depends on permeability and, as mentioned above, permeability is often influenced primarily by the mode of occurrence of the authigenic minerals formed. At present, one must rely on empirical evidence to speculate on the mode of occurrence the various minerals will assume. However, much useful information can be gained by combining empirical data with the results of simulations such as those described above.

SUMMARY

Our aim in this paper has been to demonstrate the potential for applying computer simulations using codes based on coupled reaction-transport, nonequilibrium models to investigate physicochemical controls on reservoir quality. As an introduction to this process, we chose a simple

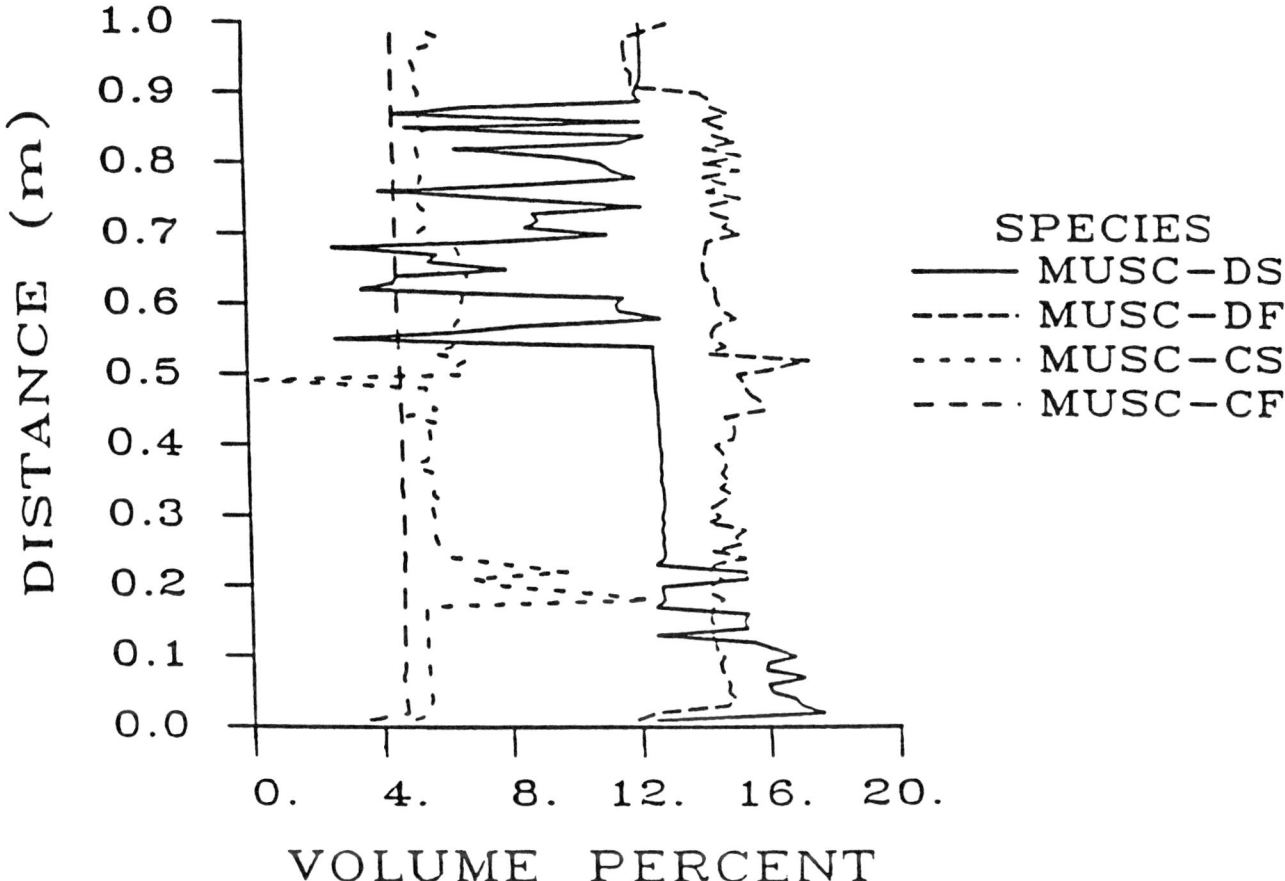

Figure 21. Comparison of the amount of muscovite* in each of the four systems simulated after 9.52 m.y. of alteration. Dirty arkoses originally contained 6% muscovite, clean arkoses contained none. Labels are as in Figure 17.

isothermal scenario and four generalized systems to investigate. We have briefly discussed the temporal and spatial evolution of the mineralogy and porosity distribution in the systems and have shown the controls exerted on them by rock and fluid compositions. The ability to predict both the temporal and spatial evolution of diagenetic alteration of sediments is a key to predicting reservoir quality. It is important to know not only if, but where and when, favorable porosity/permeability regions might develop in a reservoir, in order to gauge the potential for substantial hydrocarbon migration and accumulation. We believe that the REACTRAN code (and the more powerful code GENEX—see Chen et al., this volume) can be very effective in predicting these features when specific data on the initial mineralogy and the time/temperature history of the reservoir are included in the simulations.

ACKNOWLEDGMENT

This work was supported in part by a grant from the U.S. Department of Energy, Basic Energy Sciences Program.

REFERENCES CITED

Bear, J., 1972, Dynamics of fluids in porous media: Amsterdam, Elsevier. 764 pp.

Chen, W., A. Ghaith, A. Park, and P. Ortoleva, 1990, this volume.
Lasaga, A. C., 1984, Chemical kinetics of water-rock interactions: Journal of Geophysical Research, v. 89, no. B6, p. 4009–4025.
Ortoleva, P. J., E. Merino, C. Moore, and J. Chadam, 1987, Geochemical self-organization I: Feedback mechanisms and modeling approach: American Journal of Science, v. 287, p. 979–1007.
Ortoleva, P., W. Chen, P. Henq, T. Park, and C. Wang, 1990. GENEX user's manual: Bloomington, Geo-Chemical Research Associates.
Tigert, V., and Z. Al-Shaieb, 1990, Pressure seals: their diagenetic banding structure patterns, *in* P. Ortoleva, B. Hallet, A. McBirney, I. Meshri, R. Reeder, and P. Williams, eds., Proceedings of the Workshop on Self-Organization in Geological Systems, Santa Barbara, 1988: Earth Science Reviews (Amsterdam, Elsevier).
Wolery, T. J., 1984, EQ6, a computer code for reaction-path modeling of aqueous geochemical systems: Lawrence Livermore National Laboratory.

Interaction of Reaction, Mass Transport, and Rock Deformation During Diagenesis: Mathematical Modeling of Intergranular Pressure Solution, Stylolites, and Differential Compaction/Cementation

Thomas Dewers
Peter J. Ortoleva
Departments of Geology and Chemistry
Indiana University
Bloomington, Indiana, U.S.A.
Geo-Chemical Research Associates
Bloomington, Indiana, U.S.A.

> The recognition and description of the physical processes responsible for pressure solution and associated cementation—including aspects of thermodynamics, reaction kinetics, solute transport, and rock mechanics—are important facets of research on diagenesis. The interplay between these processes is studied by means of mathematical models coupling water-rock interaction with rock deformation. Model predictions are compared with observed pressure solution/cementation features. Examples discussed include the grain-size dependence of intergranular volume loss, the transition between stylolitic and intergranular pressure solution, an assessment of the role of clay in influencing pressure solution, and the evolution and correlation of spatially discrete domains of heightened compaction and cementation in sandstones and argillaceous carbonates.

INTRODUCTION

The operation of pressure solution during diagenesis is a topic of continued interest for researchers. Recent advances in the theoretical, experimental, and observational treatments of pressure solution have enabled us to construct mathematical reaction-transport models which describe the spatio-temporal development of various pressure solution features. These are useful in that they afford direct comparison between theoretical predictions and observations. We present an overview of our attempts to capture the evolution of several pressure solution phenomena quantitatively, and then compare the results with examples of intergranular pressure solution, stylolites, patterns of cementation in sandstones, and marl/limestone alternations.

QUANTIFICATION OF PRESSURE SOLUTION

Theoretical Considerations

Two different dynamics are thought to control the dissolution reactions of minerals under the action of applied stress during diagenesis. Following the notation in Tada et al. (1987), they are water-film diffusion, or WFD (Weyl, 1959), and free-face pressure solution, or FFPS (Bathhurst, 1958; Tada and Siever, 1986). WFD is driven by the normal stress dependence of Gibb's free energy, and is limited by the diffusion of reaction products through thin films of adsorbed fluid that line grain-to-grain contacts (Paterson, 1973; Robin, 1978; Maliva and Siever, 1988). FFPS is driven by elastic and possibly inelastic strain energy or enhanced surface free energy accompanying microgranulation at contacts, and may be limited by intracontact diffusion or by detachment kinetics at fluid-solid interfaces (Paterson, 1973; Engelder, 1982; Lehner and Bataille, 1985; Tada et al., 1987). Estimates for all the parameters necessary to roughly describe pressure solution dynamics for both WFD and FFPS exist; a comparison between the two mechanisms is found in Tada et al. (1987).

In a series of studies (Dewers and Ortoleva, 1989; in press, a, b), we examined the implications of WFD and FFPS in the context of reaction-transport modeling. This approach allowed us to determine the influence of spatial heterogeneity on the spatial distribution of (chemical) compaction and cementation. Details in the construction of the rate laws governing mineral growth and dissolution by FFPS and WFD are found in Dewers and Ortoleva (1990a, b). These rate expressions provide the necessary link in the coupling of the dynamics of grain growth and dissolution, compaction, pore-fluid chemistry, solute transport, and effective stress. Together with a means for quantifying grain texture, evolution equations for the above variables constitute a mathematical model which describes the development of pressure solution features in space and time. The required geometrical relationships are accounted for by Weyl's (1959) model of truncated spheres in simple cubic (or rectangular) packing. Although this is a simplified description for well-sorted sandstones (Rittenhouse, 1971), it allows a relatively simple method for calculating the stresses across intergranular contacts, rates of dissolution at the various grain interfaces, and thus the compactional velocity. In this model, rock texture is quantified in terms of the radius of each (truncated) sphere, and the three sides of the rectangular box within which the sphere is contained. An example of an array of such grains which have undergone progressive amounts of pressure solution and attendant cementation accompanying burial in the subsurface is shown in Figure 1.

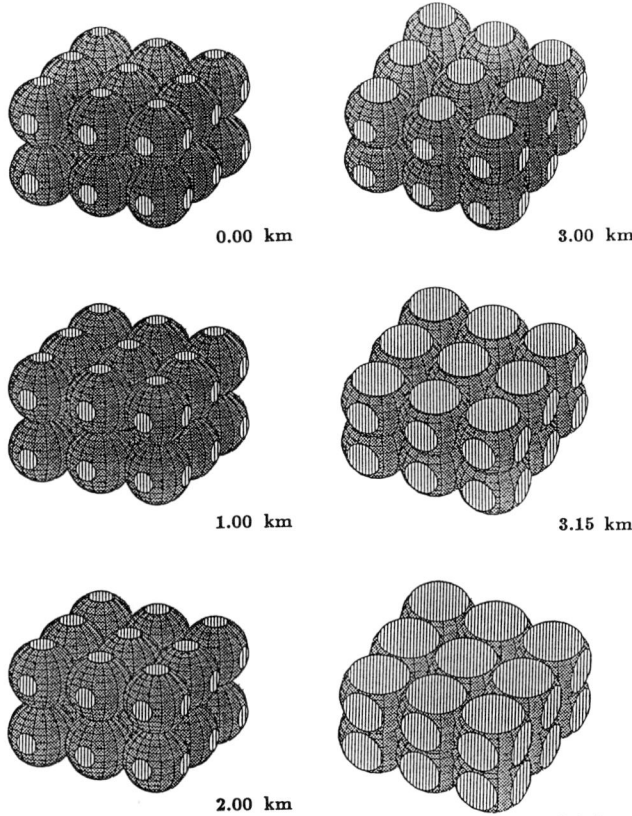

Figure 1. Three-dimensional representation of computer-simulated changes in grain shape accompanying intergranular pressure solution at grain contacts and cementation on adjacent free faces. The decrease in vertical height and increase in horizontal contact area accompanying burial become appreciable at depths below 2.5 km because of (temperature-related) enhanced reaction kinetics.

Experimental Considerations

Experimental verification of both FFPS and WFD suggests that both mechanisms may be simultaneously operating to various degrees during diagenesis. A review of recent literature is found in Tada and Siever (1989). Dissolution accompanying dislocation creep in halite in contact with a "quartz knife" prompted Tada and Siever (1986) to suggest the FFPS mechanism. They give evidence linking the dissolution to heightened elastic and inelastic strain energy at free faces adjacent to grain contacts. Brantley et al. (1986) and Schott et al. (1989) show that free-face dissolution associated with outcroppings of dislocations is the result of slightly faster reaction kinetics rather than changes in bulk solubility of a crystal due to the presence of dislocations. This effect disappears as the pore fluid approaches equilibrium with the solid (Brantley et al., 1986) and is negligible for the small supersaturations induced by dislocation-strain energy.

The existence and stability of load-supporting nanometer-thick adsorbed fluid films have been experimentally confirmed by Pashley and Kitchener (1979) for quartz-quartz contacts, and by Pashley and Israelachvili (1984) for mica-mica contacts. The structural, thermodynamic, and transport properties of these films are similar to those for interlayer water in micas (Low, 1987; Tada et al., 1987). Gratier and Guiguet (1986) give experimental evidence supporting diffusive mass transfer in compaction of quartz, and calculate a diffusion coefficient that is in order-of-magnitude agreement with that for lattice diffusion in micas (Dewers and Ortoleva, 1990b).

Experiments on quartz grains subject to varying effective stress in a flow-through apparatus made by Elias (1989) demonstrate a "reversible" increase in the pore fluid concentration of aqueous silica due to the application of a nonzero effective stress, as in the WFD mechanism; these experiments also demonstrate an "irreversible" change evidently due to some inelastic process that persisted after the removal of effective stress, possibly due to microgranulation and enhanced surface energy-related dissolution.

Observational Considerations

Recently, extensive measurements of the amount of pressure solution and overgrowth cementation in sandstones have been made (Houseknecht, 1988) by reconstructing detrital grain boundaries with the help of cathodoluminescence petrography. Amounts of pressure solution are found to be inversely related to grain size, a fact predicted by theory (Weyl, 1959; Tada et al., 1987) and experiment (Renton et al., 1969). The volume of quartz dissolved at grain contacts may or may not balance with the volume of overgrowths.

Modeling Efforts Toward Linking Theory, Experiment, and Observation

Using our quantitative reaction-transport mechanical model, we have simulated intergranular pressure solution and cementation in texturally homogeneous very-fine- to coarse-grained sandstones as a function of the burial history presented by Houseknecht (1988) for the Tuscarora (Silurian, Figure 2a) and Bromide (Ordovician, Figure 2b) sandstones. The dashed lines in Figure 2 connect simulated sandstones of different grain sizes at the same burial depth; the solid lines are regression lines for the Tuscarora (Figure 2a) and Bromide (Figure 2b) sandstones given by Houseknecht (1988). Despite the differences in thermal histories, range in uncertainty of rate coefficients, and error introduced by petrographic technique, we nonetheless obtain rather good agreement (within the uncertainty of Houseknecht's measurements) between FFPS rates and intergranular pressure solution (IPS) amounts in the Tuscarora, and between WFD rates and IPS amounts in the Bromide. Here, the FFPS rates are determined by a combination of elastic-strain energy and

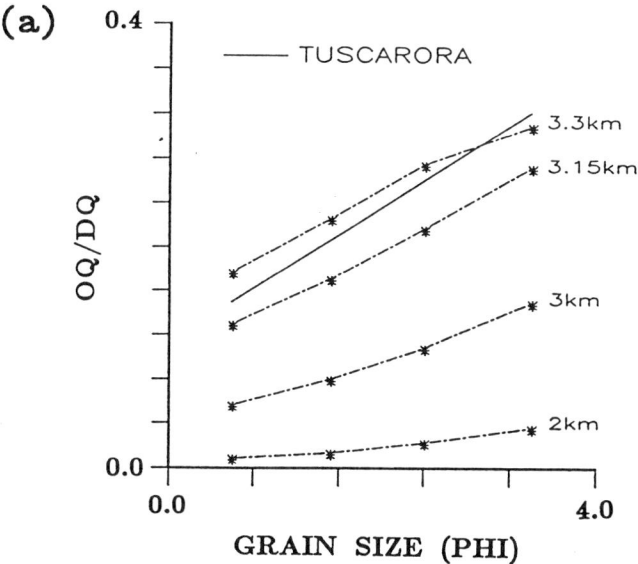

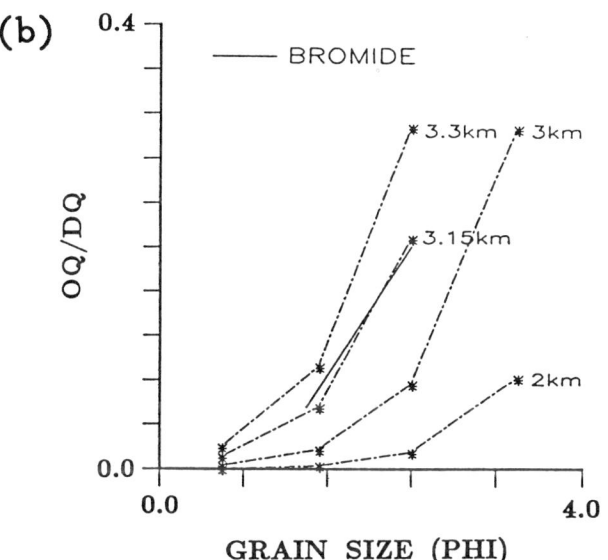

Figure 2. Results of pressure solution simulations during burial as depicted in plots of the volume of quartz dissolved at grain contacts (termed overlap quartz or "OQ") normalized by the depositional volume of grains (termed detrital quartz or "DQ") as a function of grain size in the phi scale. Shown are the predictions after burial to 2, 3, 3.15, and 3.3 km. (a) The burial path was taken to follow that postulated for the Tuscarora Sandstone (Silurian) by Houseknecht (1988), and the pressure solution mechanism controlling reaction at grain contacts was the "free-face pressure solution" mechanism. Data for coarse-, medium-, fine-, and very-fine-grained sandstones are marked by asterisks. The solid line is the regression line for the Tuscarora at maximum depth of burial given by Houseknecht (1988). (b) Same as in (a) except burial path here was that for the Bromide Sandstone (Ordovician), from Houseknecht (1988); the pressure solution mechanism here was the "water-film diffusion" mechanism. The solid line is that for the Bromide, from Houseknecht (1988).

detachment kinetics at solid/fluid interfaces. The similarity in slope between our results and those of Houseknecht in both cases show that the grain-size dependence of our rate laws is accurate, but the greater amount of pressure solution predicted by the WFD mechanism suggests that the rate coefficient in that case is too large. Houseknecht (1988) suggested that the difference in the grain-size dependency for pressure solution in the Bromide was due to the presence of clay coatings on the sand grains; we conjecture that diffusion in water films associated with clay minerals determined reaction rates in that case. It is possible that FFPS operates in clean sandstones, whereas clay in sandstones promotes the WFD mechanism.

Attempts to simulate the experiments of Elias (1989) have been successful in reproducing the pore-fluid concentration with time as a function of effective stress (Dewers and Ortoleva, 1989). We found that both the rate and level of increase of the pore-fluid concentration was consistent with WFD; the production of very fine, highly soluble particles through microgranulation at grain contacts (as advocated by Tada et al., 1987) could account for the "irreversible" response. The strain-energy-related increase in the level of aqueous silica in the pore fluid attending the FFPS mechanism was negligible and could not have accounted for the experimental behavior.

The actual concentration of pore fluid measured in the experiments of Elias (1989), and that simulated by our model, does not reflect equilibrium between the pore fluid and any quartz facet, but rather a "steady-state" concentration that reflects a compromise between the different thermodynamic and kinetic mechanisms operating at the various reaction sites around quartz grains. The assumption of equilibrium between pore fluid and quartz free-faces invoked by many authors is thereby not upheld.

The operation of either WFD or FFPS probably depends on the width of grain contacts, which differs with the deformational character of surface asperities lining the contacts; that is, whether asperities in contact on either side of a grain contact can support a sufficiently wide opening to favor FFPS, or if the contact walls are supported by the adsorbed fluid, favoring WFD. This in turn depends on the rheological nature of the minerals lining contacts, and there is probably a transition between the regions of dominance of either mechanism as the contact width changes. The rapid loading rates that characterize the experimental conditions of Elias (1989) may have "sheared" off asperities, thus promoting WFD through thin stress-supporting water films, whereas the slower loading rates that characterize diagenetic conditions may allow surface asperities (and not water films) to support the contact load. Pretreatment of quartz surfaces by etching in acid may also have influenced the mechanism of stress-induced dissolution, as Pashley and Kitchener (1979) note the stability of adsorbed aqueous films only between "clean" quartz surfaces.

The above discussion provides a basis for quantification of the grain-scale processes that operate during pressure solution. In the examples that follow, we use our quantitative model to examine the implications of textural heterogeneity on length scales larger than a single grain.

IPS TO STYLOLITE TRANSITION

The different kinetics and thermodynamic driving forces characteristic of reaction of heterogeneously stressed grains with pore fluid, inasmuch as they govern the rates of change of grain shape, can determine whether chemical compaction and cementation evolve pervasively throughout a rock, as during intergranular pressure solution, or become localized, as during stylolitization.

As was previously stated, the rates of dissolution at grain contacts may be limited by surface detachment kinetics or by diffusional transport within the contact. The latter depends on, among other factors, the typical diffusion distance (on the order of a nominal contact radius). In the transport-limiting case, reaction at grain contacts will be slower than reaction at free faces near the contact edges. Because of the heightened strain energy near grain contacts, faster dissolution at contact peripheries results in decreasing contact areas, whereas pressure solution at grain contacts otherwise leads to an increase in contact area. Dewers and Ortoleva (1990b) suggested that a decrease in the ratio of the rates of dissolution at free faces to ones in contact leads to a transition from pervasive shortening (intergranular pressure solution) to localized shortening (stylolitization). For rocks starting with even small-amplitude heterogeneity, this localization occurs from a feedback caused by a coupling of reaction with transport: chemical potential gradients can exist between spatially separate localities in a rock undergoing pressure solution at slightly different rates if the localities are sufficiently close together to communicate by diffusion. As dissolution proceeds in a region where the grains have slightly smaller contact radius, diffusion down chemical potential gradients results in reprecipitation in regions where the grains have slightly larger contact radius; this increases grain contact radius and decreases the stress acting across the contact. The free-face dissolution decreases contact radii, and thus results in a "focusing" of stress and strain energy. The contrast between the dissolving/growing regions is thereby self perpetuating. In addition, the decreased contact areas accompanying dissolution can promote plastic deformation within stylolite seams if the increasing intergranular stress accompanying free-face dissolution exceeds the strength of the contact.

In Figures 3 and 4 (after Dewers and Ortoleva, in press, c) we show the transition between pervasive IPS and stylolitization represented by the "unstable enhancement" of an initial heterogeneity in texture by the WFD mechanism in limestone buried at 2 km. These plots show the variation in the geometrical parameters that constitute our textural picture. L_z is the height of the grains along the direction parallel to the simulation domain; decreases in this variable reflect (chemical) compaction. L_f is the radius of the truncated spheres; changes in this variable reflect cementation and dissolution at free grain faces. We have taken the other two textural parameters, which measure the horizontal dimensions of the truncated sphere, to be constant, so that all compaction is in the (vertical) direction parallel to the simulation domain. In Figure 3, heightened dissolution at grain contacts accompanies dissolution at free faces; a large local increase in porosity marks the developing "stylolite;" the large differential stresses that develop there would ultimately result in localized plastic flow (not accounted for in the present model), so that the porosity maximum in Figure 3 would appear more like a fracture.

Figure 4 demonstrates that the development of localized compaction may or may not occur, depending on the kinetic parameters governing dissolution. In this simulation, the rate coefficient for calcite free-face reaction is 5 times smaller than that for Figure 3. Both compaction and cementation are relatively pervasive, and the initial textural heterogeneity does not amplify. In summary, predictions of whether or not localized rock compaction develops depend on the relative rates of grain-contact dissolution and free-face dissolution.

Depending on the parameters governing relative rates of reaction, localized compaction in the form of stylolitization may follow a sufficient porosity reduction attending intergranular pressure solution and cementation. The development of stylolites only after a sufficient degree of cementation and intergranular pressure solution is advocated by Buxton and Sibley (1981). As also exhibited in our simulation, simultaneous operation of both contact and free-face dissolution has been observed in dissolution seams in carbonates by Spang et al. (1979). Heightened cementation adjacent to stylolites is noted by Ricken (1986).

EFFECTS OF CLAY

Clay has been linked to pressure solution since Heald (1959) recognized a differential amount of pressure solution in clay-bearing beds of the St. Peter Sandstone. The enhancement of pressure solution by clay has been attributed to effects of solution chemistry (Thomson, 1959), enhanced diffusion along clay-quartz contacts (Weyl, 1959), and the preservation of relatively smaller grain contacts by the inhibition of quartz cementation by clay coatings (Sibley and Blatt, 1976).

The inhibiting effect of a thin coating of clay on the development of quartz overgrowths was recognized by Heald and Anderegg (1960) and by Heald and Larese (1974). A similar overgrowth inhibition may arise from a coating of residual hydrocarbons (Tigert and Al-Shaieb, in press). We conjecture that the attachment kinetics at quartz-fluid interfaces is physically impeded by the coating. In the case of clay, reaction at free faces may become limited by diffusion through the clay lattice if the coating is sufficiently thick.

A simple scheme for including clay coatings at free faces of a quartz grain assumes that the thickness of the clay film is equal to the volume of clay around the grain divided by the free-face surface area (Dewers and Ortoleva, in press, c). The flux of aqueous silica through the film varies with the concentration gradient across the film, the film thickness, and the clay "lattice" diffusion coefficient. This simple law yields an inverse clay dependence of the free-face growth rate that is consistent with the inverse clay-overgrowth relationship seen in sandstones (Tada and Siever, in press, their figure 2) and in carbonates (Bausch, 1968).

In a system composed of framework grains of quartz coated by illite, the rates of pressure solution at grain contacts may become limited by the slow precipitation rates at quartz-free faces. If only the amount of illite is spatially variant, with otherwise constant texture in the quartz, gradients in the concentration of aqueous silica can exist if clay inhibits quartz precipitation because free-face quartz reaction rates are then not spatially constant. This can also result in spatial nonuniformity in illite reaction rates, because quartz-water and illite-water reactions are coupled through aqueous silica and occur on similar time scales.

The influence of a small "lens" of illite-coated sand grains in an otherwise homogeneous sandstone is considered in Figure 5. The operating pressure solution mechanism here is the FFPS mechanism. The initial clay accumulation locally impedes overgrowth formation (from SiO_2 released by pressure solution at grain contacts) to a greater extent than in the surroundings; because of this, the amounts of pressure-solution-induced compaction become greater within the clay-rich region as relatively smaller contact areas are preserved there. As the system evolves, we see in Figure 5e that the initial gradients in aqueous silica are sustained over time and become amplified (the background decrease in aqueous silica concentration is related to the overall increase in contact area and loss of porosity as the stress per grain decreases). It is interesting that at the final time of the simulation, the greatest amount of intergranular volume is preserved where compaction is the greatest (smallest L_z), within the clay lens, whereas the most porosity loss is in the regions of greatest cementation (greatest L_f). The gradient in $SiO_2(aq)$ set up by the differential

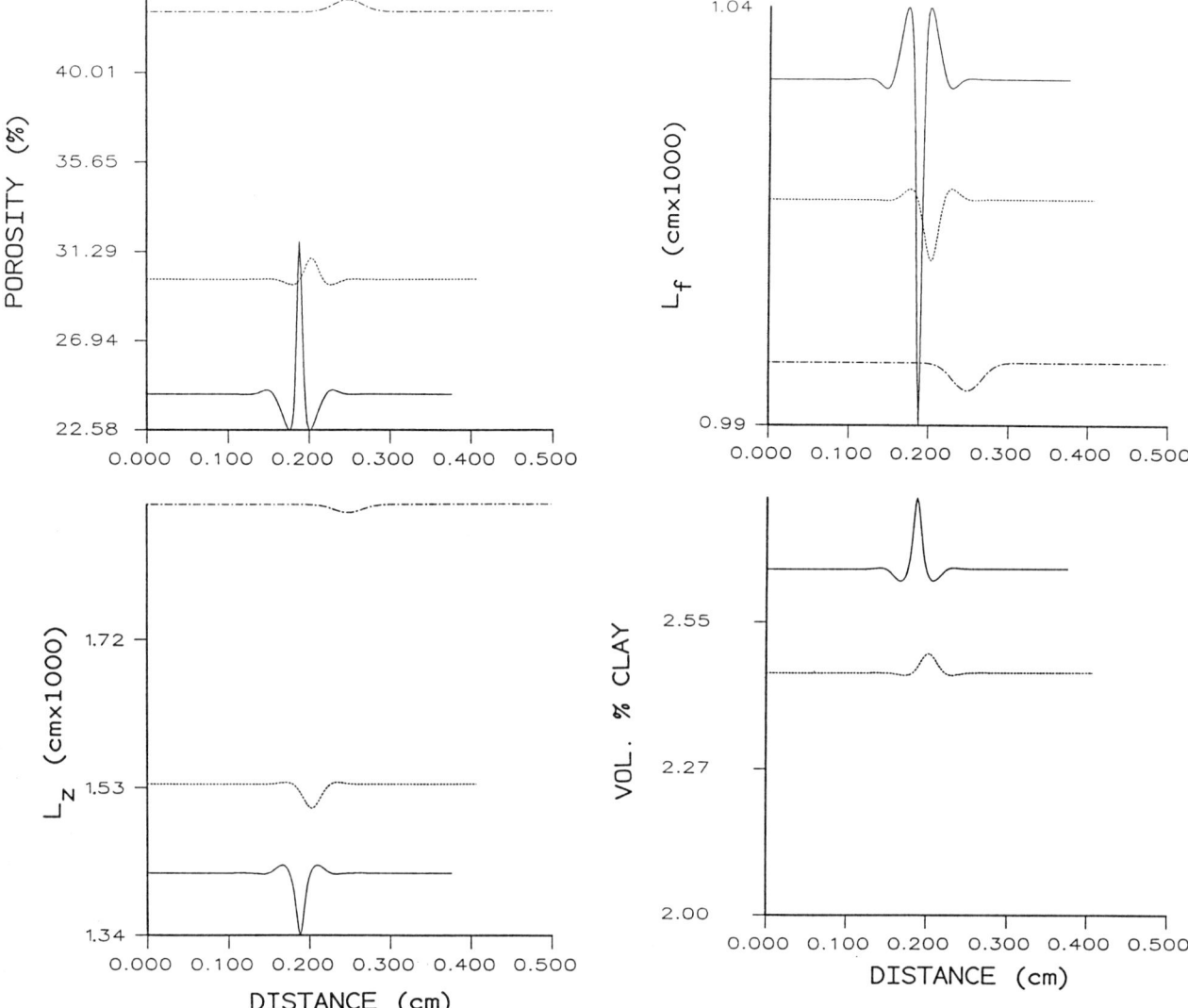

Figure 3. The transition between stylolitization and (pervasive) intergranular pressure solution accompanying a decrease in the rate of free-face reaction rates relative to contact reaction rates (after Dewers and Ortoleva, in press, c, their figure 8). For these examples, clay serves only as "insoluble" residue; burial rates are 10 m/m.y. Localized dissolution and compaction (i.e. a stylolite) develop from a small initial nonuniformity. L_f is the radius of the truncated sphere in Weyl's (1959) textural picture, and so measures amounts of overgrowth development; L_z is the height of the vertical truncation, and so measures amounts of compaction due to dissolution at grain contacts (the heightened compaction is also marked by increases in clay). Small secondary "stylolites" are seen to have emerged by the final time shown, adjacent to regions of enhanced cementation which surround the larger, primary stylolite. The broken line represents the data initializing the simulation; the solid line is the final time of the simulation run (4,470 years); the dotted line is an intermediate time.

compaction and cementation is seen to provoke gradients in the other chemical species, which in turn cause the illite to grow within the region initially more abundant in illite coatings and to dissolve from the (quartz) cementing regions.

The differential pressure solution/cementation profile, the replacement of clay in the cementing regions, and the preservation of porosity and greater pressure solution within clay-rich zones over that elsewhere in the simulation domain agree well with the observations of Heald and Anderegg (1960) on cementation patterns within the Tuscarora Sandstone. A similar feature has been described in sandstones of the Simpson Group in the Anadarko basin of Oklahoma by Tigert and Al-Shaieb (in press), except in this case, the grain coatings are chlorite. The latter occurrence is of particular interest, as the banded porous and nonporous zones consisting of heightened compaction and porosity alternating with greater cementation collectively serve as a permea-

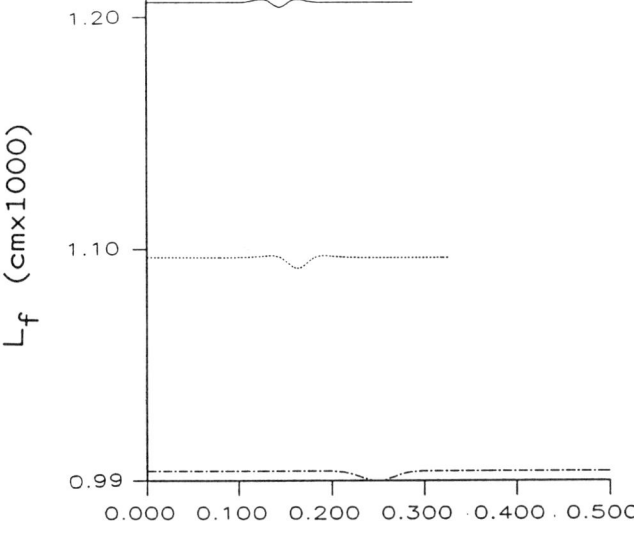

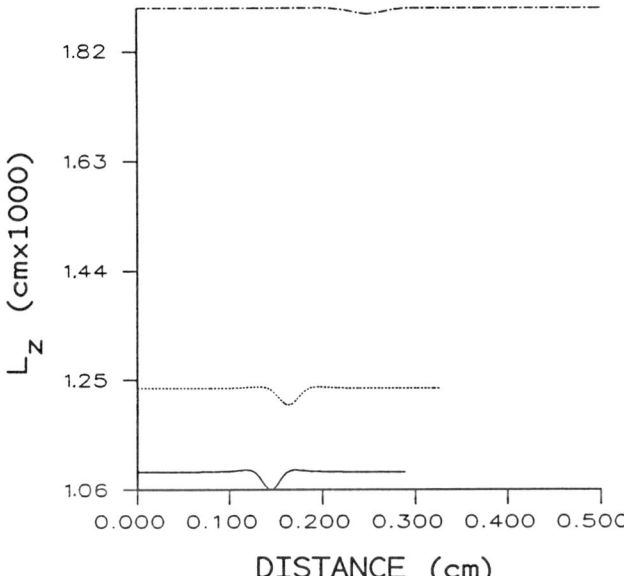

Figure 4. Same as in Figure 3, except here the calcite rate coefficient is six times smaller and the final time is after 18,300 years (after Dewers and Ortoleva, in press, c, Figure 9). Pressure solution and cementation are more or less pervasive throughout, as shown by the lack of "enhancement" of the initial heterogeneity in L_z and L_f.

bility/ pressure seal, and the porous zones allow an internal hydraulic continuity that may serve as a hydrocarbon reservoir (Powley, 1985; Tigert and Al-Shaieb, in press). Our simulation demonstrates that spatially separate regions of quartz and clay growth and dissolution develop influenced by, but not directly related to, a primary or depositional distribution of clay.

Figure 5 shows that the greatest (chemical) compaction develops on either side of the illite maximum. We liken this localized compaction to stylolitization, as shown in Figure 3. Stylolites at the transition between clay-rich and clay-poor rocks are observed in the marl-limestone alternations described by Ricken (1986, p. 141) and in the sandstones observed by Tigert and Al-Shaieb (in press).

Most field studies on stylolites and clay seams characterize the material accumulated along the dissolution seams as insoluble residue. However, as shown in Figure 5, we have seen that illite may in fact participate in reaction on the same time scale as quartz. Thomson (1959) cites evidence for the authigenic growth of illite in stylolitic seams in sandstones. Carbonates, on the other hand, have fast reaction kinetics relative to those of silicates. Clay minerals (and even quartz) may be considered to be reactively inert on the time scale of calcite reaction. We conjecture that, unless calcite reaction is strongly inhibited, clay minerals concentrated along dissolution seams in limestones may indeed be residual.

DIFFERENTIAL COMPACTION AND CEMENTATION

"Banded" Cementation in Sandstones

The differential compaction/cementation seen in Figure 5 does not necessarily require an initial, inhomogeneous distribution of clay to develop. We have demonstrated that spatially alternating regions of enhanced cementation and pressure solution may develop from initial differences in grain size in a monomineralic sandstone (Dewers and Ortoleva, 1990b). We show this in Figure 6, in a system initialized with small deviations in grain size from a uniform background. The relationship between the various textural parameters develops from small free-energy gradients that exist because of differences in texture. These small gradients set off the spatially separate zones of dissolution and reprecipitation; as the textural contrasts between adjacent regions amplify, so do the free-energy gradients between them. The textural profiles at the final time (64 m.y.) in Figure 6 demonstrate that zones of heightened cementation alternate with zones of higher porosity and pressure-solution-induced compaction. This style of alternating cementation and compaction in relation to initial depositional grain size laminations has been observed in the Tuscarora Sandstone (Heald, 1955).

By running other simulations, we have found that the amplitudes of the maxima and minima in L_z and L_f, signifying spatial differences in the amounts of compaction and cementation, respectively, increase with both amplitude of the initial "noise" (or depositional lamination) and decreases in the diffusion coefficient-diffusion path width product. Inasmuch as a transition between intergranular pressure solution-dominated to stylolite-dominated

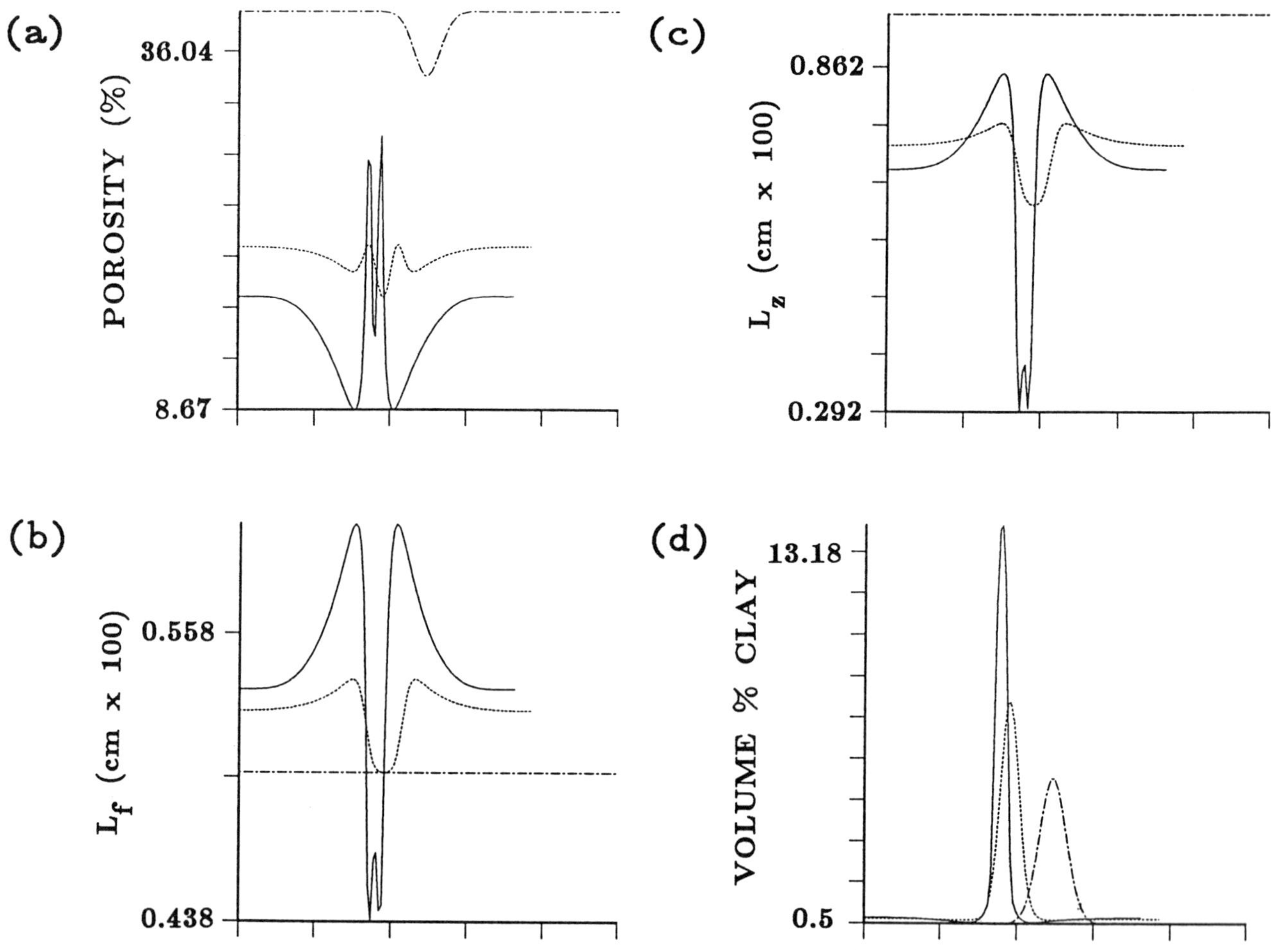

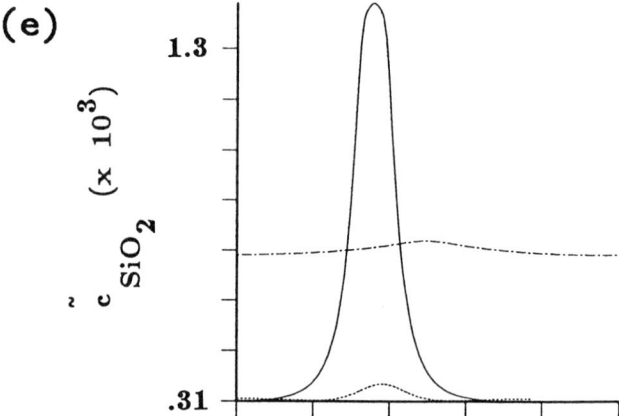

Figure 5. Evolution of an otherwise homogeneous sandstone buried at 3 km consisting of quartz grains with thin coatings of illite, initialized with a single maximum in clay volume percentage at the center of the simulation domain. Initial profiles of the variables are given by broken lines; those at 30 m.y. are given by the dotted lines; and those at the final time of 60 m.y. are given by the solid line. (a) Spatial profiles of porosity. Porosity maxima correspond to regions of greatest compaction and dissolution, which occur within the clay-rich region (seen in Figure 5d). (b) Spatial profiles of radius L_f of the truncated quartz spheres. (c) Spatial profiles of height L_z of the truncated quartz grains. (d) Spatial variations in the volume percentage of illite clay. The initial clay maximum becomes enhanced both because of the loss of intergranular volume accompanying compaction, and also because of clay growth. Zones of clay loss due to dissolution are seen in the form of small minima adjacent to the large clay maximum at the final time shown. Kinetic and thermodynamic parameters for illite were taken to be equal to those for muscovite. (e) Spatial distribution of dimensionless silica concentration (equal to the silica concentration divided by the silica concentration in equilibrium with quartz at a fluid pressure of 300 bars, minus one). Note the existence and amplification of the concentration gradients between the clay-rich and clay-poor regions. (f) Spatial profile of a combined deviatoric, dimensionless concentration variable for $Al(OH)_3$ and K^+, defined similarly as in (e), except involving illite equilibria at 300 bars and neutral pH. (g) Spatial profile of the deviatoric, dimensionless concentration of H^+ (defined by taking the true concentration, dividing it by that determined from neutral pH, and subtracting one).

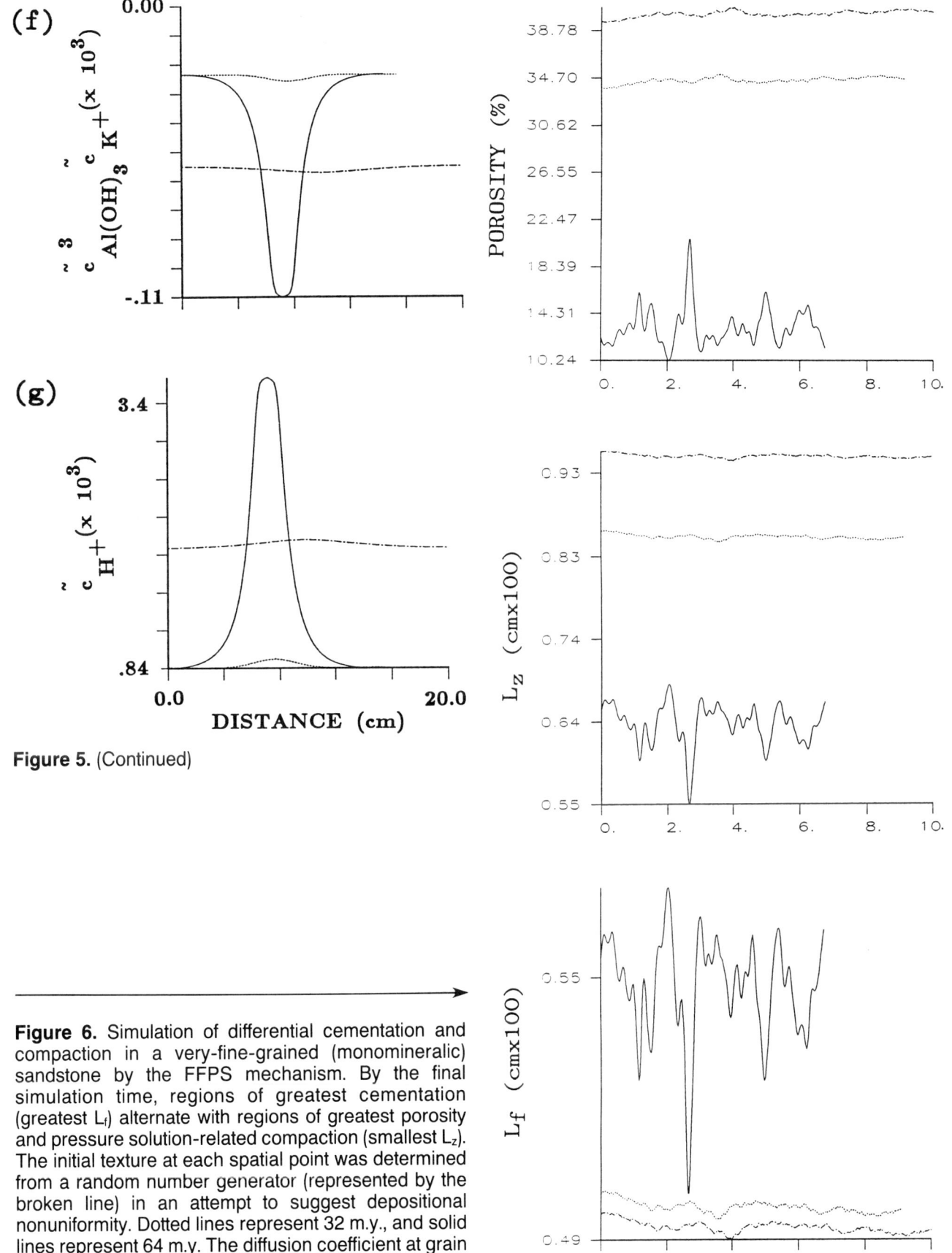

Figure 5. (Continued)

Figure 6. Simulation of differential cementation and compaction in a very-fine-grained (monomineralic) sandstone by the FFPS mechanism. By the final simulation time, regions of greatest cementation (greatest L_f) alternate with regions of greatest porosity and pressure solution-related compaction (smallest L_z). The initial texture at each spatial point was determined from a random number generator (represented by the broken line) in an attempt to suggest depositional nonuniformity. Dotted lines represent 32 m.y., and solid lines represent 64 m.y. The diffusion coefficient at grain contacts multiplied by the contact width was taken to be 10^{-11} cm^3/s.

behavior accompanies decreases in this parameter, we can state that greater amounts of differential compaction and cementation will accompany a change in parameters from the intergranular pressure solution-dominated to stylolite-dominated regime. Furthermore, the extent of differential compaction and cementation is very sensitive to the style and phase relationships of the initial variations in L_z and L_f.

Marl/Limestone Alternations

Recently, examples of marl-limestone alternations have been viewed as having arisen during diagenesis, with an emerging cyclicity that is not necessarily related to any depositional feature (Ricken, 1986; Bathurst, 1987). Ricken (1986) describes the evolution of these structures as a closed-system segregational process involving spatially alternating regions of dissolution and reprecipitation of carbonate.

Two examples of marl-limestone alternations from Ricken (1986) are shown in Figure 7. The first (Figure 7a) shows spatial differences in the texture variables of carbonate (expressed as a percentage of present-day rock volume), amount of compaction, and porosity for an uncemented foraminiferal marl (Pleistocene, Sicily; see figure 15 of Ricken, 1986, p. 35-38). Here, dominantly mechanical compaction is imposed upon a depositional bedding cyclicity that apparently is climatic in origin. In Figure 7b we show spatial textural profiles versus distance for the same three textural variables as in Figure 7a, but for an example of diagenetically enhanced bedding in Lower Cretaceous limestone-marl alternations from the French Maritime Alps. Of note are the differences in the porosity profiles between Figures 7a and 7b. If, hypothetically, the rock described in Figure 7a is taken to represent the mechanically compacted precursor to that described by Figure 7b as Ricken suggests, it appears that the diagenetic overprint on the primary bedding has resulted in the development of a rhythmicity not related to that in the initial sediment, and also in the development of correlation among the distributions of porosity, cementation, and compaction that was not present initially. Ricken (1986) states that, in cases of diagenetic bedding, "diagenesis usually generates fewer limestone layers as compared to the number of primary carbonate oscillations . . . [and the] alternations now appear . . . to be more rhythmical or cyclical than . . . in the original sediment."

In Figure 8, we show a simulation from a model developed in Dewers and Ortoleva (in press, c) that demonstrates the localization of chemical compaction of calcite in clay-rich zones and the adjacent cementation in an alternating pattern similar to the example of Ricken (1986) shown in Figure 7b. Zones of greater compactional shortening (smaller L_z), greater porosity, and greater clay content alternate with zones of greater calcite cementation (larger L_f) in a manner similar to that demonstrated in Figure 7b by the final simulation time shown. However, initially, the least porosity was associated with the zones richer in clay, as in the example in Figure 7a. This model assumes a clay-inhibition of free-face reaction similar to that used in the simulation of Figure 5.

Ricken (1986) has demonstrated an inversion of the correlation between porosity and carbonate percentage accompanying chemical compaction in pelagic sediments by plotting the porosity measured as a function of the percentage of carbonate at the same spatial locality across many limestone/marl alternations. He finds that the uncemented example of Figure 7a, having undergone only mechanical compaction, has a positive slope in the space of porosity versus carbonate; on the other hand, examples of marl/limestone alternations, which exhibit a diagenetic bedding as in Figure 7b, have negative slopes in this space. This is shown in Figure 9a. The porosity-carbonate relationship for the three consecutive times in our simulation of Figure 8 is shown in Figure 9b, which demonstrates that the correlation between cementation and porosity in our simulation is in at least qualitative agreement with Ricken's examples. The modeling supports Ricken's arguments that some types of marl/limestone alternations arise by differential compaction and cementation set off by small initial variations in texture.

SUMMARY

We agree with the assertion of Tada and Siever (1989), in that all the parameters required for the quantification of pressure solution are known, as least as well as most other geochemical processes. Furthermore, the gap between theoretical predictions and observation of pressure solution in nature may be narrowed by a mathematical reaction transport model coevolving rock texture, effective stress, rock flow, and pore-fluid chemistry in space and time. The following summarizes some results of our simulations:

Estimates of the amounts of quartz dissolved at grain contacts as a function of grain size are in good agreement with those observed in nature. The presence of clay at grain contacts in sandstones appears to favor the WFD (water-film diffusion) pressure solution mechanism, whereas monomineralic, porous sandstones undergo pressure solution by the FFPS (free-face pressure solution) mechanism.

Clay coatings at free faces may enhance pressure solution rates by inhibiting overgrowth formation. Diffusion through the clay lattice limits rates of reaction of the quartz grain free faces. Gradients in pore-fluid solutes induced by spatial differences in rock texture cause the preferential replacement of illite in regions of quartz overgrowth formation and the precipitation of illite in regions of quartz pressure solution.

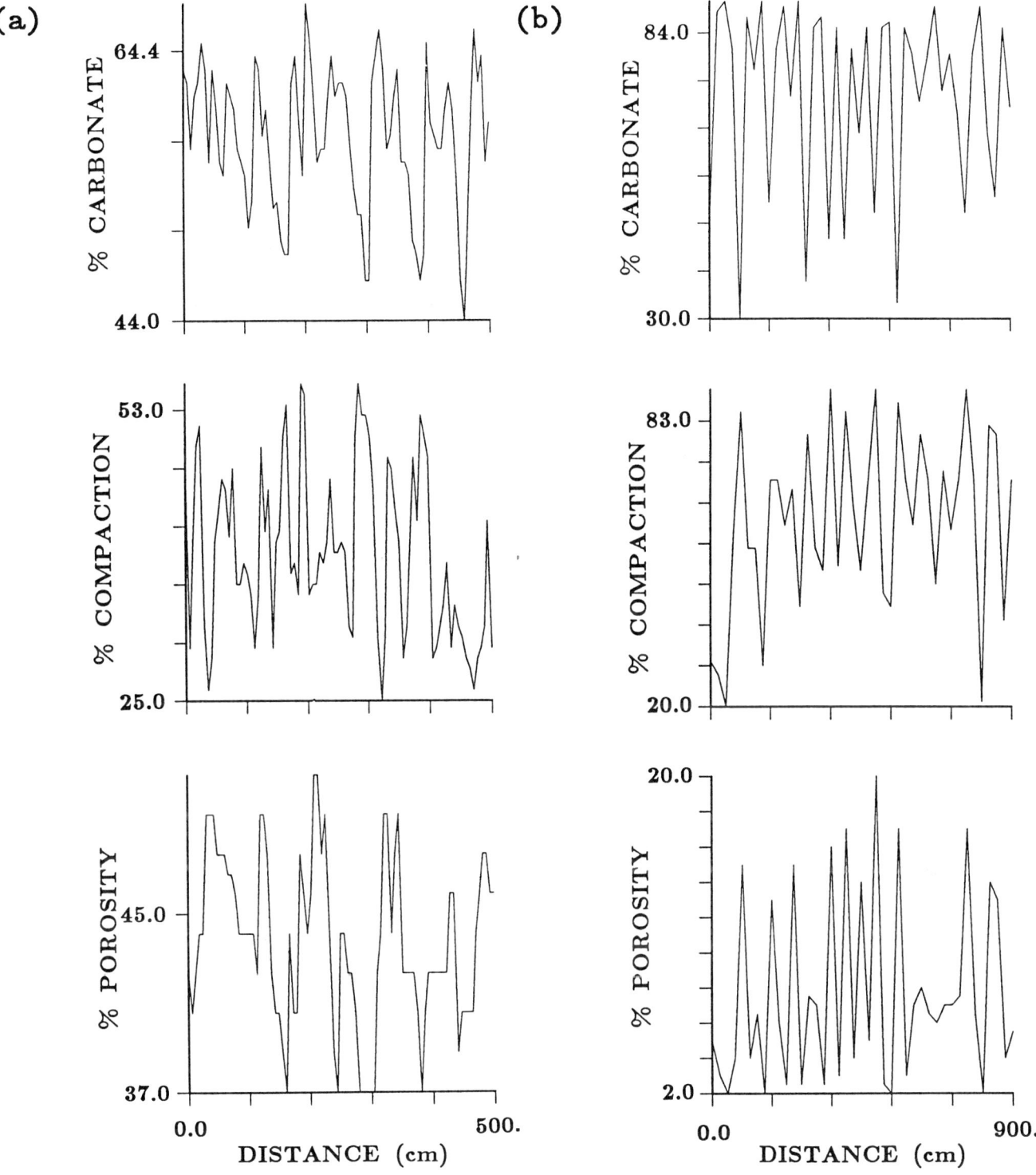

Figure 7. (a) Spatial distribution of the textural parameters of percentage of carbonate, percentage of (mechanical) compaction, and porosity, from an example of an uncemented marl given by Ricken (1986). The spatial alternations in carbonate content are apparently primary. (b) Similar to (a) except for an example of diagenetic bedding from Ricken (1986). The mechanism leading to the formation of diagenetic bedding involves the development of correlation between the textural variables of grain height due to chemical compaction, cementation, and porosity.

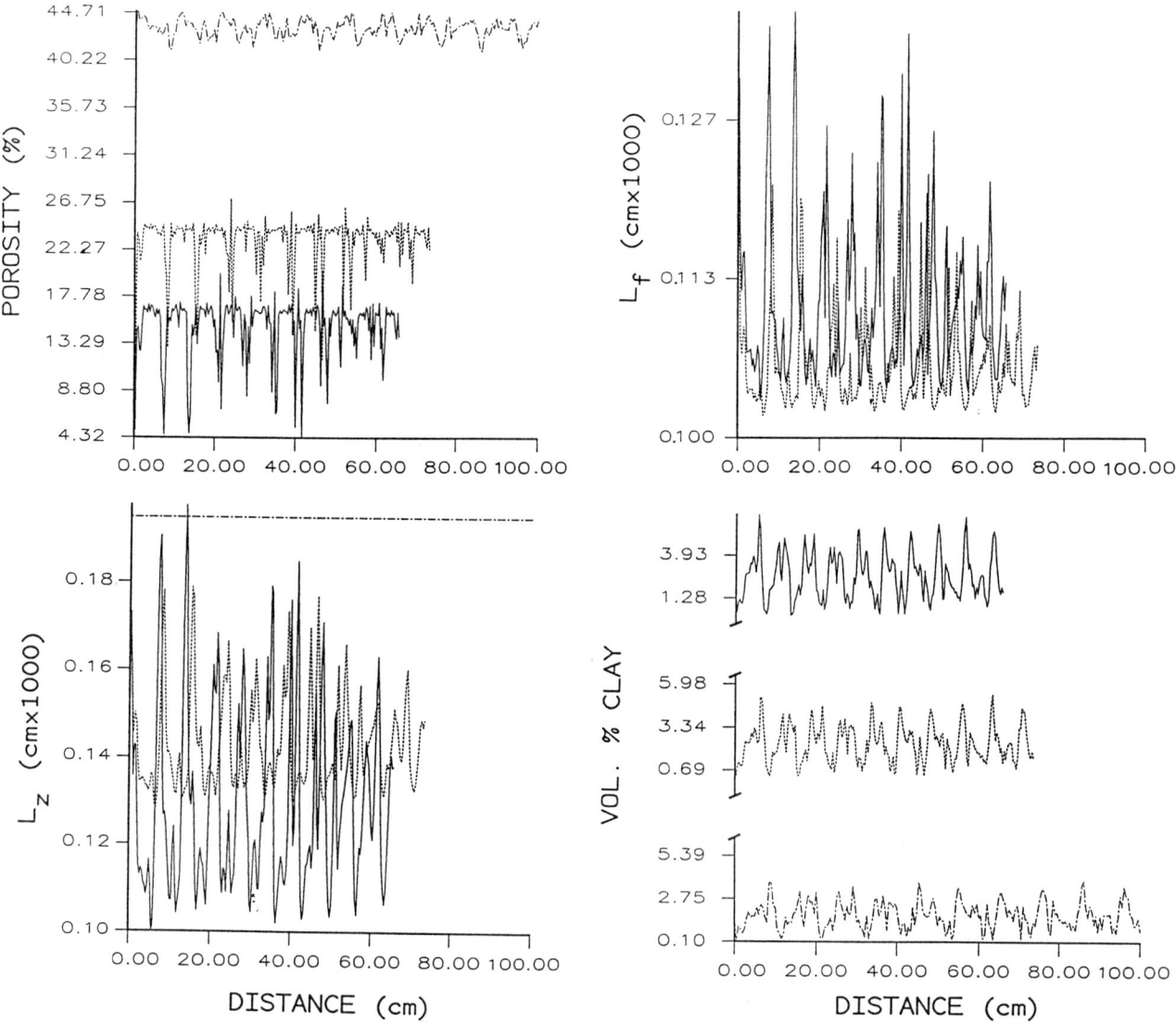

Figure 8. Simulation showing development of marl/limestone-type textural alternations. At the initial time, porosity minima correspond spatially to clay maxima; and calcite grain texture, represented by L_z and L_f, is constant in space. By the final time shown (1 m.y.), porosity maxima correspond to clay maxima, demonstrating a porosity inversion in space in a manner similar to that suggested in Figure 7. Zones of greater compaction (or lesser L_z), greater clay, and greater porosity alternate with zones of greater cementation and lower porosity. Geochemical model reflects saline pore fluid (ionic strength = .725 m), and $PCO_2 = 10^{-3}$ atm. Burial rate equals 10m/m.y.

Regions of relatively greater pressure-solution-induced compaction alternating with regions of greater cementation are shown to develop. The spatial correlation between the textural variables of cementation percentage, porosity, and compaction percentage that emerges from modeling is similar to natural examples of diagenetically enhanced bedding in marl/limestone sequences and differential compaction and cementation in sandstones.

Future modeling efforts should be directed toward including the type of processes examined here within a larger diagenetic framework. Important influences on the extent and spatial distribution of pressure solution and cementation are expected to arise from, for example, advective fluid flow, fluid overpressure, and other diagenetic reactions.

ACKNOWLEDGMENTS

This work was supported by a grant from the U.S. Department of Energy, Basic Energy Sciences

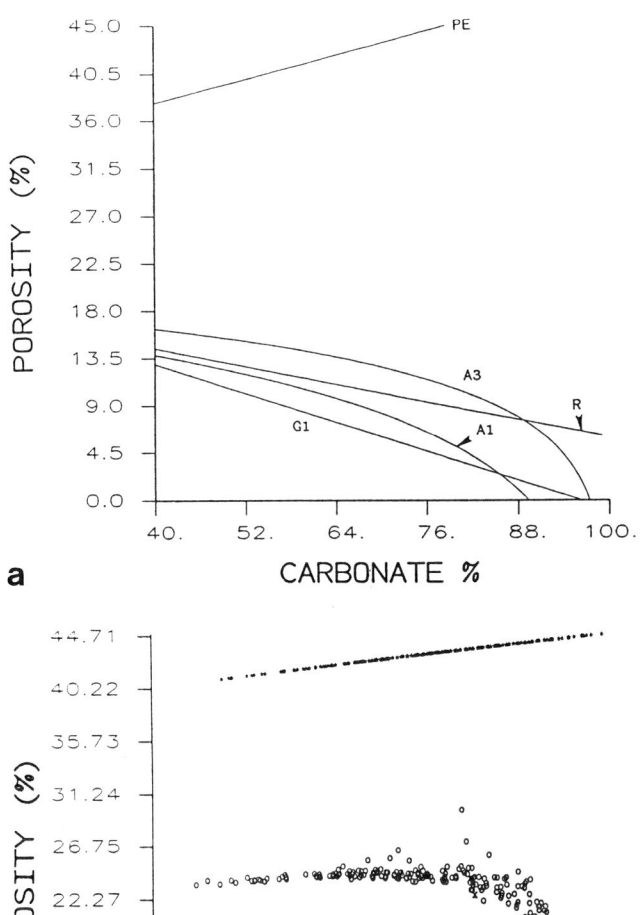

Figure 9. (a) Porosity in examples of marl-limestone alternations from Ricken (1986) plotted as a function of carbonate percentage. Labels refer to individual localities; notation is that of Ricken (1986). PE is Pleistocene marl shown in Figure 7 (a), which has undergone mechanical compaction but little cementation. The remaining examples are Jurassic or Cretaceous in age, and have experienced both chemical compaction and cementation. Clay content increases from A3, Al, Gl, to R. (b) Result of simulation of Figure 8 replotted as porosity vs. carbonate percentage, as in Figure 9. The initial time (*) slope is similar to the PE curve in (a); the final simulation time (x), representing 1 m.y., demonstrates the porosity inversion characteristic of the diagenetic differentiation between marl and limestone layers.

Program. The manuscript benefitted from reviews by R. Larese and I. Meshri.

REFERENCES CITED

Bathurst, R. G. C., 1958, Diagenetic fabrics in some British Dinantian limestones: Liverpool, Manchester Geological Journal, v. 2, p. 11-36.

Bathurst, R. G. C., 1987, Diagenetically enhanced bedding in argillaceous platform limestones: stratified cementation and selective compaction: Sedimentology, v. 34, p. 749-778.

Bausch, W. M., 1968, Clay content and calcite crystal size of limestones: Sedimentology, v. 10, p. 71-75.

Brantley, S., S. R. Crane, D. A. Crerar, R. Hellmann, and R. Stallard, 1986, Dissolution at dislocation etch pits in quartz: Geochimica et Cosmochimica Acta, v. 50, p. 2349-2361.

Buxton, T. M., and D. F. Sibley, 1981, Pressure solution features in a shallow buried limestone: Journal of Sedimentary Petrology, v. 51, p. 19-26.

Dewers, T., and P. Ortoleva, 1989, Preliminary report on the simulation of quartz "pressure solution" experiments, in Basin compartments and seals: a characterization and predictive modeling study: Gas Research Institute Progress Report IA, OKINTEX Group.

Dewers, T., and P. Ortoleva, 1990a, Geochemical self-organization III: a mechano-chemical model of metamorphic differentiation: American Journal of Science.

Dewers, T., and P. Ortoleva, 1990b, A coupled reaction-transport/ mechanical model for intergranular pressure solution, stylolites, and differential compaction and cementation in clean sandstones: Geochimica et Cosmochimica Acta.

Dewers, T., and P. Ortoleva, in press, c, Formation of stylolites, marl/limestone alternations, and clay seams accompanying chemical compaction of argillaceous carbonates, in G. V. Chilingarian and K. H. Wolf, eds., Diagenesis IV, Developments in sedimentology: New York, Elsevier.

Elias, B., 1989, The role of effective stress on silica solubility and pressure solution: an experimental investigation in a flow-through system: Masters Thesis, Texas A&M University, 36 p.

Engelder, T., 1982, A natural example of the simultaneous operation of free face dissolution and pressure solution: Geochimica et Cosmochimica Acta, v. 46, p. 69-74.

Gratier, J. P., and R. Guiguet, 1986, Experimental pressure solution-deposition on quartz grains: the crucial effect of the nature of the fluid: Journal of Structural Geology, v. 8, p. 845-856.

Heald, M. T., 1955, Stylolites in sandstones: Journal of Geology, v. 63, p. 101-114.

Heald, M. T., 1959, Significance of stylolites in permeable sandstones: Journal of Sedimentary Petrology, v. 29, p. 251-253.

Heald, M. T., and R. C. Anderegg, 1960, Differential cementation in the Tuscarora Sandstone: Journal of Sedimentary Petrology, v. 30, p. 568-577.

Heald, M. T., and R. Larese, 1974, Influence of coatings on quartz cementation: Journal of Sedimentary Petrology, v. 44, p. 1269-1274.

Houseknecht, D., 1988, Intergranular pressure solution in four quartzose sandstones: Journal of Sedimentary Petrology, v. 58, p. 228-246.

Lehner, F. R., and J. Bataille, 1985, Nonequilibrium thermodynamics of pressure solution: Pure and Applied Geophysics, v. 122, p. 53-85.

Low, P. F., 1987, Structural component of the swelling pressure of clays: Langmuir, v. 3, p. 18-25.

Maliva, R. G., and R. Siever, 1988, Diagenetic replacement controlled by force of crystallization: Geology, v. 16, p. 688-691.

Pashley, R. M., and J. N. Israelachvili, 1984, Molecular layering of water in thin films between mica surfaces and its relation to hydration forces: Journal of Colloid and Interface Science, v. 101, p. 511-523.

Pashley, R. M., and J. A. Ritchener, 1979, Surface forces in adsorbed multilayers of water on quartz: Journal of Colloid and Interface

Science, v. 71, p. 491-500.

Paterson, M. S., 1973, Non-hydrostatic thermodynamics and its geologic applications: Review of Geophysics and Space Physics, v. 11, p. 355-389.

Powley, D. E., 1985, Pressure, normal and abnormal: Lecture Notes, "Techniques of Petroleum Exploration II": AAPG School, South Padre Island, Sept. 16-19, 1985.

Renton, J. J., M. T. Heald, and C. B. Cecil, 1969, Experimental investigation of pressure solution of quartz: Journal of Sedimentary Petrology, v. 39, p. 1107-1117.

Ricken, W., 1986, Diagenetic bedding: Berlin, Springer-Verlag, 210 p.

Rittenhouse, G., 1971, Pore-space reduction by solution and cementation: AAPG Bulletin, v. 55, p. 80-91.

Robin, P. Y. F., 1978, Pressure solution at grain to grain contacts: Geochimica et Cosmochimica Acta, v. 42, p. 1383-1389.

Schott, J., S. Brantley, D. Crerar, C. Guy, M. Borcsik, and C. Willaime, 1989, Dissolution kinetics of strained calcite: Geochimica et Cosmochimica Acta, v. 53, p. 373-382.

Sibley, D. F., and H. Blatt, 1976, Intergranular pressure solution and cementation of the Tuscarora orthoquartzite: Journal of Sedimentary Petrology, v. 46, p. 881-896.

Spang, J. H., A. E. Oldershaw, and M. Z. Stout, 1979, Development of cleavage in the Banff formation at Pigeon Mountain, Front Ranges, Canadian Rocky Mountains: Canadian Journal of Earth Sciences, v. 16, p. 1108-1115.

Tada, R., and R. Siever, 1986, Experimental knife-edge pressure solution of halite: Geochimica et Cosmochimica Acta, v. 50, p. 29-36.

Tada, R., and R. Siever, 1989, Pressure solution during diagenesis: A review: Annual Review of Earth and Planetary Sciences, v. 17, p. 89-118.

Tada, R., R. Maliva, and R. Siever, 1987, A new mechanism for pressure solution in porous quartzose sandstone: Geochimica et Cosmochimica Acta, v. 51, p. 2295-2301.

Thomson, A., 1959, Pressure solution and porosity, *in* H. A. Ireland, ed., Silica in Sediments: SEPM Special Publication 7, p. 92-110.

Tigert, V., and Z. Al-Shaieb, in press, Pressure seals: their diagenetic lamination and banding patterns, *in* P. Ortoleva, B. Hallet, A. McBirney, I. Meshri, R. Reeder, and P. Williams, eds., Proceedings of the Workshop on Self-Organization in Geological Systems: Earth Science Reviews (Amsterdam, Elsevier).

Weyl, P. K., 1959, Pressure solution and the force of crystallization—a phenomenological theory: Journal of Geophysical Research, v. 64, p. 2001-2025.

Fluid-Rock Interactions in Thermal Recovery of Bitumen, Tucker Lake Pilot, Cold Lake, Alberta

Ian Hutcheon
Hugh J. Abercrombie
Department of Geology and Geophysics
The University of Calgary
Calgary, Alberta, Canada

Thermal recovery projects, in which steam-water mixtures are injected into oil-bearing rocks at high temperatures, provide a laboratory to study rock-water interactions. The temperature range of 80°-220°C is appropriate for diagenetic settings and, because injected steam that condenses to freshwater mixes with more saline water during production, the effect of mixing waters with different salinities can also be examined. In this study of the Tucker Lake pilot site in the Cold Lake heavy oil deposits of Alberta, water and gas samples were obtained at regular intervals over a 7-month period. The chemical composition of these samples was used with solution speciation models to compare the stability of waters with respect to the calculated stability of minerals known to be formed during steam injection and thermal recovery as determined from thermodynamic data.

The activity ratios of dissolved species, such as Na^+, K^+ and Mg^{2+} to H^+, tend to follow phase boundaries that represent silicate hydrolysis reactions between kaolinite, chlorite, illite, K-feldspar, smectite, and analcime. The Na concentration changes during production from approximately 500 to 4500 mg/L, and the observation that the aNa/aH ratio for the waters follows the smectite-analcime boundary during production suggests that the silicate hydrolysis reaction is buffering the pH to maintain the constant aNa/aH ratio. In contrast, the $aCa^{2+}/(aH^+)^2$ ratio follows the dissolution phase boundary for calcite, implying that the reaction rate for calcite dissolution is more rapid than that for silicate hydrolysis. Other published studies of the isotopic composition of produced gas from the Tucker Lake pilot confirm that CO_2 is produced by reaction of calcite, supporting the role of silicate hydrolysis in calcite dissolution. The buffering of fluid activity ratios by silicate hydrolysis reactions suggests that silicate hydrolysis plays an important role in dissolution of calcite and other carbonates in this thermal pilot, and potentially during natural diagenesis. Silicate hydrolysis and reactions involving silicates in general should be carefully considered in formulating diagenetic models aimed at predicting reservoir quality, particularly if prediction of dissolution porosity is considered an important factor.

INTRODUCTION

Interactions between fluids and rocks are known from a variety of natural geological environments and occur at temperatures and pressures ranging from those of the near-surface to the high P and T conditions of metamorphism. An artificial analog of natural fluid-rock interactions at elevated temperatures and pressures may be found in thermally assisted recovery of bitumen from the oil sand reservoirs of the Western Canada sedimentary basin. Because of the high viscosity of the bitumen and its occurrence at depths inaccessible to surface mining, heat (typically in the form of high-temperature water-steam mixtures) must be supplied to the reservoir to promote flow.

By analogy with natural systems, the resulting interactions of injected steam and water with reservoir rocks are expected to alter reservoir mineralogy. Studies of prerecovery and postrecovery core from thermal recovery projects (Sedimentology Research Group, 1981; Abercrombie, 1989) and experimental simulations of steam flooding (McCorriston et al., 1981; Gunter and Bird, 1988) show that mineral reactions do occur during thermal recovery and, in turn, alter the compositions of waters produced with the oil. Abercrombie and Hutcheon (1986), Cathles et al. (1987), and Hutcheon et al. (1990) used isotopic data to show conclusively that CO_2 produced during thermal recovery results from reactive destruction of calcite (Hutcheon and Abercrombie, 1987; Gunter and Bird, 1988).

This study uses the compositions of waters sampled in 1984 and 1985 from three wells at the Tucker Lake pilot site, near Cold Lake, Alberta (Figure 1). Preliminary results of a part of this study were included in Hutcheon et al. (1988). The pilot has since been shut down, but at the time was operated by Husky Oil Operations Limited. At Tucker Lake, thermal recovery was by cyclic steam stimulation or "huff 'n puff." This process involves the injection of 80% quality steam-water mixtures at about 300-320°C and 12 MPa into the formation at rates as high as several hundred m³/day for periods of as much as 90 days. The well is then capped for a short "shut-in" period before bitumen is pumped back at flow rates of about 100 m³/day (Shepherd, 1981). Each production "cycle" may last as long as 6 months and may be repeated as many as 9 times. Samples from the Husky Tucker Lake pilot are exclusively from first-cycle wells.

The water chemistry and isotope chemistry of fluid samples from the Tucker Lake pilot site give strong evidence for mineral reactions and interaction between the mineral, water, gas, and hydrocarbon phases. The reader is cautioned that the data reported in this section are from a single pilot site in the Cold Lake area and represent samples obtained from a first cycle of steam injection. In spite of this, the data are compelling, and probably can be extrapolated to other pilot sites in the Cold Lake area and to natural diagenetic settings.

COMPOSITIONS OF PRODUCED WATERS

Analyses of the produced waters from wells 6B-32-64-4W4 (6B), 10D-29-64-4W4 (10D), and 12A-28-64-4W4 (12A) are given in Table 1. Isotopic compositions were reported by Hutcheon et al. (1990). Produced fluid samples were obtained from these three wells (6B, 10D, and 12A), which were in their first cycle of steam injection and recovery. Samples of produced water, oil, and gas were collected at the well head. Water sampling procedures are substantially similar to those of Lico et al. (1982), and pressurized casing gas samples also were collected. Water samples were filtered (0.45 μm) in the field. Total alkalinity, pH, sulfide, and ammonia were measured in the field and samples were preserved for further chemical and isotopic analysis. Major element cation concentra-

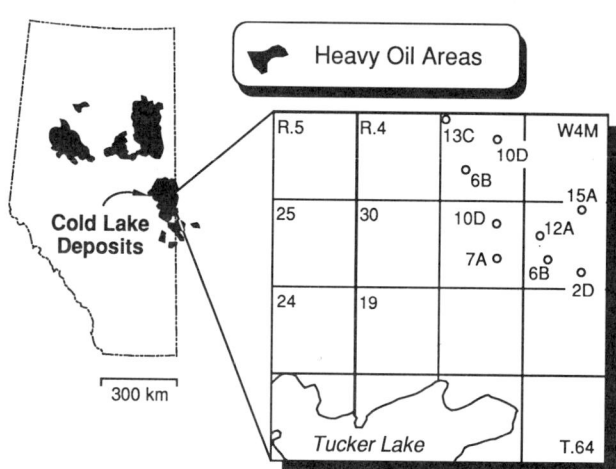

Figure 1. Map showing the location of various heavy oil and tar sand accumulations in Alberta, the location of the Tucker Lake pilot, and the locations of wells examined in this study.

tions were measured by atomic absorption; anion concentrations, including C_1-C_7 organic acids, were measured by ion liquid chromatography. Organic acid concentrations generally were less than 1 ppm and were not reported. Details of the sampling and analysis procedures are given in Abercrombie (1989).

Because steam is necessarily fresh water, the early stages of production are represented by samples that have relatively low concentrations of dissolved salts. As production proceeds, the salinity rises. Also, samples collected early in a production cycle are relatively hot, and successive samples are progressively cooler. The changes in concentration of Na^+, Cl^-, and K^+ are shown in Figure 2. The relatively continuous rise of Na^+ and Cl^- can be explained in terms of mixing of formation water with the condensed injected steam as production proceeds. The rise of K^+ concentration stops after approximately 50 days of production, and thereafter the concentration remains relatively constant. This trend is difficult to explain by simple mixing of two waters and implies that K^+ is not behaving as a conservative species, but is reacting with the rock matrix. Na^+ appears to be a conservative species and it is important to note that Na^+ concentrations generally are more than an order of magnitude greater than K^+ concentrations.

The cooling of the wells during production is reflected by the progressive decrease in the concentration of silica (Figure 3), which indicates a decrease in silica solubility with decreasing temperature and suggests that silica might be useful as a geothermometer (Arnórsson, 1975). Because the silica concentration in the injected water is low, the trend of initially high concentrations that decrease with production does not reflect mixing of waters. More problematic is the similar trend for SO_4^{2-} concentration. At present there is no obvious explanation for

Table 1. Chemical analyses of produced waters from the Tucker Lake steam pilot. Analytical values are in mg/L. Total organic acid concentrations were <1 mg/L and are not reported. Alkalinity is approximately equal to HCO_3^-

Days	T (C)*	pH (T °C)**	Alk**	S--**	NH_3**	Na	K	Ca	Mg	Sr	Si	Fe	Mn	Li	Cl	F	Br	SO_4	PO_4
Well 10D																			
20	178	7.85 (31)	906	0.6	0	720	59	7.8	1.3	0.1	207	0.3	<0.1	1.42	620	2.2	4	230	1
35	179	7.71 (27)	592	1.0	nd	1100	91	8.8	0.6	0.5	257	0.1	<0.1	3.84	1000	1.6	<0.1	109	<1
43	176	7.60 (30)	680	1.7	0	1495	121	14.6	1.3	0.3	242	0.4	<0.1	5.68	2105	1.2	13	114	1
50	172	7.56 (35)	559	0.7	0	1795	139	14.3	0.8	0.4	194	0.3	<0.1	6.26	2464	1.2	11	119	<1
63	156	7.35 (34)	760	0.8	45	2185	145	20.3	1.1	0.7	191	0.1	<0.1	6.86	2919	1.4	22	176	nd
71	159	7.50	814	0.5	320	2140	145	31.1	2.0	0.8	200	0.1	<0.1	6.89	3077	nd	9	150	<1
78	147	6.81 (40)	880	0.7	50	2185	129	27.5	1.6	0.9	177	0.1	0.1	6.22	2959	1.5	8	157	nd
88	142	7.40 (43)	1081	0.7	12.5	2375	132	29.6	2.6	0.9	164	<0.1	<0.1	6.14	3150	1.5	17	177	3
123	110	7.66 (32)	1904	<0.1	0	3815	140	19.9	8.8	1.6	124	0.3	0.1	5.47	5010	1.5	32	87	2
134	106	7.38 (43)	1982	<0.1	138	4120	142	41.4	13.4	3.2	102	0.3	0.1	4.84	5585	1.3	28	88	2
151	95	7.33 (24)	1769	<0.1	26.8	4400	129	84.5	23.8	5.2	82	0.9	0.2	4.02	5796	1.3	44	32	4
Well 6B																			
47	172	7.52 (26)	434	<0.1	nd	1266	98	9.2	1.8	0.1	238	0.1	0.1	5.92	1827	nd	4	155	2
56	186	7.32 (53)	477	1.4	nd	1220	110	23.6	4.3	0.5		4.5	0.1	4.78	4011	nd	nd	181	2
62	165	6.28 (24)	614	0.3	nd	1965	141	27.2	1.0	1.3	152	0.2	0.2	6.66	1541	1	6	137	2
70	160	7.10 (36)	772	0.8	nd	2120	145	26.3	1.3	1.0	232	0.2	<0.1	6.64	2679	nd	22	188	2
77	126	6.58 (35)	771	0.5	nd	2640	120	34.2	6.3	2.0	198	0.3	<0.1	6.52	2886	nd	17	138	nd
Well 12A																			
63	149	7.27 (33)	942	<0.1	nd	2155	131	28.9	1.2	0.8	189	0.1	0.0	5.42	3941	nd	13	144	2
74	146	7.34 (32)	1033	0.1	149	2420	142	33.5	1.7	0.8	168	0.2	0.0	5.98	3221	nd	42	112	4
91	136	7.34 (35)	1354	<0.1	14.5	2740	141	37.1	3.6	1.1	150	0	<0.1	6.03	3558	nd	33	102	nd
100	121	7.30 (30)	1552	nd	12	3060	129	32.4	4.6	1.3	138	0.2	0.0	5.36	3983	nd	47	95	<1
133	111	6.71 (32)	1998	<0.1		4120	151	53.7	14.4	3.6	110	0.6	0.2	4.81	5180	nd	107	69	nd
217		6.93 (26)	1397	<0.1	28.5	5860	104	344	93.7	17.5	45	3.51	1	2.48	8957	nd	113	8	2

* Temperature from Na/K geothermometer.
** Value measured in the field.
nd—Not detected.

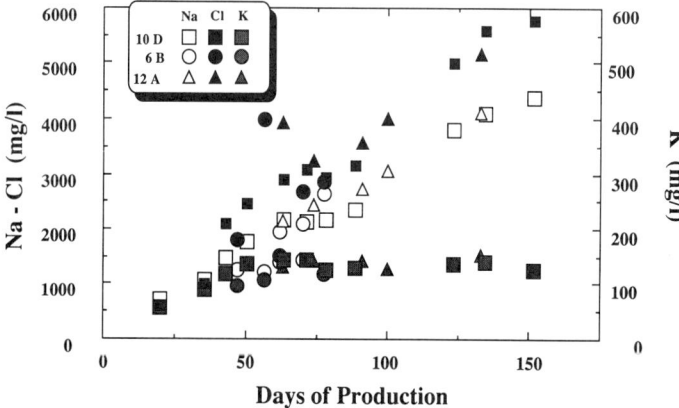

Figure 2. Trend of Na^+, Cl^-, and K^+ concentration with time. Time is measured in days; day 0 represents the day the well first yields fluids to the surface.

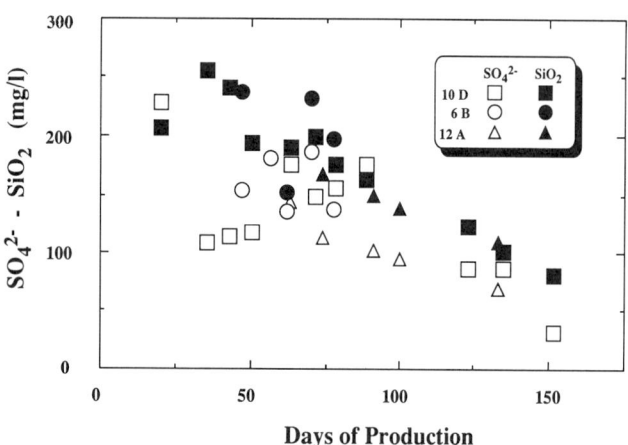

Figure 3. Silica and sulfate concentrations with time. The concentrations of silicate and sulfate continuously decrease with production. The decrease in silica concentration reflects cooling of the formation and the produced waters. The trend in sulfate concentration is unexplained.

the progressive decrease in concentration of this species with temperature.

Complete water analyses are presented in Table 1, but the general picture of changes in concentration with time is given by the species described above. Hutcheon et al. (1990) have noted the rise of Ca^{2+} and HCO_3^- during production and, using ^{13}C isotopic data, have interpreted this change to reflect the reactive dissolution of calcite to produce CO_2. Hutcheon et al. (1980) suggest that clay-carbonate reactions may generate CO_2. They also suggest that the relatively low Ca^{2+} concentration implies that this species is participating in a reaction with silicates to produce Ca-smectite, as noted in the experiments of Gunter and Bird (1988).

MEASUREMENT OF IN-SITU TEMPERATURE

The decrease in silica as production proceeds suggests that aqueous geothermometers may be useful in obtaining in-situ temperatures. Na/K and silica geothermometers (Arnórsson et al., 1983) were used by Hutcheon et al. (1990) to obtain temperatures for the waters examined in this study. Their data show that temperatures estimated for these water samples using the Na/K activity ratio and the silica concentration in equilibrium with chalcedony are similar over most of the range of temperatures observed. It is significant that the Na/K geothermometer is an empirical scale that has been calibrated by comparing Na/K ratios of produced geothermal waters to temperatures measured by other methods, and the calibration does not rely on the assumption that equilibrium is attained between rocks and their associated waters. We chose the version of the Na/K geothermometer that uses activity, rather than concentration, ratios so that temperatures would not be subject to ionic strength effects. The coincidence of the Na/K and chalcedony geothermometers suggests that they can be used to obtain the temperature of produced waters and that silica concentration can be considered to be approximately equal to that set by saturation with chalcedony. This is confirmed by considering the silica activity for each sample in terms of the temperature determined by the Na/K geothermometer (Figure 4). The silica concentration is in equilibrium with cristobalite at low temperatures (below 80°-100°C), and with chalcedony over most of the temperature range (100° to 170°C), and approaches silica saturation with respect to quartz only at temperatures above 170°C. All activity diagrams that show mineral stabilities in the following section are calculated by assuming that the silica concentration is set by equilibrium with chalcedony. Use of the chalcedony geothermometer and assuming saturation with respect to chalcedony is indicated by the presence of opaline cristobalite (chalcedony) in oil sands of the Clearwater Formation (Sedimentology Research Group, 1981).

INDICATIONS OF MINERAL REACTION

Figure 2 shows that the chloride content of the produced water increases consistently for all the wells sampled. This general rise in chloride content is thought to represent the saline formation water, mixed with progressively lower proportions of condensed steam, returning to the well during the production cycle. In contrast, the K^+ content rises fairly rapidly during a production cycle (Figure 2) and then levels off. This can be interpreted to indicate

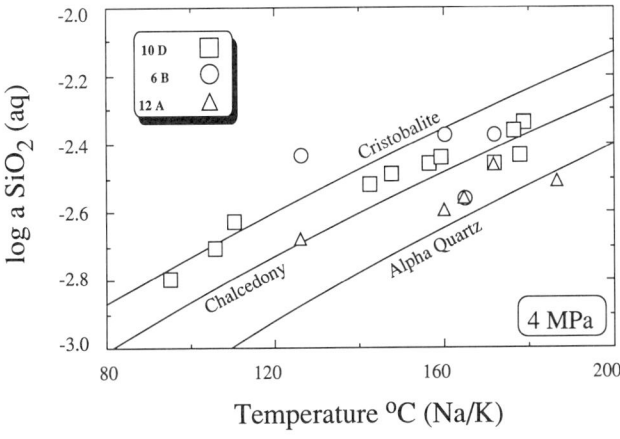

Figure 4. Activity of silica as a function of temperature, measured using the Na/K geothermometer. Over most of the temperature range, waters are saturated with respect to chalcedony.

control of the K$^+$ concentration by a mineral reaction. Hutcheon et al. (1989) noted that the rocks of the Clearwater Formation at Tucker Lake contain extensive pore-filling kaolinite. The reaction of Na$^+$ and K$^+$, either in the formation or in the injected water, with aluminous silicates could control the composition of produced waters by reactions that can be generalized as:

Illite + K$^+$ ⇔ K-feldspar + H$^+$ (1)

or:

Na-smectite + Na$^+$ ⇔ Analcime + H$^+$ (2)

No poststeam core has been recovered from Tucker Lake, and these reactions cannot be confirmed by comparison of samples from before and after steam injection. However, the Sedimentology Research Group (1981) has reported similar reactions in samples from the Clearwater Formation that were subjected to steam injection for a 2-year period. Detailed analysis of the chemistry of the produced water, using various computer programs, indicates the nature of mineral reactions which may be taking place during injection and recovery. The reactions postulated from the water chemistry are consistent with reactions observed in other steam pilots (Sedimentology Research Group, 1981).

GEOCHEMICAL MODELING

Geochemical modeling of produced fluids was used to identify and monitor the progress of mineral-water interactions occurring within the reservoir during production. Three geochemical modeling programs—SOLMNEQF (Aggarwal and Gunter, 1986), PTA-SYSTEM (Brown, et al. 1988), and EQCALC (Flowers, 1986)—have been used. SOLMNEQ-F, a FORTRAN version of the program SOLMNEQ (Kharaka and Barnes, 1973), was used to calculate activities of aqueous species in produced waters. A program in BASIC, which utilizes the geothermometers developed by Arnórsson et al. (1983), was used to estimate the temperature at which the sample last equilibrated with minerals in the reservoir. EQCALC uses the mineral and aqueous species data of Helgeson and Kirkham (1974) and Helgeson et al. (1978, 1981), combined with the equation of state for water of Haar et al. (1979), to calculate equilibrium constants for mineral-water reactions. PTA-SYSTEM was also used to calculate equilibrium constants for mineral-water reactions and, in general, the results obtained with PTA-SYSTEM and EQCALC are in close agreement for the reactions examined.

Analyzed concentrations of the major elements present in produced waters do not account for the distribution of each element between aqueous species. In order to determine mineral saturation indices accurately, the total amounts of a particular element present in solution must be distributed between all species that contain the element. An example of this is an alkalinity titration where the total alkalinity represents contributions from a variety of species, including the carbonate and bicarbonate ions and possibly the acetate or other titratable weak acid anions. SOLMNEQ-F is designed to "distribute" each element between a large number of possible species so that the state of saturation of the produced fluid with respect to minerals within the reservoir can be determined. SOLMNEQ-F also allows the user to simulate heating the sample to reservoir temperatures to assess saturation indices under more realistic conditions.

In-field measurements of pH, temperature, and other analytical data were entered into SOLMNEQ-F. The first run on SOLMNEQ-F is carried out at approximately 25°C (in fact, the actual temperature at which field measurements were made is used, and this is slightly different, see Table 1) and speciation and activities of ions in solution are calculated. Measured values for total SiO$_2$, Na$^+$, K$^+$, Mg^{2+}, Ca^{2+}, and Li$^+$ and calculated activities of SiO$_2$, Na$^+$ and K$^+$ are used with the published geothermometers of Fournier and Truesdell (1973), Arnórsson et al. (1983), Kharaka et al. (1985), Truesdell (1985), and Kharaka and Mariner (1989) to estimate temperatures within the reservoir. The Na/K geothermometer, using activity ratios as calibrated by Arnórsson et al., was found to give the most consistent results, but in general there was good agreement between various geothermometers. To be consistent, only the activity ratio of the Na/K geothermometer of Arnórsson et al. was used in the subsequent calculations.

After an in-situ temperature is estimated, SOLMNEQ-F is used to calculate the distribution of species at the estimated reservoir temperature. The new distribution of species and the corresponding ion activities, including pH, may be plotted alone or as ratios against temperature, allowing the effects of decreasing temperatures throughout the production cycle to be examined. As a test for the accuracy of pH measurements, SOLMNEQ-F was used to calculate how much $CO_2(g)$ was required at reservoir temperatures to saturate the solution with calcite. This was useful in confirming the pH at reservoir temperatures calculated using the dissociation constant for water from pH measurements near 25°C.

MINERAL REACTIONS

The results of computer modeling of produced water chemistries from wells 10D, 6B, and 12A are shown in Figures 5 through 8. In all figures the indicated log of the activity or activity ratio is plotted against temperatures given by the Na/K geothermometer. Time since production increases to the left, because points at lower temperature represent samples collected later in the production cycle. The phase boundaries between various minerals were calculated using the thermodynamic data for minerals and aqueous species in EQCALC or PTA-SYSTEM. The points plotted on the phase diagrams are derived using the water analyses and the thermodynamic data for the aqueous species only. Agreement between the positions of phase boundaries and water analyses, indicated by points, can be considered to be an indication that the waters are in equilibrium with the mineral assemblage specified by the particular phase boundary.

Phase relations in thermal waters may be portrayed at either quartz or amorphous silica saturation. Because produced waters from wells 10D, 6B, and 12A are close to equilibrium with chalcedony, Figures 5 to 8 have been constructed at chalcedony saturation using EQCALC. Portrayal of produced waters on diagrams at chalcedony saturation will remain valid as long as chalcedony is present in the reservoir. Once this more soluble silica polymorph is exhausted, silica activities should decrease to levels consistent with quartz saturation.

Phase relations between kaolinite, pyrophyllite, illite, and microcline at chalcedony saturation are shown in Figure 5 in terms of temperature and the aK^+/aH^+ ratio at 4 MPa (40 bars). In this phase diagram the thermodynamic component muscovite is used to calculate the stability field of illite. The effects of decreasing activities of muscovite in illite are shown as dashed lines. Muscovite activity in illite may be estimated from measured illite compositions using an ideal multi-site mixing model. These activities, however, do not account for differences in crystallinity between muscovite and illite. An estimated "best-fit" activity for muscovite in illite is about 0.04. If compositional data were available for illite, the method of Giggenbach (1983) could be used to estimate illite activities based on a three-

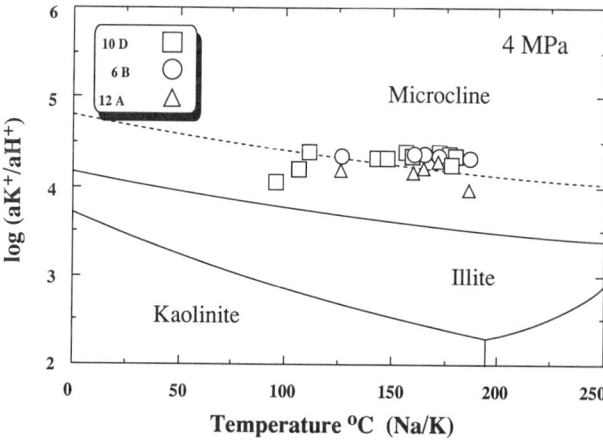

Figure 5. Activity ratio of K^+/H^+ vs. temperature. The activity ratio K^+/H^+ of the produced fluids tends to parallel the phase boundary between illite and K-feldspar. The diagram is calculated at 4 MPa total pressure, and SiO_2 is set by chalcedony saturation.

end-member solid-solution model involving muscovite, celadonite, and pyrophyllite.

The calculated K^+/H^+ activity ratios clearly indicate that the evolution of fluid compositions parallels the illite/microcline boundary. This strongly suggests that the activity of K^+ in produced waters is controlled by reactions involving microcline and illite. Because aluminum was not measured, SOLMNEQ-F could not be used to evaluate saturation with respect to either illite or microcline. The low slope of the illite/microcline boundary on Figure 5 indicates that the plateau in K^+ analytical values noted earlier may be a result of control of fluid compositions by a reaction involving illite and microcline.

The Na^+/H^+ activity ratio versus temperature for kaolinite, pyrophyllite, Na-smectite, and analcime are shown in Figure 6. Paragonite was used as the thermodynamic component to describe Na-smectite, and a reduced activity of 0.5, estimated for paragonite in smectite, is shown as a dashed line. Analcime has been chosen instead of albite as the stable sodium aluminum silicate mineral. This choice was made because analcime has been noted in poststeam core from the Clearwater Formation (Sedimentology Research Group, 1981). Secondly, with only minor reduction of analcime activity, say by substitution of Ca^{2+} for $2Na^+$, analcime is stabilized with respect to albite. The observation that analcime forms rather than albite may also imply that in fluids supersaturated with respect to both analcime and albite, analcime is kinetically favored because of sluggish precipitation kinetics for albite at these temperatures. Matthews (1980) has shown that, at 250°C, precipitation rates for albite are on the order of years. Thus the kinetic data are consistent with petrographic observations and water compositional data, all of which indicate that albite is not precipitated during steam injection. As with potassium, the evolution

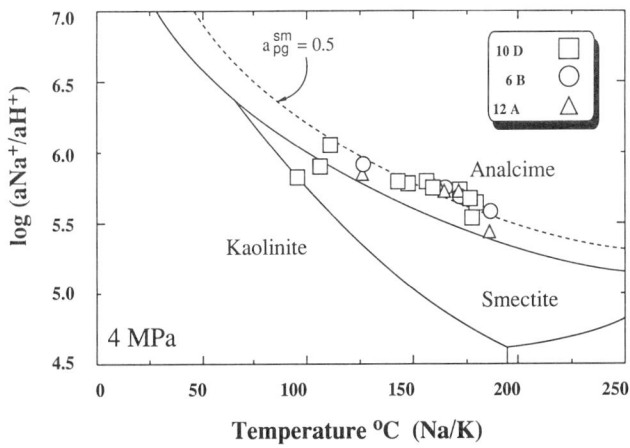
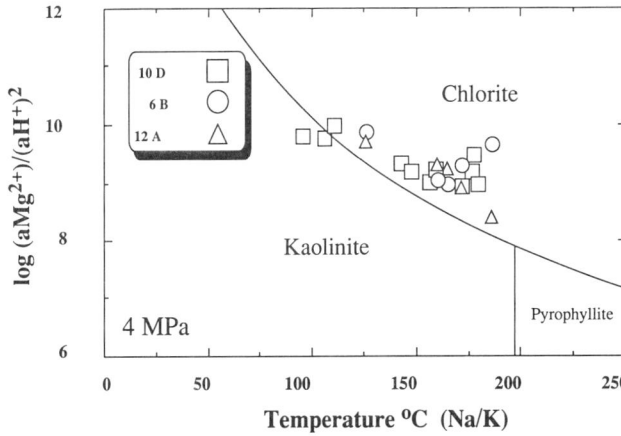

Figure 6. Activity ratio of Na$^+$/H$^+$ vs. temperature. The activity ratio Na$^+$/H$^+$ of the produced fluids tends to parallel the phase boundary between smectite and analcime. The diagram is calculated at 4 MPa total pressure, and SiO$_2$ is set by chalcedony saturation.

Figure 7. Activity ratio of Mg^{2+}/(H$^+$)^2 vs. temperature. The activity ratio Mg^{2+}/(H$^+$)^2 of the produced fluids tends to parallel the phase boundary between kaolinite and chlorite. The diagram is calculated at 4 MPa total pressure, and SiO$_2$ is set by chalcedony saturation.

of sodium activities with time follows the reaction boundary between Na-smectite and analcime. Unlike potassium, however, concentrations of sodium rise through the production cycle because of the progressive mixing of formation water with steam condensate. It might appear to be fortuitous that sodium, which increases nearly linearly with time because of return of formation waters, should show Na$^+$/H$^+$ activity ratios that follow the Na-smectite/analcime phase boundary. If we consider that aNa$^+$ is 8 or 9 orders of magnitude greater than aH$^+$, it is obvious that very small changes in pH will allow the aNa$^+$/aH$^+$ ratio to follow the smectite/analcime phase boundary. In other words, the hydrolysis of the silicates, as implied in previously described reactions (1) and (2), may be capable of buffering the pH under conditions where aNa$^+$ is changing quite rapidly in response to fluid mixing, even at temperatures in the range of 100°C or lower.

The ratio aMg^{2+}/(aH$^+$)^2 versus temperature (Figure 7) shows phase relationships between kaolinite, pyrophyllite, and 7Å-clinochlore, a magnesium chlorite. Although 7Å-clinochlore is metastable, it was chosen over 14Å-clinochlore because its thermodynamic properties more accurately reproduce phase relations in hydrothermal, geothermal, and diagenetic environments (Helgeson et al., 1978). Compositional data on chlorites present in the Clearwater Formation indicate that they are Fe-Mg chlorites. Although it is not shown in Figure 7, the chlorite/kaolinite and chlorite/pyrophyllite boundaries are not significantly affected by 50% substitution of Fe for Mg in chlorite. Unlike Figures 5 and 6, for K$^+$ and Na$^+$ respectively, Figure 7 shows a migration of early Mg^{2+}/(H$^+$)^2 activity ratios from a position within the Mg-chlorite stability field toward the kaolinite field. The Mg^{2+}/(H$^+$)^2 activity ratio for injection water, calculated by SOLMNEQ-F, is 11.4 at 300°C. The observed decrease in Mg^{2+}/(H$^+$)^2 activity ratios is consistent with a decrease from the activity ratio in injection fluids to values buffered by a reaction involving kaolinite and chlorite.

When compared with the Na$^+$ and K$^+$ systems, the buffering of Mg^{2+} activities by mineral reactions is less effective. Both Na$^+$ and K$^+$ appear to be buffered throughout the production cycle, whereas Mg^{2+} does not reflect activities consistent with equilibrium until after about 60 days of production. This is probably related to the kinetics of reactions involving Mg-chlorite and kaolinite, which in turn depend on both the temperature and the surface area of the reactant minerals. In this case, surface area may be most important, because chlorite is the least abundant clay present, rarely making up more than 10% of the clay fraction (Sedimentology Research Group, 1981; Hutcheon et al., 1989).

The reversal from decreasing to increasing Mg^{2+}/(H$^+$)^2 activity ratios after about 60 days, and the coincidence of this trend with the Mg-chlorite/kaolinite boundary, strongly suggest equilibration between produced waters and minerals present in the reservoir. This is not to say that both kaolinite and Mg-chlorite are stable; rather, it implies that reactions involving the two are controlling the Mg^{2+}/(H$^+$)^2 ratio, presumably by metastable equilibria.

The phase relationships for Na$^+$, K$^+$, and Mg^{2+} might lead us to expect that the Ca^{2+}/(H$^+$)^2 activity ratio would coincide with an alumino-silicate equilibrium. However, as Figure 8 shows, the Ca^{2+}/(H$^+$)^2 activity ratio is in equilibrium with calcite dissolution into the vapor phase according to the reaction:

$$CaCO_3 + 2H^+ \Leftrightarrow Ca^{2+} + H_2O + CO_2 \quad (3)$$

Hutcheon et al. (1990) have shown that CO$_2$ is produced in this pilot site by reactive dissolution of calcite, and the Ca^{2+}/(H$^+$)^2 activity ratio of the fluids

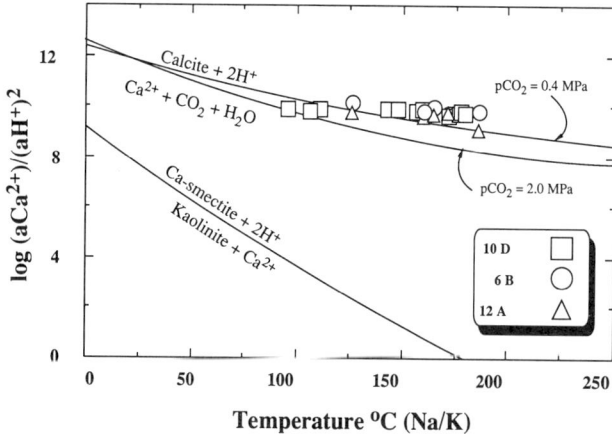

Figure 8. Activity ratio of $Ca^{2+}/(H^+)^2$ vs. temperature. The activity ratio $Ca^{2+}/(H^+)^2$ of the produced fluids tends to parallel the phase boundary specified by dissolution of calcite into the vapor phase. These data imply that calcite dissolution proceeds more rapidly than silicate hydrolysis. The diagram is calculated at pressures set by the boiling curve for pure water, and the activity of SiO_2 is set by chalcedony saturation. The pCO_2 is estimated from field data to be between 0.4 MPa (one-tenth the total pressure) and 2 MPa (one-half the total pressure).

appear to be in equilibrium with calcite, rather than a silicate reaction. To calculate the position of reaction (3) requires that the fugacity of CO_2 be known. The points shown on Figure 8 were calculated by taking the concentration of CO_2 in the vapor, reported by Hutcheon et al. (1990), assuming that there is vapor-liquid equilibrium for CO_2, and applying the fugacity coefficient at the appropriate temperature and pressure. The range of fCO_2 was found to be small and close to 1.0 MPa. Reaction (3), as shown on Figure 8, is calculated for $fCO_2=1$ MPa with pressures determined by the boiling curve for pure water.

DISCUSSION

The coincidence of the activity ratios Na^+/H^+, K^+/H^+ and $Mg^{2+}/(H^+)^2$ of the produced fluids with phase boundaries among minerals observed to form during thermal recovery is a strong indication that metastable equilibria control the activity ratios via water-rock interaction. The boundaries, and the corresponding reactions, observed undoubtedly represent a series of metastable states of the system, not the final equilibrium state. However, the evidence strongly suggests that these metastable equilibria are capable of moderating activity ratios of some constituents of the fluid. At the temperatures and times observed in steam flooding, the reaction rates for these silicate equilibria are rapid enough for equilibrium of the silicates with Na^+/H^+, K^+/H^+, and $Mg^{2+}/(H^+)^2$ to be attained. However, as we might anticipate, the reaction rate for calcite dissolution is more rapid than the reaction rates for the silicates, and thus the $Ca^{2+}/(H^+)^2$ activity ratio is set by calcite dissolution, not a silicate equilibrium. The combination of silicate equilibria for Na- and K-bearing minerals controlling fluid activity ratios suggests that silicate hydrolysis equilibria may act as pH buffers in water-rock systems at the conditions of thermal recovery or conditions appropriate for diagenesis.

Consider that the Na^+/H^+ activity ratio is set by the silicate equilibria, even though the bulk concentration of Na^+ is apparently controlled by the mixing of formation waters and condensed steam. The concentration of Na^+ varies from less than 500 to more than 4500 mg/L, or by about 1 order of magnitude. Over that same time span the aNa^+/aH^+ ratio varies more than half an order of magnitude, suggesting that pH should decrease by more than half a unit (order of magnitude) over the same period. Figure 9 shows that pH does decrease more than half an order of magnitude over the sampling period, suggesting that the silicates are buffering the pH. Various authors (Surdam et al., 1984; Means and Hubbard, 1987; Surdam and Crossey, 1987; Surdam et al., 1989) have suggested that in diagenetic systems pH is buffered by the organic acids. Hutcheon and Abercrombie (1990) have calculated buffering potentials for acetate, dissolved carbonate, and silicates and generally have found that the pH is most strongly buffered by silicate equilibria. The form of the equation used to calculate buffering potential (see Butler, 1964, for example) suggests that the buffer capacity of silicate equilibria should be infinite compared with the other potential pH buffers. The pH buffer capacity of a range of mineral and organic reactions in a range of feasible water compositions should be calculated by using appropriate mass transfer considerations, including allowing variation of the solution composition at each stage of the calculation. The calculations would be complex but are ideally suited to EQ3/6 (Wolery, 1983) or some similar computer program that can track solution compositions, including speciation of organic constituents, while maintaining saturation with respect to a number of mineral species.

If pH is buffered by the silicates and Na^+ is the dominant cation in formation waters, the mixing of more saline with less saline waters should cause progressive reductions in pH. This should be reflected by a decrease in the stability of various minerals, especially carbonates. One would therefore expect that dissolution might be prevalent in the vicinity of salt domes or at salt dissolution edges, such as in the subcrop of the Paleozoic in the eastern part of the Western Canada sedimentary basin.

Increasing temperature tends to favor the stability of the more K^+- and Na^+-rich mineral species, as shown in Figures 5 and 6, respectively. In other

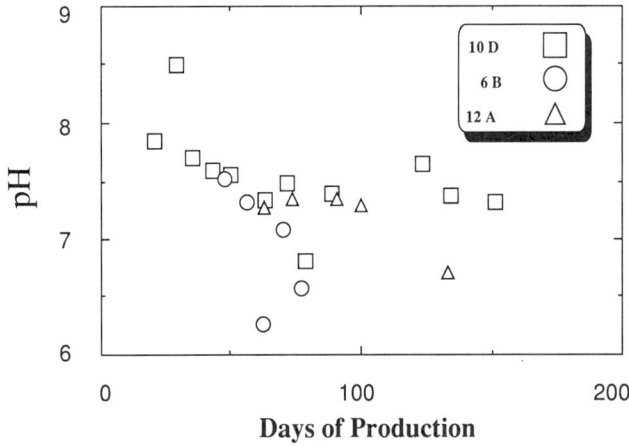

Figure 9. Production time vs. pH. The trend of pH with time is as would be expected if silicate hydrolysis reactions were buffering pH over the range of change in concentration observed for Na^+ during production.

words, as diagenesis proceeds, K^+ and Na^+ will be exchanged for H^+ in hydrolysis reactions, and pH (relative to neutral pH) will continuously decrease. In this respect, aluminous silicates such as kaolinite can be considered to be "acids." Hydrolysis should result in progressive dissolution, particularly of carbonates, during burial and diagenesis. The requirement of the illite-kaolinite reaction for K^+ in natural diagenetic settings can be expected to be reflected by dissolution of K-feldspar.

CONCLUSIONS

Injection of steam to assist recovery of viscous oils is a suitable laboratory to consider the effects of changing salinity and temperature on diagenetic water-rock interactions. The range of temperatures and changes in salinity are appropriate for most natural systems, but the time scale is considerably shorter and the water-rock ratio is much higher. The effect of shorter times and higher water-rock ratios should be to make approach to equilibrium less likely. However, even under these extreme conditions, metastable silicate reactions apparently are rapid enough to attain equilibrium with the water. The exact metastable equilibria that govern water-rock evolution during diagenesis may be different from those considered here, but because the form of the equilibrium constants will be the same, the overall processes will be similar.

The progress of silicate hydrolysis reactions is such that they can buffer the activity ratios of metal ions, such as Na^+ and K^+ to H^+, and this functions as a means of producing H^+ as salinity or temperature increase. The hydrolysis process is the rate-limiting step in silicate carbonate reactions, allowing the $Ca^{2+}/(H^+)^2$ ratio to be determined by the carbonates, in this case calcite. Although the rate of increase of Ca^{2+} is less than the alkalinity change, in a nearly closed system, with only minor addition or removal of major elements in the water, the result would be progressive decreases in Na^+ and K^+ relative to Ca^{2+}. Many basin waters (Carpenter, 1978; Boles, 1984) have been noted to be relatively enriched in Ca^{2+} and silicate-hydrolysis-driven calcite dissolution in combination with albitization of plagioclase could provide the source of Ca^{2+}. The hypothesis that silicate hydrolysis drives calcite dissolution is not new, having been suggested by Carpenter (1978) in a study of brines from central Mississippi.

Although it is not entirely clear that the organic-inorganic-water system is governed by any single reaction or class of reactions, it is clear from the data presented here that any model attempting to predict diagenetic process, such as dissolution, requires silicate equilibria to be an integral part of that model. With the possible exception of the clay minerals, zeolites, and association constants for organic acids, most of the requisite thermodynamic data, formation water compositions, and computational tools (EQ3/6; Wolery, 1983) are available to evaluate the water-rock-organic system and its interactions. A clear concept of how fluid flows in the subsurface during burial is the most enigmatic part of the diagenetic problem.

ACKNOWLEDGMENTS

We gratefully acknowledge assistance in the field and laboratory by Cynthia Nahnybida, William Tang Kong, Maurice Shevalier, and Nenita Lozano. We thank Husky Oil Operations Ltd., and particularly A. F. S. Bau and Ed Klovan, for providing core material and obtaining permission to sample producing wells. Peter Putnam, formerly of Husky, was instrumental in making the arrangements to start this project. The Alberta Oil Sands Technology Research Authority (AOSTRA) provided funding and permission to publish. Graduate research support from the Petroleum Aid to Education Fund and scholarships provided by AOSTRA and the Killam Trust are acknowledged by H. J. Abercrombie.

REFERENCES CITED

Abercrombie, H. J., 1989, Water-rock interaction during diagenesis and thermal recovery, Cold Lake Alberta: Ph.D. dissertation, Department of Geology and Geophysics, The University of Calgary, Calgary, Alberta.

Abercrombie, H. J., and I. E. Hutcheon, 1986, Remote monitoring of water-rock-bitumen interactions during steam-assisted heavy oil recovery: Reykjavik, Iceland, International Association of Geochemistry and Cosmochemistry Fifth International Symposium on Water-Rock Interactions Proceedings, p. 1–4.

Aggarwal, P. K., and W. D. Gunter, 1986, SOLMNEQF: a FORTRAN-77 version of SOLMNEQ: Alberta Research

Council/Alberta Oil Sands Technology and Research Authority, Joint Agreement Report 8687-4.

Arnórsson, S., 1975, Application of the silica geothermometer in low temperature hydrothermal areas in Iceland: American Journal of Science, v. 275, p. 763-784.

Arnórsson, S., E. Gunnlaugsson, and H. Svavarsson, 1983, The chemistry of geothermal waters in Iceland. III. Chemical geothermometry in geothermal investigation: Geochimica et Cosmochimica Acta, v. 47, p. 567-577.

Boles, J. R., 1984, Secondary porosity reactions in the Stevens Sandstone, San Joaquin Valley, California, in D. A. McDonald and R. C. Surdam, eds., Clastic diagenesis: AAPG Memoir 37, p. 217-224.

Brown, T. H., R. G. Berman, and E. H. Perkins, 1988, GEOCALC II: PTA-SYSTEM: software for the calculation and display of pressure-temperature-activity phase diagrams: University of British Columbia, Department of Geological Sciences, Vancouver, British Columbia, manual for computer program, 35 p.

Butler, J. N., 1964, Ionic equilibrium: a mathematical approach: Reading, Massachusetts, Addison-Wesley, 547 p.

Carpenter, A. B., 1978, Origin and chemical evolution of sedimentary brines: Oklahoma Geological Survey Circular 79, p. 60-77.

Cathles, L. M., M. Schoell, and R. Simon, 1987, CO_2 generation during steamflooding: a geologically based kinetic theory that includes carbon isotope effects and application to high-temperature steamfloods: San Antonio, Texas, Society of Petroleum Engineers International Symposium on Oilfield Chemistry Proceedings, Paper Number SPE 16267, p. 255-270.

Flowers, G. C., 1986, Computation of thermodynamic properties of reactions involving minerals and aqueous solutions with the aid of a personal computer: Computers and Geosciences, v. 12, p. 361-380.

Fournier, R. O., and A. H. Truesdell, 1973, An empirical Na-K-Ca geothermometer for natural waters: Geochimica et Cosmochimica Acta, v. 37, p. 1255-1275.

Giggenbach, W. F., 1983, Construction of thermodynamic stability diagrams involving alkali clay minerals: Misasa, Japan, Fourth International Symposium on Water-Rock Interaction Proceedings, p. 156-159.

Gunter, W. G., and G. W. Bird, 1988, CO_2 production in tar sand reservoirs under in situ steam temperatures: reactive calcite dissolution: Chemical Geology, v. 70, p. 301-311.

Haar, L., J. S. Gallagher, and G. S. Kell, 1979, Thermodynamic properties of fluid water, in Water and steam: their properties and current industrial applications: New York, Hemisphere Publishing Company, 122 p.

Helgeson, H. C., and D. H. Kirkham, 1974, Theoretical prediction of the thermodynamic behavior of aqueous electrolytes at high pressures and temperatures. I. Summary of the thermodynamic and electrostatic properties of the solvent: American Journal of Science, v. 274, p. 1089-1198.

Helgeson, H. C., D. H. Kirkham, and G. C. Flowers, 1981, Theoretical prediction of the thermodynamic properties of aqueous electrolytes at high temperatures and pressures. IV. Calculation of activity coefficients, osmotic coefficients, and apparent molal and standard and relative partial molar properties to 600°C and 5Kb: American Journal of Science, v. 281, p. 1249-1516.

Helgeson, H. C., J. M. Delany, H. W. Nesbitt, and D. K. Bird, 1978, Summary and critique of the thermodynamic properties of rock-forming minerals: American Journal of Science, v. 278A, p. 1-229.

Hutcheon, I., and H. Abercrombie, 1987, Monitoring mineral reactions by fluid compositions from heavy oil pilots (abs.): Saskatoon, Saskatchewan, GAC/MAC Joint Annual Meeting, Program with Abstracts, v. 12, p. 57.

Hutcheon, I., and H. J. Abercrombie, 1990, Carbon dioxide in clastic rocks and silicate hydrolysis: Geology, v. 18, p. 541-544.

Hutcheon, I., A. Oldershaw, and E. Ghent, 1980, Diagenesis of sandstones of the Kootenay Formation at Elk Valley (southeastern British Columbia) and Mount Allan (southwestern Alberta): Geochimica et Cosmochimica Acta, v. 44, p. 1425-1435.

Hutcheon, I., H. J. Abercrombie, and H. R. Krouse, 1990, Carbon dioxide production in thermal recovery: Chemical and isotopic evidence and applications: Geochimica et Cosmochimica Acta, v. 59, p. 165-171.

Hutcheon, I., H. Abercrombie, M. Shevalier, and C. Nahnybida, 1988, A comparison of formation reactivity in quartz-rich and quartz-poor reservoirs during steam assisted recovery: Edmonton, Alberta, United Nations Institute for Training and Research/United Nations Development Program Fourth International Conference, v. 3, paper 235, p. 1-12.

Hutcheon, I., H. J. Abercrombie, P. Putnam, H. R. Krouse, and R. Gardner, 1989, Diagenesis and sedimentology of the Clearwater Formation at Tucker Lake: Bulletin of Canadian Petroleum Geology, v. 37, p. 83-97.

Kharaka, Y. K., and I. Barnes, 1973, Solution mineral equilibrium computations: USGS NTIS Document PB 215 899.

Kharaka, Y. K., and R. H. Mariner, 1989, Chemical geothermometers and their application to formation waters from sedimentary basins, in N. D. Naeser and T. H. McCulloh, eds., Thermal history of sedimentary basins: New York, Springer-Verlag, p. 99-118.

Kharaka, Y. K., R. W. Hull, and W. W. Carothers, 1985, Water-rock interactions in sedimentary basins, in Relationship of organic matter and mineral diagenesis: SEPM Short Notes, Course 17, p. 79-176.

Lico, M. S., Y. K. Kharaka, W. W. Carothers, and V. A. Wright, 1982, Methods of collection and analysis of geopressured geothermal and oil field waters: USGS Water-Supply Paper 2194, 21 p.

Matthews, A., 1980, Influence of kinetics and mechanism in metamorphism: a study of albite crystallization: Geochimica et Cosmochimica Acta, v. 44, p. 387-402.

McCorriston, L. L., R. A. Demby, and E. C. Pease, 1981, Study of reservoir damage produced in heavy oil formations due to steam injection: San Antonio, Texas, Society of Petroleum Engineers 56th Annual Fall Technical Conference SPE Paper 10077.

Means, J. L., and N. Hubbard, 1987, Short-chain aliphatic acid anions in deep subsurface brines: a review of their origin, occurrence, properties, and importance and new data on their distribution and geochemical implications in the Palo Duro basin, Texas: Organic Geochemistry, v. 11, p. 177-191.

Sedimentology Research Group, 1981, The effects of in situ steam injection on Cold Lake oil sands: Bulletin of Canadian Petroleum Geology, v. 29, p. 447-478.

Shepherd, D. W., 1981, Steam stimulation recovery of Cold Lake bitumen, in R. F. Meyer and C. T. Steele, eds., The future of heavy crude oils and tar sands: New York, McGraw Hill, p. 349-360.

Truesdell, A. H., 1985, Chemical geothermometers for geothermal exploration, in R. W. Henley, A. H. Truesdell, and P. B. Barton, Jr. (with J. A. Whitney), eds., Fluid-mineral equilibria in hydrothermal systems: Reviews in Economic Geology, 1.

Surdam, R. C., and L. J. Crossey, 1987, Integrated diagenetic modeling: a process-oriented approach for clastic systems: Annual Review of Earth and Planetary Sciences, v. 15, p. 141-170.

Surdam, R. C., S. W. Boese, and L. G. Crossey, 1984, The chemistry of secondary porosity, in D. A. MacDonald and R. C. Surdam, eds., Clastic diagenesis: AAPG Memoir 37, p. 127-149.

Surdam, R. C., L. G. Crossey, E. S. Hagen, and H. P. Heasler, 1989, Organic-inorganic interactions and sandstone diagenesis: AAPG Bulletin, v. 73, p. 1-23.

Wolery, T. J., 1983, EQ3NR, a program for geochemical aqueous speciation-solubility calculations: User's guide and documentation: Lawrence Livermore Laboratory Report UCRL-53414, 191 p.

Index

Activity ratios, of fluid constituents 166-168
Africa, South, erosional surfaces 5
Alaskan North Slope
    dissolution 12
    erosional surfaces 5-6
    porosity enhancement 5, 12
Algeria
    erosional surfaces 7
    porosity enhancement 7
Alkalinity, of formation waters 57
Anoxic conditions, effect of depositional environment 42-43
Arkose
    see also subarkose
    altered by fresh water 138-140
    altered by saline water 134-138
    clean 136-140
    dirty 134-136, 138
    muscovite in 145
    porosity in diagenetic simulation 143
    quartz distribution in diagenetic simulation 143
    simulation of diagenesis 133-144
Australia
    erosional surfaces 4, 9
    karst 8
    porosity enhancement 9-10
    sandstones 8
Banding, by compaction/cementation 105, 153-156
Basins
    compartmentalization 105-106
    overpressurized compartments in 106-107, 120-127
    Red Desert 55-67
    sealed compartments in 106-107, 120-127
    self-organized petroleum distributions in 104-107
    subsiding compartments in 126-127, 128
    Washakie 55-67
Benard flow cells 104
Brent Sand Formation, North Sea 91-92
Bulk solution, concentration of ions in 75
Bulk solution disequilibrium
    in aqueous fluids 71-82
    in diagenetic carbonates 71-82
    versus equilibrium 72-74
Buoyancy-driven convection
    by methane 106, 112-117
    reaction mediated 105, 106
    reaction-transport modeling 112
    in shales 118
Calcite
    concentration profiles 75, 76
    concentration ratios of diagenetic waters 81
    diagenesis 74
    dissolution 75-76, 117-120, 121-122
    distribution in arkose diagenesis 143
    high-Mg versus low-Mg 79, 80
    isotopic composition data 72
    precipitation 76
    at reaction fronts 117-120, 121-122
    recrystallization 75-77, 79
    thermodynamic equilibrium 77
    trace element data 72, 74
Canada, Cold Lake oil deposits, Alberta 161-169
Carbonates
    cementation of 156
    diagenesis 72-74, 77-81
    marl-limestone alternations 156
    precipitation 76
    recrystallization of 75-81
Caves
    in China 7
    in Venezuela 9
Cementation
    authigenic, banded 104-105
    banded 153-156
    at reaction fronts 104-105
    in sandstones 153-156
    simulation 155
    of Upper Almond Sandstone 67
Chemical modeling
    applications 49-51
    classification 46
    computer codes 48-51, 109-111, 131-132, 165
    coupling 108
    of diagenesis 46, 131-145
    effect of fluid and rock composition 131-145
    evolution equations 108-111
    feedback 108
    geologic data 52
    input data 48-52, 132
    input parameters 62-64
    kinetic data 51-52
    limitations 49-51
    mathematical modeling of overpressurized compartment 122-126
    of mineral reactions 166
    open system 63-64
    output data 48-51, 132
    overview 47
    of produced fluids by steam injection 165-166
    rate laws 108-111
    reaction-path modeling 49-50
    reaction-transport modeling 50-51
    in reservoir prediction 103-129
    rock description 107-108
    rock-water interactions 55-67
    speciation-solubility modeling 48-49, 161-169
    of subarkose diagenesis 67, 93-99
    theory 48-51
    thermodynamic data 51-52
    Upper Almond formation waters 62-67
    use in porosity prediction 45-52
Chemical weathering, of sandstones 2-4
China
    carbonates 7
    entrapment of oil in karst facies 13
    erosional surfaces 4, 9, 13
    karst 6, 7
    porosity enhancement 9
Clays
    in arkose 134-136
    authigenic 34, 38, 43
    composition in depositional facies 33
    diagenesis of kaolinite 85-99
    effect on pressure solution 151-152
Climate
    rock dissolution 5
    subaerial exposure 4
Closed system, Upper Almond formation waters 64-65, 68, 69
Cold Lake oil deposits
    compositions of produced waters 162-164
    geochemical modeling 165-166
    mineral reactions in 164-165, 166-168

in situ temperature measurement	164

Compaction
- and cementation banding 105, 153-156
- differential 153-156
- evolution in sandstones 154
- simulation in sandstones 155

Compartmentalization, of basins 105-106

Computer codes
- available reaction-transport codes 111-112
- for chemical modeling of diagenesis 48-51, 109-111, 165-166
- for mineral reactions 166-168

Computers
- hardware 111
- simulations with 111

Concentration profiles
- of ions in solution 75
- of magnesium and calcite 76

Continental shelf environments 39, 40

Convection
- buoyancy-driven 112-117
- cellular 113-117
- by methane in basin 106
- by methane in basins 112-117
- ocsillatory 113-117
- in presence of hydraulic heterogeneity 115-117
- from random kerogen distributions 114-117
- simulation of 113-117

Crevasse-splay delta environments 28, 36

Delta-front environments
- continental shelf 39, 40
- distal bar 38, 39
- distributary-mouth bar 37-38, 39
- prodelta 38-39, 40

Delta-plain environments
- crevasse-splay delta 28, 32, 36
- diagrammatic representation 28
- distributary channel 27-28
- lake 28
- marine bay 28, 35
- point count modal analyses 29-31

Depositional environments
- control of diagenesis 27-39
- deltaic 27-39
- effect on pore-water chemistry 41
- effect on sediment composition 40-41
- effect on sediment texture 41
- geochemical classification 43
- shoreline 32-37
- Texas Gulf Coast 25-43

Deposition environments, delta-front environments 37-39

Diagenesis
- of arkose 133-144
- boundary conditions in simulation 94, 112-113
- of carbonates 72-74, 77-81
- chemical modeling 46, 48-51, 94-99, 131-145
- computer codes for chemical modeling 48-51
- effect of fluid and rock composition 131-145
- effect of pore-water chemistry 41-43
- facies control 25-43
- input data in simulation 51-52, 132
- interaction of reaction, mass transport, and deformation 147-158
- numerical methods of simulation 93-94
- output data in simulation 132
- reaction-transport modeling of 104-107, 131-145
- of sandstones 85-99
- simulations of 55-67, 93-99, 131-145
- of subarkose 93-99
- through coupled processes 103-129
- of Upper Almond Sandstone 65-66, 67

Diagenetic carbonates
- bulk solution disequilibrium 71-82
- distribution coefficients 78
- molar concentration ratios 81
- recrystallization 77-81
- starting composition 78

Diagenetic modeling, see chemical modeling

Dissolution
- effect of climate 4, 5
- effect of framework composition 4
- Alaskan North Slope 12
- of calcite 117-120, 121-122
- in diagenetic modeling of subarkose 98
- during subaerial exposure 2-4
- experiments on kaolinite 86-90, 93
- factors controlling 2-4
- finger-shaped reaction front 119
- at grain contacts 150-151
- by meteoric waters 2
- North Sea, Norwegian 17, 18
- in open system 2-3
- rate of calcite 98
- rate for imperfect crystal 92
- vs. solution saturation in kaolinite 92

Distal bar environments 38, 39
Distributary channel environments 27-28
Distributary-mouth bar environments 37-38, 39
Dolomitization, at reaction fronts 117-120

Erosional surfaces
- Alaskan North Slope 5-6
- Algeria 7
- ancient 5-10
- Australia 4, 9
- China 4, 9, 13
- Libya 7
- modern 4-5
- The Netherlands 7, 13
- Norwegian Sea 6, 12
- in sandstones 10
- South Africa 5
- Texas 10, 14
- Venezuela 5, 10

Erosional unconformities, in sandstones 1-23
Exposure, duration of subaerial 4

Facies
- clay compositions 33
- control of diagenesis 27-43

Fluid injection, oscillatory 106-107

Fluid-rock interactions
- in Cold Lake oil deposits, Alberta 161-169
- compositions of produced waters 162-164
- mineral reactions in 164-165, 166
- in situ temperature measurement 164
- speciation-solubility modeling of 161-169
- in thermal recovery of bitumen 161-169
- of Upper Almond Sandstone 55-67

Formation waters
- activity ratios of constituents 166-168
- alkalinity 57
- bromide content 59, 63
- classification 57
- dilution 56-57
- HCO_3/Cl ratio 61, 66
- meteoric mixing 59-61, 64, 65
- mineral reactions in 164-165, 166-168
- $\delta^{18}O$ versus total dissolved solids 58-59, 61
- pH 64-65, 68, 69
- present-day composition 56-62
- pressure-depth profiles 59, 62

Fractures
- produced by steam injection 162-164
- salinity 57
- sulfate depletion 61
- Upper Almond Sandstone 56-62

Fractures
- in overpressurized compartments 106-107
- subaerial exposure 3-4

Framework composition, subaerial exposure 4
Free-face pressure solution (FFPS) 148-150, 156
Geochemical depth profiles, subaerial exposure 73
Geochemical modeling, see chemical modeling
Geothermometers, use of 164, 165
Grain shape, computer-simulated changes in 148-150
Homogeneous Distribution Law 78
Hydrochemical processes, in soil zone 3
Intergranular pressure solution (IPS), see pressure solution
Isotopic composition data, calcite 72
Jurassic, sandstones in North Sea 91-92

Kaolinite
- dissolution experiments 86-90
- mineral stability diagram 91
- occurrence in sandstone reservoirs 90-93
- precipitation experiments 86-90

Karst
- in Australia 8
- in China 6

Kinetics
- dissolution/precipitation of kaolinite 85-99
- input data 51-52

Lagoon environments 36, 37
Lake environments 28, 35
Late Cretaceous, of Wyoming 56

Libya
- erosional surfaces 7
- porosity enhancement 7

Limestone
- see also carbonates
- and marl diagenesis 156

Marine bay environments 28, 35

Marl-limestone alternations
- cementation 156
- differential compaction 156
- porosity during diagenesis 159
- textural parameters 157, 158

Mathematical modeling
- see also chemical modeling; chemical modeling
- of overpressurized compartments 122-126
- of pressure solution 147-158

Meteoric waters
- carbonate diagenesis 80
- subaerial exposure effects 2
- Upper Almond Sandstone 59-61

Methane
- buoyancy-driven convection by 106, 112-117
- contour map of convection 113

Muscovite, in arkose diagenesis 145

Netherlands, The
- erosional surfaces 7, 13
- porosity enhancement 7, 13

North Sea, Jurassic sandstones 91-92

North Sea, Norwegian
- dissolution 17, 18
- porosity observed 23
- porosity prediction 10-12, 15, 17, 18-23
- stratigraphy 16

Norwegian Sea
- see also North Sea, Norwegian
- erosional surfaces 6, 12
- porosity enhancement 6, 12

Oil recovery, by steam injection 161-169

Open system
- subaerial exposure 2-3
- Upper Almond formation waters 63-64, 67

Oscillatory intracrystalline zoning 104

Overpressurized compartments (in basins)
- mathematical modeling of 122-126
- oscillation in 126-127
- oscillatory fluid ejection in 106-107
- oscillatory fluid release 120-127
- permeability of 125, 126
- in petroleum basins 106-107, 120-127

Paleocene-Eocene, Texas Gulf Coast 25-43
Permeability, of sealed compartments 125, 126

pH
- in diagenetic modeling of subarkose 98
- of formation waters 64-65, 68, 69
- production versus time 169

Pore-water chemistry
- see also formation waters
- effect of depositional environment 41-43
- effect on diagenesis 41-43
- marine versus fresh water 42
- oxic versus anoxic conditions 42-43
- Wilcox Group, Texas Gulf Coast 44-43

Porosity
- in arkose diagenesis 143
- in diagenetic modeling of subarkose 98
- primary, in sandstones 70

Porosity enhancement
- Alaskan North Slope 5, 12
- Algeria 7
- Australia 9-10
- beneath ancient erosional surfaces 5-10
- China 9
- Libya 7
- The Netherlands 7, 13
- Norwegian Sea 6, 12
- Texas 10, 13
- Venezuela 10

Porosity evolution
- of sandstones 85-99
- of subarkose 85-99

Porosity observed, North Sea, Norwegian 23

Porosity prediction
- chemical modeling 45-52
- North Sea, Norwegian 10-12, 15, 17, 18-23
- in sandstones 1-2, 10-12
- using erosional unconformities 1-23

Precipitation
- calcite 76
- carbonates 76
- in diagenetic modeling of subarkose 98
- experiments on kaolinite 86-90, 93
- rate of quartz 98
- vs. solution saturation in kaolinite 92

Pressure solution
- chemical modeling 149
- effect of clay 151-153
- free-face 148-150
- mathematical modeling of 147-158
- quantification 148-150
- simulations during burial 149
- theory of dynamics 148-150
- transition to stylolites 150-151, 152

Primary porosity, Upper Almond Sandstone 70
Prodelta environments 38-39, 40
Provenance, Wilcox Group, Texas Gulf Coast 40-41

Reaction front
- calcite dissolution at 117-120, 121-122
- dolomitization...117-120
- fingering and scalloping 105, 106
- morphology..117-120
- triply fingered 120

Reaction-path modeling 46, 49-50
- applications 50
- computer codes...49-50
- input and output...49
- limitations 50
- theory..49
- of Upper Almond formation water 64-65

Reaction-transport modeling............................46, 50-51
- applications 51
- boundary conditions 112-113
- computer codes................50-51, 109-111, 131-132
- feedback in 104-107
- importance in petroleum exploration........127-129, 144-145
- input and output..50, 132-133
- limitations 51
- of methane buoyancy-driven flows 112
- self-organization of......................................104-107
- theory..50
- of Upper Almond formation water 62-64

Recrystallization
- of calcite...75-77, 79
- of carbonates 71, 75-81
- of high-Mg calcite 79
- and thermodynamic equilibrium 77

Reservoir prediction, chemical modeling in 103-129

Rock dissolution, see dissolution
Rock-fluid interactions, see fluid-rock interactions
Rock-water interactions, see fluid-rock interactions
Salinity, of formation waters......................................57

Sandstones
- in Australia..8
- banded cementation......................................153-156
- chemical weathering...2-4
- compaction evolution of 154
- deltaic 27-39
- diagenesis of calcite cement in 117-120
- differential compaction 153-156
- dissolution rate constants for minerals in 95
- effect of clay on pressure solution of...........151-152
- equilibrium constants for minerals in 95
- erosional unconformities in 1-23
- facies control of diagenesis..............................27-39
- kaolinite in..90-93
- North Sea Jurassic..91-92
- petrology..32-40
- point modal count analyses 29-31
- porosity evolution 85-99
- porosity prediction 1-23, 10-12
- reservoir quality..10
- reservoirs...90-93
- Tertiary Texas Gulf Coast 93

Sedimentary environments, see depositional environments
Sediment composition
- effect of depositional environments 40-41
- Wilcox Group, Texas Gulf Coast 40-41

Sediment texture, effect of depositional environments.......41
Shales, buoyancy-driven flows in 118
Shoreface environments 32, 35, 37

Shoreline environments
- lagoon 36, 37
- shoreface..32, 35, 37
- tidal channel 36-37, 38
- tidal flat 36, 37

Silica, activity as a function of temperature 165

Soil zone
- hydrochemical processes in..................................3
- subaerial exposure...2

Speciation-solubility modeling............................46, 48-49
- applications 49
- computer codes...48-49
- of fluid-rock interactions 161-169
- input and output...48
- limitations 49
- theory..48

Stylolites...105
- transition to pressure solution 150-151, 152

Subaerial exposure
- climate 4
- duration of 4
- fractures..3-4
- framework composition 4
- geochemical depth profiles.................................73
- of limestone 73
- meteoric waters effects..2
- open system 2-3
- rock dissolution during.....................................2-4
- soil zones 2
- topography 4
- weathering...3

Subaerial unconformity, see subaerial exposure
Subarkose
- see also arkose
- chemical modeling 67
- diagenesis...55-67
- porosity evolution 85-99
- simulation of diagenesis.................................93-99
- volume percentages of minerals in 97

Temperature, measurement of in situ 164
Tertiary, Texas Gulf Coast sandstones 93

Texas
- collapse breccias...14
- erosional surfaces..10, 13
- porosity enhancement 10, 13

Texas Gulf Coast
- depositional environments............................25-43
- facies control of diagenesis............................25-43
- regional setting 26
- Tertiary Sandstones 93
- Wilcox Group (see also Wilcox Group, Texas Gulf Coast).....25-43

Thermal recovery, of bitumen............................161-169

Thermodynamic disequilibrium
- in aqueous fluids 71-82
- versus equilibrium in carbonate diagenesis 72-74

Thermodynamic equilibrium
- concentration of ions in solid solution 75
- and recrystallization of calcite............................77
- of silicate dissolution 85-86, 168
- versus disequilibrium in bulk solution 72-74

Thermodynamics
- equilibrium theory..48, 49
- input data 51-52

Tidal channel environments...........................36-37, 38
Tidal flat environments....................................36, 37
Topography, subaerial exposure 4

Trace element data
- calcite...72
- depth profiles for calcite......................................74
- use of Homogeneous Distribution Law 78

Tucker Lake pilot site, Alberta..............................161-169
Unsaturated zone, see open system

Upper Almond formation waters
- alkalinity 57

bromide content 59, 63
chemical modeling 62-67
classification 57
closed system 64-65, 68, 69
HCO$_3$/Cl ratio 61, 66
meteoric mixing 59-61, 64, 65
open system 63-64, 67
δ^{18}O versus TDS 58
present-day composition 56-62
pressure-depth profiles 59
reaction-path modeling 64-65
reaction transport modeling 62-64
salinity 57
sampled wells 59
sampling locations 58
sulfate depletion 61

Upper Almond Sandstone
   see also Upper Almond formation waters
   cementation of 67
   depositional setting 56
   diagenesis 65-66, 67, 70
   paragenesis 65-66, 70
   petrography 65, 70
   primary porosity 70
   rock-water interaction 55-67

stratigraphy 56
Venezuela
   caves 9
   erosional surfaces 5, 10
   porosity enhancement 10
Water-film diffusion (WFD) 148-150, 156
Weathering, subaerial exposure 3
Wilcox Group, Texas Gulf Coast
   delta-front environments 37-40
   delta-plain environments 27-32
   depositional environments 27-39
   diagenesis 27-43
   location map 26
   occurrence of kaolinite 93
   pore-water chemistry 44-43
   provenance 40-41
   regional setting 26
   sediment composition 40-41
   shoreline environments 32-37
Wyoming
   paleogeography 56
   regional geology 56, 57
   stratigraphy 56
   Upper Almond Sandstone 55-67
Zoning, oscillatory intracrystalline 104